AF358723

Novel Electrochemical Biosensors for Clinical Assays

Novel Electrochemical Biosensors for Clinical Assays

Editor

Antonio Guerrieri

MDPI • Basel • Beijing • Wuhan • Barcelona • Belgrade • Manchester • Tokyo • Cluj • Tianjin

Editor
Antonio Guerrieri
Dipartimento di Scienze
Università della Basilicata
Potenza
Italy

Editorial Office
MDPI
St. Alban-Anlage 66
4052 Basel, Switzerland

This is a reprint of articles from the Special Issue published online in the open access journal *Sensors* (ISSN 1424-8220) (available at: www.mdpi.com/journal/sensors/special_issues/biosensors_Clinical).

For citation purposes, cite each article independently as indicated on the article page online and as indicated below:

LastName, A.A.; LastName, B.B.; LastName, C.C. Article Title. *Journal Name* **Year**, *Volume Number*, Page Range.

ISBN 978-3-0365-2344-6 (Hbk)
ISBN 978-3-0365-2343-9 (PDF)

Contents

About the Editor

Antonio Guerrieri

Researcher from June 1990 to October 1992 at the Faculty of Science, University of Bari, from November 1992 to now Antonio Guerrieri (Ph.D.) is Associate Professor in Analytical Chemistry at the Faculty of Science, now Department of Sciences, at University of Basilicata (Potenza, Italy). He teaches analytical chemistry in the chemistry and biotechnologies courses of his University, as well as in Ph.D. in Chemical Sciences.

Antonio Guerrieri was Head of Chemistry Courses, Head of Biotechnology Courses as well as Member of the Board of Governors and Vice Rector for Triennial Planning at University of Basilicata.

His research interests mainly deal with bioelectrochemistry and the application of biological electron transfer processes in bioelectronic devices, the interaction between electropolymerised films and biological systems, the development of amperometric biosensors based on electropolymerised thin polymer films and the study of protein/biomaterial interactions.

Antonio Guerrieri is co-author of about 100 scientific publications in international journals and congress meetings.

Preface to "Novel Electrochemical Biosensors for Clinical Assays"

Biosensors, i.e., devices where biological molecules or bio(mimetic)structures are intimately coupled to a chemo/physical transducer for converting a biorecognition event into a measurable signal, have recently gained a wide (if not huge) academic and practical interest for the multitude of their applications in analysis, especially in the field of bioanalysis, medical diagnostics, and clinical assays. Indeed, thanks to their very simple use (permitting sometimes their application at home), the minimal sample pretreatment requirement, the higher selectivity, and sensitivity, biosensors are an essential tool in the detection and monitoring of a wide range of medical conditions from glycemia to Alzheimer's disease, as well as in the monitoring of drug responses. Soon, we expect that their importance and use in clinical diagnostics will expand rapidly so as to be of critical importance to public health in the coming years.

In this context, electrochemical biosensors definitely play an innovative and quite promising role, particularly due to their clear advantages over, e.g., the spectroscopic methods, since they can be used even when the clinical samples are turbid or coloured. More importantly, electrochemical biosensors and the relevant instrumentations are cheap, easy to use, and usable in field analysis with a minimum or no sample pretreatment. Of course, to be effective in clinical assays, the biorecognition event of the desired analyte needs to be developed and optimized, so further efforts are required in choosing/designing the required biomolecule/structure and its efficient coupling to the transducer; no less important, the flow of the bio/chemical information starting from the biorecognition event to the chemo/physical detection needs to be studied and optimized as well. On the other hand, electrochemical detection is not without its drawbacks, since it is barely selective, so novel and more effective electrochemical detection approaches need to be studied and developed to assure an interferent-free detection of analytes in clinical samples without sample pretreatment.

This Special Issue would like to focus on recent research and development in the field of biosensors as analytical tools for clinical assays and medical diagnostics.

It was really a pleasure for me to collaborate with MDPI for this Special Issue. All the contributors are fully acknowledge for the realization of this Special Issue. A special thanks goes to Janetta Li (Section Managing Editor) for all the editorial work she did and for her huge patience.

Antonio Guerrieri
Editor

Review

Determination of Folic Acid Using Biosensors—A Short Review of Recent Progress

Alessio Di Tinno [1,2,†], **Rocco Cancelliere** [1,3,†] **and Laura Micheli** [1,4,*]

[1] Department of Chemical Sciences and Technologies, University of Rome Tor Vergata, Via della Ricerca Scientifica 1, 00133 Rome, Italy; alessio.ditinno@alumni.uniroma2.it (A.D.T.); rocco.cancelliere@uniroma2.it (R.C.)

[2] Department of Electrical and Information Engineering, University of Cassino and Southern Lazio, Via Gaetano di Biaio 1, 03043 Cassino, Italy

[3] CNR—National Research Council of Italy, Institute of Crystallography (IC), Via Salaria Km 29,300, 00015 Rome, Italy

[4] INBB—Consorzio Interuniversitario Istituto Nazionale di Biostrutture e Biosistemi, Viale Medaglie d'Oro 305, 00136 Rome, Italy

[*] Correspondence: laura.micheli@uniroma2.it

[†] These authors contributed equally to this work.

Abstract: Folic acid (FA) is the synthetic surrogate of the essential B vitamin folate, alternatively named folacin, pteroylglutamic acid or vitamin B_9. FA is an electroactive compound that helps our body to create and keep our cells healthy: it acts as the main character in a variety of synthetic biological reactions such as the synthesis of purines, pyrimidine (thus being indirectly implied in DNA synthesis), fixing and methylation of DNA. Therefore, physiological folate deficiency may be responsible for severe degenerative conditions, including neural tube defects in developing embryos and megaloblastic anaemia at any age. Moreover, being a water-soluble molecule, it is constantly lost and has to be reintegrated daily; for this reason, FA supplements and food fortification are, nowadays, extremely diffused and well-established practices. Consequently, accurate, reliable and precise analytical techniques are needed to exactly determine FA concentration in various media. Thus, the aim of this review is to report on research papers of the past 5 years (2016–2020) dealing with rapid and low-cost electrochemical determination of FA in food or biological fluid samples.

Keywords: folic acid; real samples; analytical methods; electrochemical tools

Citation: Di Tinno, A.; Cancelliere, R.; Micheli, L. Determination of Folic Acid Using Biosensors—A Short Review of Recent Progress. *Sensors* **2021**, *21*, 3360. https://doi.org/10.3390/s21103360

Academic Editor: Antonio Guerrieri

Received: 31 March 2021
Accepted: 8 May 2021
Published: 12 May 2021

1. Introduction

Folic acid (FA) or pteroylglutamic acid is a water-soluble B-complex vitamin and, due to its extremely important functions, represents an essential constituent of the human diet. It is the synthetic substitute of the essential B vitamin folate, also known as pteroylglutamic acid, folacin or vitamin B9. The IUPAC name of FA is (2S)-2-[[4-[(2-amino-4-oxo-1H-pteridin-6-yl)methylamino]benzoyl]amino]pentanedioic acid.

Beginning in the 2000s, FA and folate derivatives received increasing interest due to their importance for human well-being and due to growing understanding of the consequences of deficiency [1,2]. These molecules are considered essential compounds, because precursors of fundamental coenzymes are needed in many crucial biochemical reactions. FA molecular structure is made up of three components: a pteridine portion linked by through p-aminobenzoic acid to L-glutamic acid (Figure 1). The acyl group coming from the pteroic acid is a pteroyl group [3].

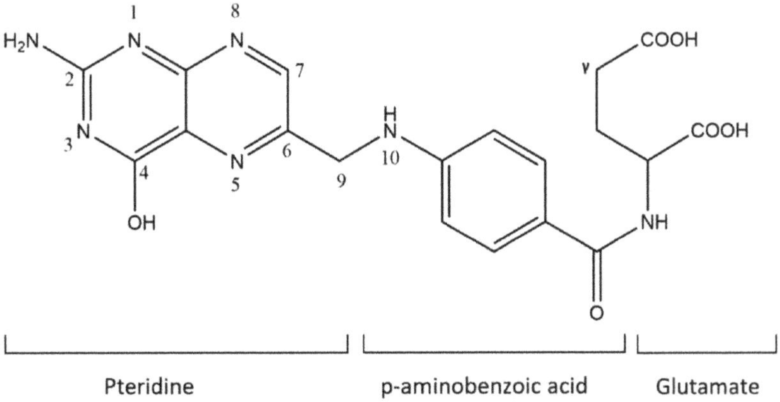

Figure 1. Schematic representation of folic acid (mono-glutamate derivate).

FA, though, has no biological activity itself but acts as fundamental precursor of a group of crucial coenzymes. Furthermore, FA is not produced by human body, and thus it must be obtained through diet (in the folate form), including liver, yolk, kidney beans, green leafy vegetables and fresh fruits [4]. Due the extreme importance of FA in cell proliferation, its assimilation is more and more frequently achieved by taking food supplements [5]. Indeed, all cells need folate in its reduced form in order to renew their cellular components. Tetrahydrofolate, for instance, acts as a cofactor in some essential metabolic pathways, such as DNA synthesis and biological methylation [6]. Different organisms exploit different strategies to obtain folate: plants and some microorganisms can produce folate from scratch with minimal variations of the same biosynthetic pathway [7–9]. Conversely, mammals, being auxotrophs, obtain folate through their diet or by exploiting intestinal bacteria capable of synthesizing it [10,11]. Despite all the different functions of the folate coenzyme, the main one is to transfer one-carbon groups in a variety of synthetic reactions. This capability varies depending upon the state of oxidation of the transferred group. Furthermore, FA can participate in a plethora of key reactions for fundamental cell functions: among the most important reported are the synthesis of purines, pyrimidines (and thus, indirectly, in the synthesis of DNA) and methionine, and the repair and methylation of DNA. Folate deficiency may result in degenerative conditions, such as neural tube defects in developing embryos and megaloblastic anaemia at any age [12]. Being water-soluble, folate cannot be stored in the human body and consequently is continuously lost. Therefore, its deficiency is one of the most commonly found vitamin deficits. The effective folate lack in the world is not well understood even though it appears a usual condition for many vulnerable classes. The use of so-called FA antagonists in certain disease (such as cancer [13], leukaemia [14], psoriasis [15], rheumatoid arthritis [16,17], polymyositis [18,19], dermatomyositis [20,21] and so on) may seem to suggest that a folate surplus in the diet would be harmful. Moreover, there have recently been rising concerns that FA supplementation could actually increase the risk of cancer frequency [22], as animal and human studies have indicated that high folate status may promote the progression of preneoplastic and undiagnosed neoplastic lesions [23,24]. There is little evidence to support such a view, nor it is well understood if FA supplements hamper the therapeutic effectiveness of these medications. Due to FA's ability to itself act as dihydrofolate reductase inhibitor, it could quite possibly be not only reliable but even advantageous in the treatment of these disorders [25,26]. For this reason, FA supplements are widely used to prevent and handle folate deficiency in at-risk groups and also to prevent adverse events associated with antifolate medications [27]. A daily intake of FA, through the commercially available supplements, is recommended to fertile and pregnant women in order to limit possible neural disorders in developing fetuses [28]. Folate is highly recommended also in subjects with heart failure due to its ability to lower the blood-homocysteine level, which has been linked to increased risk of cardiovascular events [29,30]. Moreover, FA promotes the formation of vigorous and healthy red blood cells [31,32]. However, some countries decided to not adopt FA fortification, being afraid of the possible negative consequences [33,34]. All these different benefits and disadvantages have led the research world to develop and optimize analytical methods, which can dependably and accurately monitor the FA concentration in natural sources, fortified foods, and multivitamin dietary supplements. In Europe, the EFSA (European Food Safety Authority) balances the fortification of flour (wheat and maize), establishing a minimum and maximum of 140 and 220 µg of FA per 100 g. For women with a history of congenital malformation, the recommended daily dose is 5 mg to reduce the risk of recurrence of the problem. A plethora of different analytical methods have been applied to determine FA in natural sources, using laboratory instrumentation as thermogravimetry [35], spectrophotometry [36,37], high performance liquid chromatography (HPLC) [38,39] with mass spectroscopy [40,41], colorimetry [42,43], fluorescence [44], electrophoresis [45,46] and so on (Table 1). Recently, an interesting overview about the future trends in the market for electrochemical biosensing has been proposed in [47], in which the current outline of the sensors and biosensors market is summarized. Some of the most recent advances are discussed, along with future prospects for biosensing development that could make an impact on the future global market. This short review reports on research papers of the past 5 years (2016–2020) dealing with the electrochemical determination of FA in food or biological fluid samples.

Table 1. Summary of analytical methods for the determination of FA in various samples.

Analyte	Technique	LR *	LOD	Sensitivity	RSD%	Ref
FA in pharmaceutical preparations	Spectrophotometric determination by coupling reaction	0.1–8.0 µg mL^{-1}	0.0469 µg mL^{-1}	0.0066 µg cm^{-2}	0.2805	[36]
FA in vegetables	HPLC-UV-Vis	0.3–100 ng mL^{-1}	0.1 ng mL^{-1}	/	0.3	[39]
FA in beer	LC-MS/MS	/	0.3 µg L^{-1}	1.2 µg L^{-1}	/	[40]
FA in egg yolks	Modified EMR-lipid method combined with HPLC-MS/MS	0.1–100 ng mL^{-1}	18.3 ng mL^{-1}	/	3.9 (HC) 8.1 (MC) 10 (LC)	[41]
FA in commercial preparations	Chemiluminometric procedure	6.0–114 µg mL^{-1}	2.0 µg mL^{-1}	/	1	[44]
	Fluorimetric procedure	0.02–1.1 µg mL^{-1}	0.002 µg mL^{-1}		0.7	
FA in human urine	Capillary electrophoresis	0.5–6.0 mg L^{-1}	0.30 mg L^{-1}	/	0.4–0.7(MT) 2.0–3.9 (PA) 1.2–1.7 (PH)	[45]
FA in pharmaceutical tablets	Capillary electrophoresis with chemiluminescence determination	5.0×10^{-8}–10^{-5} M	2.0×10^{-8} M	/	1.1 (MT) 1.5 (PA) 4.9 (PH)	[46]
FA in pharmaceutical preparations	Flow-injection/chemiluminescence determination	2.5×10^{-5}–3×10^{-7} M	2.3×10^{-8} M	/	3.5	[48]

* LR (linear range), LOD (limit of detection), RSD (relative standard deviation), HC (high-concentration), MC (medium-concentration), LC (low-concentration), MT (migration time), PA (peak area) and PH (peak height).

2. Electrochemical Determination

In recent decades, the growing interest in FA, due to its physiological importance, has involved the development of a plethora of methods for its determination [49]. Thus, a sensitive, specific and easy-to-use way to quantify FA is crucial. In this section, a collection of the most relevant and original electrochemical sensors of the past 5 years for FA determination is reported. In particular, two different subsections are mentioned: traditional and screen-printed based sensors.

2.1. Traditional Sensors

In 2018, Mohammadi et al. [50] aimed to fabricate an FA-sensitive platform based on manganese ferrite nanoparticles modified by means of trimethoxy silane (3-amynopropyl). In particular, in this study, core shell magnetic nanoparticles (CMNP) were involved in order to produce 2FTNE ((2-(4-Ferrocenyl-[1,2,3]triazol-1-yl)-1-(naphthalen-2-yl) ethanone), successively used in the modification of CMPE paste electrodes (2FTNE-modified CMNP paste electrodes), whereby this platform has been carefully characterized. The main goal of this research study was the concurrent measurements of epinephrine (EP), uric acid (UA) and FA. For this purpose, the concentrations of these analytes were simultaneously changed over time. The analysis was conducted using 2FTNE-modified CMNP paste electrodes (2FTNEMCPPE) and square wave voltammetry (SWV) as electrochemical technique, as reported in Figure 2. It is possible to observe that three well-distinguished anodic peaks at defined potentials of 430, 730 and 930 mV, corresponding to the oxidation peaks of EP, UA and FA, respectively, were obtained, which confirmed that the concurrent measurements of these analytes was possible (Figure 2). With the use of bare CPE, an overlapping voltammogram for the analytes was obtained. The sensitivity of the 2FTNEMCNPPE toward EP was 1.813 μA μmol^{-1} L, while without UA and FA, it was found to be 1.839 μA μmol^{-1} L. As a consequence, it is possible to have the independent or the concurrent measurements and quantification of EP, UA and FA. Finally, 2FTNEMCNPPE was applied to measure EP, UA and FA in an EP ampoule, an FA tablet and urine samples. All the different outcomes are summarized in Table 2. In addition, the recovery of EP, UA and FA of the samples spiked with known amounts of EP, UA and FA was assessed.

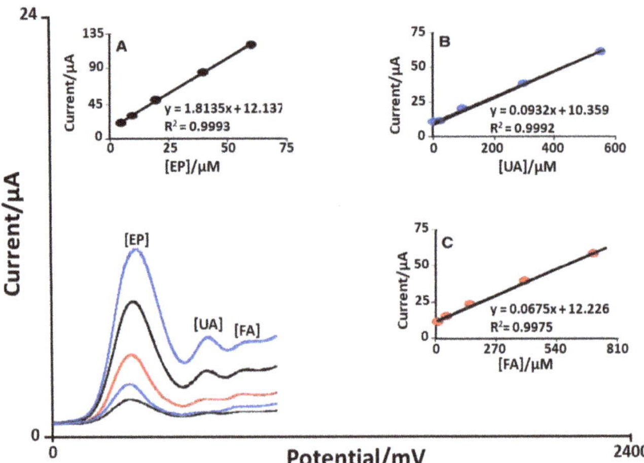

Figure 2. SWVs of 2FTNEMCNPPE in 0.1 M PBS (pH 7.0) with various concentrations (μmol L^{-1}) of EP, UA FA in a mixed solution: (1) 5.0 + 5.0 + 10.0, (2) 50.0 + 25.0 + 10.0, (3) 150.0 + 100.0 + 20.0, (4) 400.0 + 300.0 + 40.0, (5) 700.0 + 550.0 + 60.0. (**A–C**): plots of peak currents as a function of EP, UA and FA concentration, respectively [50].

Back in 2016, Lavanya et al. [51] developed a selective electrochemical sensor based on Mn doped SnO$_2$ nanoparticles (NPs) modified glassy carbon electrode (Mn-SnO$_2$/GCE). In particular, this platform was employed for the simultaneous determination of ascorbic acid (AA), uric acid (UA) and FA. These nanoparticles were synthesized by microwave irradiation and fully characterized using different spectroscopical and morphological techniques: X-ray diffraction (XRD), transmission electron microscopy (TEM), X-ray photoelectron spectroscopy (XPS), vibrating sample magnetometer (VSM), EIS, CV and SWV, respectively. AA is an important interfering analyte, which coexists with FA and UA in real body fluids. For this reason, the concentrations of UA and FA were determined in presence of a very high concentration of AA. The SWVs, reported in Figure 3, were obtained from the mixture of UA and FA in the presence of 200 μM AA.

Table 2. The use of 2FTNEMCNPPE in simultaneous quantification of EP, UA and FA in an EP ampule, an FA tablet and urine samples (n = 5). The amounts of EP and FA in ampoules and tablets were found to be equal to 0.98 mg mL^{-1} and 1.01 mg/tablet, respectively. These outcomes show no significant difference between the results of the 2FTNEMCNPPE and the nominal value on the ampoule label and tablet label (1.00 mg mL^{-1} and 1.00 mg/tablet, respectively) [50].

Sample	Spiked (μM)			Found (μM)			Recovery (%)			R.S.D. (%)		
	EP	UA	FA	EP	UA	EP	EP	UA	FA	EP	UA	FA
EP ampoule	0	0	0	9.0	-	-	-	-	-	2.7	-	-
	2.5	15.0	17.5	11.4	15.5	17.1	99.1	103.3	97.7	3.2	1.9	2.8
	5.0	25.0	27.5	14.3	24.8	27.9	102.1	99.2	101.4	3.1	2.3	2.7
	7.5	35.0	37.5	17.1	35.1	37.3	103.6	100.3	99.5	1.9	3.3	2.4
	10.0	45.0	47.5	18.5	45.6	48.8	97.3	101.3	102.7	2.2	1.8	3.1
FA tablet	0	0	0	-	-	17.0	-	-	-	-	3.4	-
	5.0	17.5	2.5	5.1	17.1	19.7	102.7	97.7	101.0	2.3	1.9	3.2
	10.0	2.5	5.0	9.8	22.9	21.9	98.0	101.8	99.5	3.1	2.3	1.9
	15.0	27.5	7.5	15.1	27.1	24.3	100.7	98.5	99.2	1.7	2.8	2.7
	20.0	32.5	10.0	19.8	33.5	27.5	99.0	103.1	101.8	2.8	3.1	1.8
	0	0	0	-	10	-	-	-	-	-	-	-
	7.5	10.0	30.0	7.4	20.2	30.9	98.7	101.0	103.0	2.9	3.2	1.6
	12.5	20.0	40.0	12.7	29.5	39.1	101.6	98.3	97.7	3.4	2.7	2.6
	17.5	30.0	50.0	18.1	41.2	49.5	103.4	103.0	99.0	1.6	2.6	3.1
	22.5	40.0	60.0	22.4	49.8	61.5	99.5	99.6	102.4	2.2	1.8	2.9

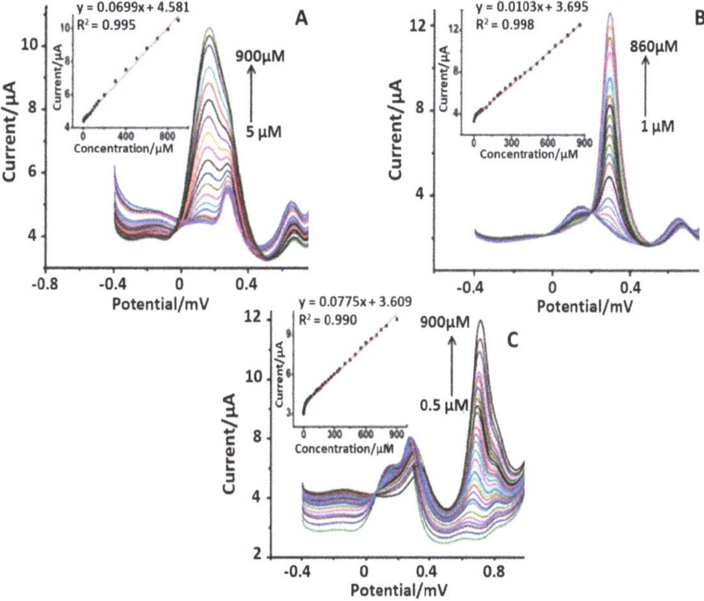

Figure 3. SWVs obtained for various concentrations of UA (5 to 500 μM) (**A**) and FA (1 to 800 μM) (**B**) at Mn SnO$_2$/GCE in the presence of 200 μM AA in 0.1 M PBS (pH 6.0) and inreal samples (**C**). (insert of **A**–**C**): plots of the oxidation peak currents as a function of various concentrations of UA and FA, respectively [51].

It can be easily noticed (Figure 3A) that simultaneously increasing the FA and UA concentrations a corresponding linear increase in the anodic peak currents have been obtained. Thus, a linear relationship between analytes concentration and their anodic peak currents is established. Furthermore, using different concentrations of UA and FA (in the range 5.0 to 500 µM for UA and 1.0 to 500 µM for FA), detection limits of 0.025 and 0.038 µM, respectively, have been obtained. To verify the applicability of this platform (Mn-SnO$_2$/GCE sensor) to real samples, AA and FA have been determined in pharmaceutical products. A recovery value corresponding to 99.4% (labeled content of FA mg/tablet 5, observed content of FA mg/tablet 4.88) was found using 3 wt% Mn-SnO$_2$/GCE. Thus, the method studied could be reliably applied for the determination of FA and AA in commercial samples.

Another important work, published in 2020 by Sadeghi et al. [52], proposed an electrochemical amplified FA sensor based on paste electrode (PE) modified with CuO-CNTs and 1-butyl-2,3-dimethylimidazolium hexafluorophosphate (BDHFP). Specifically, this device is based on the FA oxidation current registered by means of DP voltammograms, which increased 2.8 times using PE/M/CuO-CNTs-BDHFP with respect to PE. The modification produced an increase in the active surface area of PE (from 0.11 cm^2 to 0.18 cm^2 after modification with CuO-CNTs and BDHFP). DP voltammograms of different concentrations of FA were recorded on the surface of PE/M/CuO-CNTs/BDHFP with a linear dynamic range between 3.0 nM and 250 µM with a detection limit of 0.8 nM using PE/M/CuO-CNTs/BDHFP as an electrochemical sensor (Figure 4).

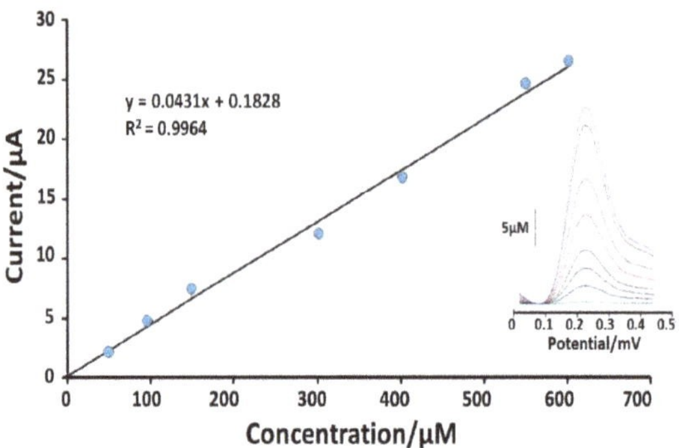

Figure 4. Current-concentration curve for the oxidation of FA in the range of 3.0 nM–250 µM. Inset: relative DP voltammograms of FA at surface of PE/CuO-CNTs/BDHFP [52].

The ability of PE/CuO-CNTs/BDHFP was also tested in real samples. To this purpose, it is employed in the determination of FA in orange and apple juices by applying the standard addition method (data repeated in Table 3). The results, summarized in Table 3, confirmed the powerful performances of this sensors for real sample analysis.

Table 3. Real sample analysis of FA using PE/CuO-CNTs/BDHFP (n = 4) [52].

Sample	FA Added (µM)	FA Expected (µM)	FA Found (µM)	Recovery %
Orange Juice	/	/	9.89 ± 0.54	/
	10.00	19.89	20.21 ± 0.87	101.6
Apple Juice	/	/	8.51 ± 0.34	/
	10.00	18.51	18.38 ± 0.65	99.29

Mollaei et al. [53] in 2019 reported an electrochemical sensor for FA detection based on the adsorption of a cationic surfactant (n-dodecylpyridinium chloride, DPC), at the surface of carbon paste electrode (CPE). The electrochemical performances in FA detection were compared with cetyltrimethylammonium bromide (CTAB). This quantitative analysis is performed using different voltammetric techniques: the differential pulse voltammetry (DPV), cyclic voltammetry (CV) and chronocoulometry (CC). Moreover, the determination of FA in urine and pharmaceuticals were performed in order to demonstrate the capability of the improved method and its reproducibility in real sample matrix (Figure 5). To overcome the matrix effect drawback, in both cases, the conventional standard addition method was utilized. The results are summarized in Table 4, indicating acceptable recoveries (ranging between 94.1 and 103.5) and working range.

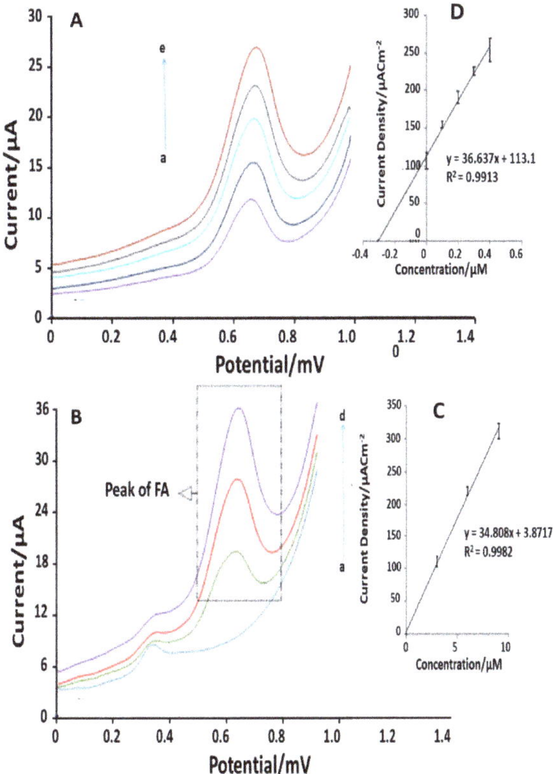

Figure 5. (**A**): Differential pulse voltammograms of CPE in 0.1 M PB (pH = 8.0) and drug tablet (3.0 µM FA) containing different concentrations of pure FA: a–e corresponds to 0.0–4.0 µM of pure FA. (**B**): Differential pulse voltammograms of CPE in 0.1 M PB (pH = 8.0) and 0.2 mL of a urine sample containing a different concentration of FA: the letters a–d correspond to 0.0–9.0 µM of FA. (**C,D**): Standard addition curve of the peak current density J (µM cm^{-2}) (extrapolated from respectively voltammograms A and B) vs. the concentration of FA [53].

Table 4. Determination of FA in drug tablet and urine samples by applying differential pulse voltammetry at the surface of CPE [53].

Sample	Initial Found (µM)	Added (µM)	Found (µM)	Recovery (%)
	3.0	0.0	2.82	94
	3.0	1.0	4.14	104
TABLET	3.0	2.0	5.11	102
	3.0	3.0	6.07	101
	3.0	4.0	6.86	98
	0.0	0.0	0.00	-
	0.0	3.0	3.07	102
URINE	0.0	6.0	6.20	103
	0.0	9.0	8.84	98

Looking at Figure 5, it can be noticed that CPE in the presence of DPC has good operating properties such as selectivity, sensitivity, stability, repeatability, low detection limit (2.9 nM) and wide linear concentration range (0.01–10.69 mM). Finally, Tahernejad-Javazmi et al. [54] in 2019, using previous experience with electrochemical sensor employing CuO nanoparticle decorated on single wall carbon nanotubes (CuO/SWCNTs) nanocomposite and 1-butyl-3-methylimidazolium hexafluorophosphate [55], developed an electroanalytical sensor based on reduced graphene oxide (GO) modified carbon paste electrode (CPE). In particular, the GO modified platform were functionalized with $FeNi_3$ ($FeNi_3$)/rGO-ionic liquid n-hexyl-3-methylimidazolium hexafluoro phosphate ($HMPF_6$). This platform was then employed for the simultaneous determination of FA and TBHQ (the antioxidant additive tertbutylhydroquinone), respectively. The modification process of CPE with $FeNi_3$/rGO and $HMPF_6$ was followed by EIS measurements (Figure 6A) and using 1.0 mM $[Fe(CN)_6]^{3/,4-}$ as an electrochemical probe. Using this technique and going from bare electrode to FeNi3-modifed platform ($FeNi_3rGO/HMPF_6/CPE$), an important decrease in the charge transfer resistance (Rct) value (from 6480 Ω to 870 Ω) was observed. This confirmed that the $FeNi_3$/rGO and $HMPF_6$ based modification improved conductivity and the electron transfer process at CPE's interface. The $FeNi_3$/rGO/$HMPF_6$/CPE performances have been compared to previous published methods [56,57]. The results, displayed in Table 5, show the applicability of this modified CPE for real sample analysis.

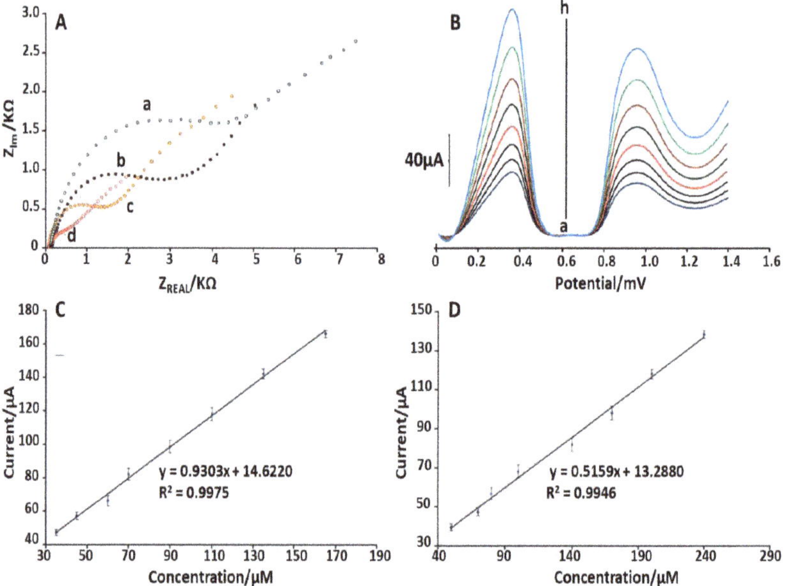

Figure 6. (**A**): Nyquist diagrams of (a) CPE, (b) $FeNi_3$/rGO/CPE, (c) $HMPF_6$/CPE and (d) $FeNi_3$/rGO/ $HMPF_6$/CPE in 1.0 mM $[Fe(CN)_6]^{3-,4-}$. (**B**): SW voltammograms of solution containing TBHQ and FA at the $FeNi_3$/rGO/$HMPF_6$/CPE; (a) 35.0 + 50.0; (b) 45.0 + 70.0; (c) 60.0 + 80.0; (d) 70.0 + 100.0; (e) 90.0 + 140.0; (f) 110.0 + 170.0; (g) 135.0 + 200.0 and (h) 165.0 + 240.0 µM. (**C**): The Ipa vs. TBHQ concentration obtained from SW voltammograms. (**D**): The Ipa vs. folic acid concentration obtained from SW voltammograms [54].

Table 5. Determination of TBHQ and FA in real samples (n = 4) [54]. The different results are compared to those from previously published methods for TBHQ [57] and FA [56], respectively.

Sample	TBHQ Added	FA Added	Found TBHQ Proposed Method	Found TBHQ Published Method	Found FA Proposed Method	Found FA Published Method
Soybean oil	-	-	2.6 ± 0.2	2.6 ± 0.2	-	-
	5.00	-	7.5 ± 0.3	7.7 ± 0.4	-	-
Sesame oil	-	-	5.7 ± 0.4	5.8 ± 0.6	-	-
	10.00	-	15.4 ± 0.5	15.3 ± 0.6	-	-
Apple juice	-	-	-	-	10.4 ± 0.7	10.2 ± 0.8
	-	10.00	-	-	20.6 ± 0.7	20.1 ± 0.8
Drinking water	10.00	10.00	9.8 ± 0.7	10.5 ± 0.6	10.5 ± 0.5	9.8 ± 0.8

2.2. Screen Printed Electrodes (SPEs)-Based Sensors

Screen printed electrodes (SPEs)-based sensors are an improvement of traditional sensors for performing rapid and accurate in situ analyses and for the development of portable devices. In the past 5 years, few biosensors based on SPEs were developed for the detection of FA in pharmaceutical and biological products, as summarized in Table 6.

Table 6. Summary of biosensors based on SEPs for the determination of FA in various samples.

Analyte	Technique	LOD	Working Range	Sample	Ref.
Vitamin B9 in real specimens.	SPE modified with La+3/Co3O4 nano-cubes.	0.3 µM	1–600 µM	Human urine samples, tablet.	[58]
N-acetylcysteine in the presence of paracetamol and folic acid	CPE modified with Pt-Co nanoparticles and 2-(3,4 dihydroxy phenethyl) isoindoline-1,3-dione	0.04 µM	0.08–650 µM	Human urine samples, tablet.	[59]
Simultaneous determination of sulfisoxazole and folic acid.	CuO Nanoparticles decorated on SWCNT nanocomposite modified CPE.	0.8 µM	0.07–500 µM	Human urine and tablet.	[59]
Folic acid in real specimens.	SPE modified with Graphene Oxide Nanoribbons	0.02 µM	0.1–1600 µM	Human urine samples, tablet	[60]
Folic acid in real specimens.	Mn-zeolite/Graphite modified Screen-printed Carbon Electrode	0.003 µM	0.004–1 µM	Pharmaceutical samples	[61]
Folic acid in real specimens.	SPE modified using GNS-MoS2-AuNPs	38.5 nM	50 nM–1150 µM	Human urine	[62]
Simultaneous determination of folic acid and epinephrine	Graphite SPE modified with Fe3O4@SiO$_2$	1 µM	5–1000 µM	Human blood serum and urine	[63]
Folic acid in real specimens.	SPE modified with NiFe$_2$O$_4$ nanoparticles	0.034 µM	0.1–500 µM	Human urine	[64]

The great results have been achieved in the application of SPEs for the detection of FA and the main results are described below in detail.

In 2016, Mani et al. [62] developed a facile method to determine FA in real samples. In this work, the preparation of a ternary nanocomposite made of graphene nanosheets (GNS), molybdenum disulfide (MoS_2) and gold nanoparticles (AuNPs) were detailed and its electrochemical sensing suitability studied. This ternary nanocomposite, GNS-MoS_2-AuNPs, characterized by particle sizes of about 10–50 nm, is deposited onto the surface of screen-printed electrodes (SPE) and applied for the quantification of FA. Due to the good synergic effect between GNS, MoS_2 and AuNPs, the composite shows excellent electrocatalytic ability. Impedance and electrochemical attributes of the nanocomposite were carefully studied: charge transfer resistance (R_{ct}) compared (Figure 7) for unmodified SPE, MoS_2/SPCE, GNS–MoS_2/SPCE and GNS–MoS_2–AuNPs/SPE. The R_{ct} values are reported in the following order: unmodified SPE > MoS_2/SPE > GNS–MoS_2/SPE > GNS–MoS_2–AuNPs/SPE, respectively. In particular, R_{ct} obtained at GNS–MoS_2–AuNPs/SPE is the lowest compared with control electrodes, indicating lowest resistance at this electrode interface. Thus, the electrochemical impedance spectroscopy (EIS) results revealed that the GNS–MoS_2–AuNPs composite interface presents a better electrical conductivity than the other platforms. Furthermore, this EIS study was followed by a potentiometric investigation in amperometry. Using this technique, the determination of FA in human urine sample occurs in a wide linear range of 50 nM–1150 μM and displays low detection limit of 38.5 nM. To apply these platforms in real samples, a recovery study was conducted, obtaining satisfactory results (values range from 97.16 to 98.55%). The sensor performance of the GNS–MoS$_2$–AuNPs is either superior or comparable to the previously reported electrodes [65–67].

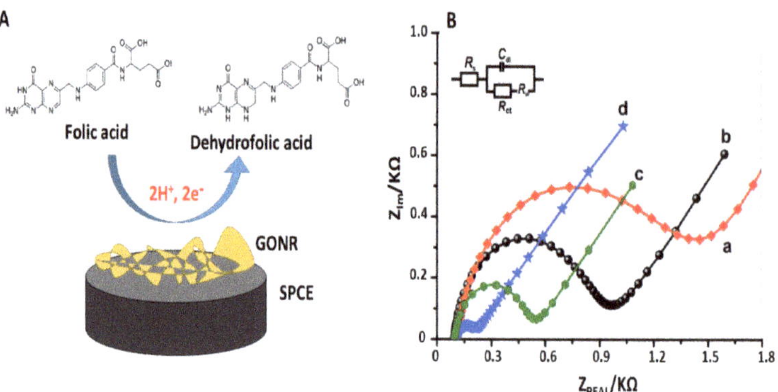

Figure 7. (**A**): Scheme of the electrochemical oxidation of FA on GONR/SPE. (**B**): EIS curves of unmodified SPE (a) MOS$_2$-SPE (b), GNS-MoS$_2$ /SPE (c), MoS$_2$-AuNPs/SPE, (d) in 0.1 M KCl containing 5 mM Fe(CN)$_6$ $^{3-/4-}$. Frequency: 0.1 Hz to 100 kHz [63].

Safaei and co-workers [63] in 2019 proposed a device for the simultaneous determination of FA and epinephrine (EP). This sensitive and convenient electrochemical sensor is based on Fe$_3$O$_4$@SiO$_2$/GR nanocomposite modified graphite SPE using cyclic voltammetry as detection tools. In particular, cyclic voltammograms (CVs) were recorded in the presence of analytes after having cycled the potential 20 times at a scan rate of 50 mV s^{-1}. The peak potentials were unchanged and the currents decreased by less than 2.3%. Therefore, at the surface of Fe$_3$O$_4$@SiO$_2$/GR /SPE, not only the sensitivity increases, but the fouling effect of the analyte and its oxidation product also decreases. To ascertain the analytical applicability of the proposed method, real samples matrices were evaluated. In particular, the simultaneous determination of epinephrine and FA, using the standard addition method and DPV as analytical technique, in human blood serum, urine, epinephrine injection and folic acid tablets samples, were analyzed in depth. In Table 7, the relative results are reported.

Satisfactory recoveries were found for epinephrine and FA, as reported in the above table. Reproducibility is reported as mean relative standard deviations (RSD%). The sensor exhibited notable electrochemical activity towards the oxidation of epinephrine and FA, and solved the overlapping anodic peak outcomes of EP and FA into two well-defined peaks.

Safaei et al. [64], in 2019, successfully synthesized and used NiFe$_2$O$_4$ (NFO) nanoparticles in order to develop a modified novel voltammetric sensor for determination of FA in urine samples. The morphological characterization of NFO nanoparticles was examined by means of scanning-electron microscopy (SEM), and it was observed that all the particles are nearly spherical, not agglomerated and less than 10 nm. In Figure 8C, cyclic voltammograms (CVs) are depicted, obtained using NFO-modified SPE. Analyzing equal concentration of substrate, NFO/SPE shows much higher anodic peak currents for the oxidation of folic acid compared to the unmodified SPE. This indicates that the modification of bare SPE with NiFe$_2$O$_4$ nanoparticles has significantly improved the performance in terms of electron transfer process between the electrode and the folic acid. In Figure 8C, the effect of potential scan rates on the oxidation currents of FA is described.

Table 7. Determination of epinephrine and FA in human blood serum, urine, epinephrine injection and folic acid tablet samples. All the concentrations are in µM (n = 5) [63].

Sample	Spiked		Found		Recovery, %		Rsd %	
	Epinephrine	FA	Epinephrine	FA	Epinephrine	FA	Epinephrine	FA
Human blood serum	0	0	-	-	-	-	-	-
	10.0	5.0	10.3	4.9	103.0	98.0	3.2	2.4
	20.0	60.0	19.8	61.6	99.0	102.7	17	2.7
Urine	0	0	-	-	-	-	-	-
	12.5	45.0	12.3	45.3	98.4	100.7	2.4	3.1
	22.5	55.0	23.1	53.7	102.6	97.6	1.8	2.8
Epinephrine Injection	0	0	10.5	-	-	-	3.2	-
	2.5	30.0	12.7	30.3	97.7	101.0	1.9	2.6
	5.0	39.7	15.9	40.3	102.6	99.2	2.4	3.3
Folic Acid Tablet	0	0	-	15.0	-	-	-	2.7
	5.0	25.0	4.9	0.9	98.0	102.2	2.4	1.6
	10.0	35.0	10.1	49.2	101.0	98.4	2.7	3.0

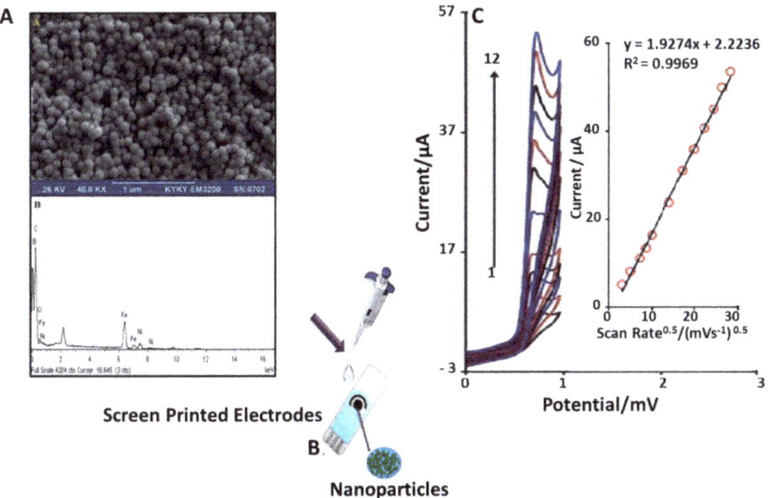

Figure 8. (**A**): SEM Micrograph of NiFe$_2$O$_4$ and its EDX. (**B**): deposition of NPs on SPE with drop casting. (**C**): CVs of NFO/SPE in 0.1 M PBS (pH 7) containing 150 µM of FA at various scan rates; numbers 1–12 correspond to 10, 25, 50, 75, 100, 200, 300, 400, 500, 600, 700 and 800 mV s^{-1}, respectively. Inset: variation of anodic peak current vs. square root of scan rate [64].

Since differential pulse voltammetry (DPV) has the advantage of having an increase in sensitivity and better characteristics for analytical applications, DPV technique was performed in order to determine various FA concentrations; a dynamic range between 1.0×10^{-7} and 5.0×10^{-4} M and a detection limit of 3.4×10^{-8} M were found. This protocol has been applied to real samples—FA tablets and urine samples—by using standard addition method. The results for the determination are summarized in Table 8.

Table 8. Determination of FA in FA tablet and urine samples. All the concentrations are in μM (n = 5) [64].

Sample	Spiked	Found	Recovery (%)	Rsd (%)
	0	15.0	-	3.2
	2.5	17.8	101.7	1.7
Folic acid tablet	5.0	19.5	97.5	2.8
	7.5	23.3	103.5	2.2
	10.0	24.8	99.2	2.4
	0	-	-	-
	10.0	10.3	103.0	3.4
Urine	20.0	19.9	99.5	1.7
	30.0	29.1	97.0	2.3
	40.0	40.5	101.2	2.8

Satisfactory recovery and reproducibility values of the experimental were found for FA, as demonstrated by the mean relative standard deviation (RSD%).

3. Conclusions

FA has a crucial role in some extremely important biochemical processes including synthesis of nucleic acid, cell division, growth and development of fetuses. For this reason, FA supplements are widely used to prevent and handle folate deficiency in at-risk groups and also to prevent adverse events associated with antifolate medications. Due to the critical rule of FA in human health, during the past years, many efforts have been made in the analytical field in order to develop reliable and precise sensors for its determination.

In more detail, the aim of this short review is to report the most interesting and original electrochemical tools proposed in the past 5 years for FA determination in pharmaceutical preparations, food supplements and other real samples. In conclusion, all the proposed analytical platforms can be applied to the determination of FA in a plethora of real samples: human urine, blood serum, pharmaceutical or commercial preparations. Remarkable recovery values and relative standard deviations have been found in all the proposed research papers. Among all, particularly interesting is the device based on $Fe_3O_4@SiO_2/GR$ nanocomposite modified graphite SPE, which shows good recovery values and RSD% in the simultaneous determination of FA by means of convenient electrochemical sensor. Among traditional sensors, noteworthy is the electrochemical behavior of 2FTNE-modified CMNP paste electrodes (2FTNECMNPPE). Using SWV, the concurrent measurement of epinephrine, uric acid and FA has been realized and no significant differences between the results of the 2FTNECMNPPE and the nominal values of real samples have been observed.

Author Contributions: Conceptualization, L.M. and A.D.T.; methodology, R.C. and A.D.T.; data curation, A.D.T. and R.C.; writing—original draft preparation, A.D.T. and R.C.; writing—review and editing, L.M.; supervision, L.M.; project administration, L.M. All authors have read and agreed to the published version of the manuscript.

Funding: This research received no external funding.

Institutional Review Board Statement: Not applicable.

Informed Consent Statement: Not applicable.

Data Availability Statement: Not applicable.

Conflicts of Interest: The authors declare no conflict of interest.

References

1. Andlid, T.A.; D'Aimmo, M.R.; Jastrebova, J. Folate and bifidobacteria. In *the Bifidobacteria and Related Organisms*; Elsevier: Amsterdam, The Netherlands, 2018; pp. 195–212.
2. Langston, W.C.; Darby, W.J.; Shukers, C.F.; Day, P.L. Nutritional Cytopenia (Vitamin M Deficiency) in the Monkey. *J. Exp. Med.* **1938**, *68*, 923–940. [CrossRef]
3. Talwar, G.P. *Textbook of Biochemistry, Biotechnology, Allied and Molecular Medicine*, 4th ed.; Place of Publication Not Identified; Prentice-Hall of India: New Delhi, India, 2015.
4. Bandžuchová, L.; Šelešovská, R.; Navrátil, T.; Chýlková, J. Electrochemical behavior of folic acid on mercury meniscus modified silver solid amalgam electrode. *Electrochim. Acta* **2011**, *56*, 2411–2419. [CrossRef]
5. Akbar, S.; Anwar, A.; Kanwal, Q. Electrochemical determination of folic acid: A short review. *Anal. Biochem.* **2016**, *510*, 98–105. [CrossRef] [PubMed]
6. Ebara, S. Nutritional role of folate. *Congenit. Anom.* **2017**, *57*, 138–141. [CrossRef] [PubMed]
7. Maynard, C.; Cummins, I.; Green, J.; Weinkove, D. A bacterial route for folic acid supplementation. *BMC Biol.* **2018**, *16*, 1–10. [CrossRef] [PubMed]
8. Gorelova, V.; Ambach, L.; Rébeillé, F.; Stove, C.; Van Der Straeten, D. Folates in Plants: Research Advances and Progress in Crop Biofortification. *Front. Chem.* **2017**, *5*, 21. [CrossRef] [PubMed]
9. Levin, I.; Giladi, M.; Altman-Price, N.; Ortenberg, R.; Mevarech, M. An alternative pathway for reduced folate biosynthesis in bacteria and halophilic archaea. *Mol. Microbiol.* **2004**, *54*, 1307–1318. [CrossRef] [PubMed]
10. Strozzi, G.P.; Mogna, L. Quantification of Folic Acid in Human Feces After Administration of Bifidobacterium Probiotic Strains. *J. Clin. Gastroenterol.* **2008**, *42*, S179–S184. [CrossRef]
11. Aufreiter, S.; Gregory, J.F.; Pfeiffer, C.M.; Fazili, Z.; Kim, Y.-I.; Marcon, N.; Kamalaporn, P.; Pencharz, P.B.; O'Connor, D.L. Folate is absorbed across the colon of adults: Evidence from cecal infusion of 13C-labeled [6S]-5-formyltetrahydrofolic acid. *Am. J. Clin. Nutr.* **2009**, *90*, 116–123. [CrossRef] [PubMed]
12. Wien, T.N.; Pike, E.; Wisløff, T.; Staff, A.; Smeland, S.; Klemp, M. Cancer risk with folic acid supplements: A systematic review and meta-analysis. *BMJ Open* **2012**, *2*, e000653. [CrossRef]
13. Berlin, N.I.; Rall, D.; Mead, J.A.R.; Freireich, E.J.; Van Scott, E.; Hertz, R.; Lipsett, M.B. Folic Acid Antagonists. *Ann. Intern. Med.* **1963**, *59*, 931–957. [CrossRef] [PubMed]
14. Mehranfar, S.; Zeinali, S.; Hosseini, R.; Mohammadian, M.; Akbarzadeh, A.; Feizi, A.H.P. History of Leukemia: Diagnosis and Treatment from Beginning to Now. *Galen Med. J.* **2017**, *6*, 2017. Available online: https://gmj.ir/index.php/gmj/article/view/702 (accessed on 12 February 2017).
15. Bronckers, I.M.G.J.; Seyger, M.M.B.; West, D.P.; Lara-Corrales, I.; Tollefson, M.; Tom, W.L.; Hogeling, M.; Belazarian, L.; Zachariae, C.; Mahé, E.; et al. Safety of Systemic Agents for the Treatment of Pediatric Psoriasis. *JAMA Dermatol.* **2017**, *153*, 1147–1157. [CrossRef]
16. Essouma, M.; Noubiap, J.J.N. Therapeutic potential of folic acid supplementation for cardiovascular disease prevention through homocysteine lowering and blockade in rheumatoid arthritis patients. *Biomark. Res.* **2015**, *3*, 1–11. [CrossRef] [PubMed]
17. Nogueira, E.; Gomes, A.C.; Preto, A.; Cavaco-Paulo, A. Folate-targeted nanoparticles for rheumatoid arthritis therapy. *Nanomed. Nanotechnol. Biol. Med.* **2016**, *12*, 1113–1126. [CrossRef] [PubMed]
18. Arnett, F.C.; Whelton, J.C.; Zizic, T.M.; Stevens, M.B. Methotrexate therapy in polymyositis. *Ann. Rheum. Dis.* **1973**, *32*, 536–546. [CrossRef] [PubMed]
19. Metzger, A.L.; Bohan, A.; Goldberg, L.S.; Bluestone, R.; Pearson, C.M. Polymyositis and Dermatomyositis: Combined Methotrexate and Corticosteroid Therapy. *Ann. Intern. Med.* **1974**, *81*, 182–189. [CrossRef] [PubMed]
20. Cobos, G.A.; Femia, A.; Vleugels, R.A. Dermatomyositis: An Update on Diagnosis and Treatment. *Am. J. Clin. Dermatol.* **2020**, *21*, 339–353. [CrossRef] [PubMed]
21. Papadopoulou, C.; Wedderburn, L.R. Treatment of Juvenile Dermatomyositis: An Update. *Pediatr. Drugs* **2017**, *19*, 423–434. [CrossRef]
22. Kim, Y.-I. Folate and colorectal cancer: An evidence-based critical review. *Mol. Nutr. Food Res.* **2007**, *51*, 267–292. [CrossRef]
23. Cole, B.F.; Baron, J.A.; Sandler, R.S.; Haile, R.W.; Ahnen, D.J.; Bresalier, R.S.; Mckeown-Eyssen, G.; Summers, R.W.; Rothstein, R.I.; Burke, C.A.; et al. Folic Acid for the Prevention of Colorectal Adenomas. *JAMA* **2007**, *297*, 2351–2359. [CrossRef] [PubMed]
24. Ebbing, M.; Bønaa, K.H.; Nygård, O.; Arnesen, E.; Ueland, P.M.; Nordrehaug, J.E.; Rasmussen, K.; Njølstad, I.; Refsum, H.; Nilsen, D.W.; et al. Cancer Incidence and Mortality After Treatment With Folic Acid and Vitamin B12. *JAMA* **2009**, *302*, 2119–2126. [CrossRef] [PubMed]
25. Moazzen, S.; Dolatkhah, R.; Tabrizi, J.S.; Shaarbafi, J.; Alizadeh, B.Z.; De Bock, G.H.; Dastgiri, S. Folic acid intake and folate status and colorectal cancer risk: A systematic review and meta-analysis. *Clin. Nutr.* **2018**, *37*, 1926–1934. [CrossRef] [PubMed]
26. Cheung, A.; Bax, H.J.; Josephs, D.H.; Ilieva, K.M.; Pellizzari, G.; Opzoomer, J.; Bloomfield, J.; Fittall, M.; Grigoriadis, A.; Figini, M.; et al. Targeting folate receptor alpha for cancer treatment. *Oncotarget* **2016**, *7*, 52553–52574. [CrossRef]
27. Bailey, L.B.; Rampersaud, G.C.; Kauwell, G.P.A. Folic Acid Supplements and Fortification Affect the Risk for Neural Tube Defects, Vascular Disease and Cancer: Evolving Science. *J. Nutr.* **2003**, *133*, 1961S–1968S. [CrossRef] [PubMed]
28. Shlobin, N.A.; LoPresti, M.A.; Du, R.Y.; Lam, S. Folate fortification and supplementation in prevention of folate-sensitive neural tube defects: A systematic review of policy. *J. Neurosurg. Pediatr.* **2021**, *27*, 294–310. [CrossRef] [PubMed]

29. Peng, Y.; Ou, B.-Q.; Li, H.-H.; Zhou, Z.; Mo, J.-L.; Huang, J.; Liang, F.-L. Synergistic Effect of Atorvastatin and Folic Acid on Cardiac Function and Ventricular Remodeling in Chronic Heart Failure Patients with Hyperhomocysteinemia. *Med. Sci. Monit.* **2018**, *24*, 3744–3751. [CrossRef]

30. Pieroth, R.; Paver, S.; Day, S.; Lammersfeld, C. Folate and Its Impact on Cancer Risk. *Curr. Nutr. Rep.* **2018**, *7*, 70–84. [CrossRef]

31. Herbert, V.; Zalusky, R. Interrelations of vitamin b12 and folic acid metabolism: Folic acid clearance studies. *J. Clin. Investig.* **1962**, *41*, 1263–1276. [CrossRef]

32. Quinlivan, E.; McPartlin, J.; McNulty, H.; Ward, M.; Strain, J.; Weir, D.; Scott, J. Importance of both folic acid and vitamin B12 in reduction of risk of vascular disease. *Lancet* **2002**, *359*, 227–228. [CrossRef]

33. Mason, J.B. Diet, folate, and colon cancer. *Curr. Opin. Gastroenterol.* **2002**, *18*, 229–234. [CrossRef]

34. Hirsch, S.; Sanchez, H.; Albala, C.; De La Maza, M.P.; Barrera, G.; Leiva, L.; Bunout, D. Colon cancer in Chile before and after the start of the flour fortification program with folic acid. *Eur. J. Gastroenterol. Hepatol.* **2009**, *21*, 436–439. [CrossRef] [PubMed]

35. Vora, A.; Riga, A.; Dollimore, D.; Alexander, K.S. Thermal stability of folic acid. *Thermochim. Acta* **2002**, *392–393*, 209–220. [CrossRef]

36. Nagaraja, P.; Vasantha, A.R.; Yathirajan, H.S. Spectrophotometric determination of folic acid in pharmaceutical preparations by coupling reactions with iminodibenzyl or 3-aminophenol or sodium molybdate–pyrocatechol. *Anal. Biochem.* **2002**, *307*, 316–321. [CrossRef]

37. Matias, R.; Ribeiro, P.R.S.; Sarraguça, M.; Lopes, J.A. A UV spectrophotometric method for the determination of folic acid in pharmaceutical tablets and dissolution tests. *Anal. Methods* **2014**, *6*, 3065–3071. [CrossRef]

38. Osseyi, E.S.; Wehling, R.L.; Albrecht, J.A. HPLC Determination of Stability and Distribution of Added Folic Acid and Some Endogenous Folates During Breadmaking. *Cereal Chem. J.* **2001**, *78*, 375–378. [CrossRef]

39. Jastrebova, J.; Witthöft, C.; Grahn, A.; Svensson, U.; Jägerstad, M. HPLC determination of folates in raw and processed beetroots. *Food Chem.* **2003**, *80*, 579–588. [CrossRef]

40. Bertuzzi, T.; Rastelli, S.; Mulazzi, A.; Rossi, F. LC-MS/MS Determination of Mono-Glutamate Folates and Folic Acid in Beer. *Food Anal. Methods* **2018**, *12*, 722–728. [CrossRef]

41. Sun, D.; Jin, Y.; Zhao, Q.; Tang, C.; Li, Y.; Wang, H.; Qin, Y.; Zhang, J. Modified EMR-lipid method combined with HPLC-MS/MS to determine folates in egg yolks from laying hens supplemented with different amounts of folic acid. *Food Chem.* **2021**, *337*, 127767. [CrossRef]

42. Yang, Z.; Gong, F.; Yu, Z.; Shi, D.; Liu, S.; Chen, M. Highly sensitive folic acid colorimetric sensor enabled by free-standing molecularly imprinted photonic hydrogels. *Polym. Bull.* **2021**, 1–15, in press. [CrossRef]

43. Peng, Y.; Dong, W.; Wan, L.; Quan, X. Determination of folic acid via its quenching effect on the fluorescence of MoS2 quantum dots. *Microchim. Acta* **2019**, *186*, 605. [CrossRef]

44. Anastasopoulos, P.; Mellos, T.; Spinou, M.; Tsiaka, T.; Timotheou-Potamia, M. Chemiluminometric and Fluorimetric Determination of Folic Acid. *Anal. Lett.* **2007**, *40*, 2203–2216. [CrossRef]

45. Flores, J.R.; Peñalvo, G.C.; Mansilla, A.E.; Gómez, M.R. Capillary electrophoretic determination of methotrexate, leucovorin and folic acid in human urine. *J. Chromatogr. B* **2005**, *819*, 141–147. [CrossRef] [PubMed]

46. Zhao, S.; Yuan, H.; Xie, C.; Xiao, D. Determination of folic acid by capillary electrophoresis with chemiluminescence detection. *J. Chromatogr. A* **2006**, *1107*, 290–293. [CrossRef] [PubMed]

47. Neves, M.M.P.D.S.; González-García, M.B.; Hernández-Santos, D.; Fanjul-Bolado, P. Future trends in the market for electrochemical biosensing. *Curr. Opin. Electrochem.* **2018**, *10*, 107–111. [CrossRef]

48. Wabaidur, S.M.; Alam, S.M.; Lee, S.H.; Alothman, Z.A.; Eldesoky, G.E. Chemiluminescence determination of folic acid by a flow injection analysis assembly. *Spectrochim. Acta Part A Mol. Biomol. Spectrosc.* **2013**, *105*, 412–417. [CrossRef]

49. Batra, B.; Narwal, V.; Kalra, V.; Sharma, M.; Rana, J. Folic acid biosensors: A review. *Process. Biochem.* **2020**, *92*, 343–354. [CrossRef]

50. Mohammadi, S.Z.; Beitollahi, H.; Khodaparast, B.; Hosseinzadeh, R. Electrochemical determination of epinephrine, uric acid and folic acid using a carbon paste electrode modified with novel ferrocene derivative and core–shell magnetic nanoparticles. *Res. Chem. Intermed.* **2019**, *45*, 1117–1129. [CrossRef]

51. Lavanya, N.; Fazio, E.; Neri, F.; Bonavita, A.; Leonardi, S.; Neri, G.; Sekar, C. Electrochemical sensor for simultaneous determination of ascorbic acid, uric acid and folic acid based on Mn-SnO2 nanoparticles modified glassy carbon electrode. *J. Electroanal. Chem.* **2016**, *770*, 23–32. [CrossRef]

52. Sadeghi, H.; Shahidi, S.-A.; Raeisi, S.N.; Ghorbani-HasanSaraei, A.; Karimi, F. Electrochemical Determination of Folic Acid in Fruit Juices Samples Using Electroanalytical Sensor Amplified with CuO/SWCNTs and 1-Butyl-2,3-dimethylimidazolium Hexafluorophosphate. *Chem. Methodol.* **2020**, *4*, 743–753. [CrossRef]

53. Mollaei, M.; Ghoreishi, S.M.; Khoobi, A. Electrochemical investigation of a novel surfactant for sensitive detection of folic acid in pharmaceutical and biological samples by multivariate optimization. *Measurement* **2019**, *145*, 300–310. [CrossRef]

54. Tahernejad-Javazmi, F.; Shabani-Nooshabadi, M.; Karimi-Maleh, H. 3D reduced graphene oxide/FeNi3-ionic liquid nanocomposite modified sensor; an electrical synergic effect for development of tert-butylhydroquinone and folic acid sensor. *Compos. Part B Eng.* **2019**, *172*, 666–670. [CrossRef]

55. Karimi-Maleh, H.; Amini, F.; Akbari, A.; Shojaei, M. Amplified electrochemical sensor employing CuO/SWCNTs and 1-butyl-3-methylimidazolium hexafluorophosphate for selective analysis of sulfisoxazole in the presence of folic acid. *J. Colloid Interface Sci.* **2017**, *495*, 61–67. [CrossRef] [PubMed]

56. Jamali, T.; Karimi-Maleh, H.; Khalilzadeh, M.A. A novel nanosensor based on Pt:Co nanoalloy ionic liquid carbon paste electrode for voltammetric determination of vitamin B9 in food samples. *LWT* **2014**, *57*, 679–685. [CrossRef]
57. Rahmanpour, M.S.; Khalilzadeh, M.A. ZnO nanoparticle modified carbon paste electrode as a sensor for electrochemical determination of tert-butylhydroquinone in food samples», Anal. Bioanal. *Electrochemistry* **2016**, *8*, 922–930.
58. Aflatoonian, M.R.; Tajik, S.; Ekrami-Kakhki, M.-S.; Aflatoonian, B.; Beitollai, H. A nano-sensor based on screen printed electrode (SPE) for electro-chemical detection of vitamin B9. *Eurasian Chem. Commun.* **2020**, *2*, 609–618. [CrossRef]
59. Karimi-Maleh, H.; Hatami, M.; Moradi, R.; Khalilzadeh, M.A.; Amiri, S.; Sadeghifar, H. Synergic effect of Pt-Co nanoparticles and a dopamine derivative in a nanostructured electrochemical sensor for simultaneous determination of N-acetylcysteine, paracetamol and folic acid. *Microchim. Acta* **2016**, *183*, 2957–2964. [CrossRef]
60. Mani, V. Highly Sensitive Determination of Folic Acid Using Graphene Oxide Nanoribbon Film Modified Screen Printed Carbon Electrode. *Int. J. Electrochem. Sci.* **2017**, *12*, 475–484. [CrossRef]
61. Porada, R.; Fendrych, K.; Baś, B. Development of novel Mn-zeolite/graphite modified Screen-printed Carbon Electrode for ultrasensitive and selective determination of folic acid. *Measurement* **2021**, *179*, 109450. [CrossRef]
62. Mani, V. Determination of Folic Acid Using Graphene/Molybdenum Disulfide Nanosheets/Gold Nanoparticles Ternary Composite. *Int. J. Electrochem. Sci.* **2017**, 258–267. [CrossRef]
63. Safaei, M.; Beitollahi, H.; Shishehbore, M.R. Simultaneous Determination of Epinephrine and Folic Acid Using the $Fe_3O_4@SiO_2$/GR Nanocomposite Modified Graphite. *Russ. J. Electrochem.* **2018**, *54*, 851–859. [CrossRef]
64. Safaei, M.; Beitollahi, H.; Shishehbore, M.R. Modified Screen Printed Electrode for Selective Determination of Folic Acid. *Acta Chim. Slov.* **2019**, *66*, 777–783. [CrossRef]
65. Gross, R.L.; Reid, J.V.; Newberne, P.M.; Burgess, B.; Marston, R.; Hift, W. Depressed cell-mediated immunity in megaloblastic anemia due to folic acid deficiency. *Am. J. Clin. Nutr.* **1975**, *28*, 225–232. [CrossRef] [PubMed]
66. Cinková, K.; Švorc, Ľ.; Šatkovská, P.; Vojs, M.; Michniak, P.; Marton, M. Simple and Rapid Quantification of Folic Acid in Pharmaceutical Tablets Using a Cathodically Pretreated Highly Boron-doped Polycrystalline Diamond Electrode. *Anal. Lett.* **2015**, *49*, 107–121. [CrossRef]
67. Keeley, G.P.; O'Neill, A.; Holzinger, M.; Cosnier, S.; Coleman, J.N.; Duesberg, G.S. DMF-exfoliated graphene for electrochemical NADH detection. *Phys. Chem. Chem. Phys.* **2011**, *13*, 7747–7750. [CrossRef] [PubMed]

 *sensors*

Review

Electrochemical Nanobiosensors for Detection of Breast Cancer Biomarkers

Veronika Gajdosova [1], Lenka Lorencova [1], Peter Kasak [2,*]  and Jan Tkac [1,*]

[1]   Institute of Chemistry, Slovak Academy of Sciences, Dubravska cesta 9, 845 38 Bratislava, Slovakia;
      Veronika.Gajdosova@savba.sk (V.G.); Lenka.Lorencova@savba.sk (L.L.)
[2]   Center for Advanced Materials, Qatar University, Doha 2713, Qatar
*   Correspondence: peter.kasak@qu.edu.qa (P.K.); Jan.Tkac@savba.sk (J.T.)

Received: 27 May 2020; Accepted: 15 July 2020; Published: 20 July 2020

Abstract: This comprehensive review paper describes recent advances made in the field of electrochemical nanobiosensors for the detection of breast cancer (BC) biomarkers such as specific genes, microRNA, proteins, circulating tumor cells, BC cell lines, and exosomes or exosome-derived biomarkers. Besides the description of key functional characteristics of electrochemical nanobiosensors, the reader can find basic statistic information about BC incidence and mortality, breast pathology, and current clinically used BC biomarkers. The final part of the review is focused on challenges that need to be addressed in order to apply electrochemical nanobiosensors in a clinical practice.

Keywords: breast cancer; nanobiosensors; biomarkers; electrochemistry; impedance; immobilization; nanomaterial; nanoparticles (NPs); magnetic NPs; self-assembled monolayers (SAMs); signal amplification

1. Introduction

According to the World Health Organization, the year of 2030 should witness roughly 12 million cancer-related deaths, making cancer one of the most prominent death-causing factors around the globe. In fact, the number of new cases of cancer (cancer incidence) is currently around 439 *per* 100,000 men and women *per* year [1]. Breast cancer (BC) has been considered the most frequent type of cancer disease worldwide among women, impacting 2.1 million women each year. In 2018, it was estimated that around 627,000 women died from BC; that is approximately 15% of all cancer deaths among women [2]. The highest incidence rates were observed in the United States and Western Europe. In the US, there were 101 new cases reported *per* 100,000 women, and in Europe, there were 85 [3]. East Asia has the lowest incidence with 21 cases *per* 100,000 women [3]. In Africa, the incidence is slightly higher with 23 cases *per* 100,000 women, but this amount can be undervalued due to a lack of accurate data [3].

BC is one of the leading causes of cancer-related mortality. The disease had always been common among women. That is supported by the fact that one of the first surgical treatments ever performed was BC treatment during the first surgical revolution at the end of the 19th century. BC rates are globally increasing and are higher among women in developed regions. The incidence and fatality increase with the increasing age of women as well. It was reported that statistically, women with an age 65 and above die with higher probability due to the disease [4–6]. The probability of the disease to develop within a woman's lifetime has grown over the past few decades from 1 in 11 in 1975 to 1 in 8 in 2016 [6]. There are several risk factors behind BC, including age, geographic location (country of origin), socioeconomic status, lifestyle risk factors (smoking, alcohol, diet, obesity, and physical activity), low rates of breastfeeding, family history of BC, mammographic density, ionizing radiation, etc. [5].

If the BC is diagnosed at an early stage, a 5-year survival rate can reach up to 90% in developed countries [7]. On the other hand, once a BC is metastatic, the patient's 5-year survival rate falls down to 27.4% [8]. Early diagnosis is needed for a successful treatment and high survival rate. T1 tumors with size less than 2 cm show a 10-year survival rate of approximately 85%, while T3 tumors show a 10-year survival of less than 60% mainly as the result of delayed accurate diagnosis [9]. Nowadays, mammography is used as a gold standard for early BC screening and detection, but it is less sensitive for young women (under 40 years old) with a sensitivity of 25–59%. A factor that is limiting the diagnosis of young women is a denser breast tissue compared to older women. Other limitations of mammography are high rates of false-positive and false-negative results which lead to biopsy, high cost of treatment, and procedural discomfort for the women [10].

To avoid development of the disease into advanced stages, there is clear need for early diagnostics, efficient treatment, and post-treatment monitoring. Therefore, there is an enormous demand for efficient less-invasive diagnosis i.e., analysis of cancer biomarkers in plasma/serum samples [11].

Although several review papers have been published recently describing the electrochemical biosensing of cancer biomarkers [12–16], such studies only partly covered the biosensing of BC biomarkers or the electrochemical biosensing of BC biomarkers. There are only two review papers specifically covering the electrochemical biosensing of BC biomarkers published in 2017 [17,18], but with only a minor coverage of beneficial properties of nanoparticles within electrochemical transducing schemes. To our best knowledge, this is the first review paper comprehensively covering the use of nanomaterials for enhanced electrochemical detection of breast cancer biomarkers.

2. Breast Pathology

In humans, the breast has a number of functions. The mammary gland is a distinguishing feature of mammals, and its primary role is to produce milk to nourish offspring. The breast develops in the superficial fascia. Until puberty, the breast includes only few ducts in both men and women. In females, true breast development begins at puberty due to the effect of estrogen and progesterone [19]. The breast consists of 12–15 major breast ducts, which lead to the formation of a nipple. These are connected to ducts ending in a duct lobular unit, which is the functional milk-producing unit of a breast [20]. Breast ducts are surrounded by myoepithelial cells supported by connective tissue stroma and a variable amount of fat [19]. The terminal duct lobular units spread during pregnancy. Milk is produced due to the secretion of prolactin and oxytocin. Imbalance in estrogen and progesterone concentrations *prior* to menopause results in atrophic changes of a breast tissue.

From a clinical point of view, the lymphatic drainage of a breast has a big importance. Approximately 5% of the lymph from the breast drains through the intercostal spaces to nodes along the internal mammary vessels. The remaining 95% of the lymph drains toward the axilla in one or two larger channels. As a result of this fact, all patients with invasive BC should go through some form of auxiliary surgery to find out whether there is lymph node involvement [21].

BC is a heterogeneous disease with various subtypes accompanied by a series of genetic changes. Breast tumors originate in the anatomical structures of the mammary gland that form the mammary gland, fibrous tissue, and adipose tissue. Mammary gland tumors can be benign such as papillomas and fibroadenomas. The most common malignant tumors are carcinomas. BC is most often caused by the terminal lobes (lobular) of the mammary gland and their ducts (ductal) (Figure 1). Globally, approximately 80% of all diagnosed BC cases are of the ductal subtype [22]. These two subtypes cover 40%–75% of all diagnosed cases [23]. Other types of cancer are present in 10% of cases where inflammatory BC, male BC, Paget's disease of breast, papillary carcinoma, and others are included [24]. BC can be classified by immunohistochemical examination into four subtypes: estrogen receptor (ER), progesterone receptor (PR), human receptor tyrosine-protein kinase erbB-2 (HER2), and antigen Ki-67 dependent [25].

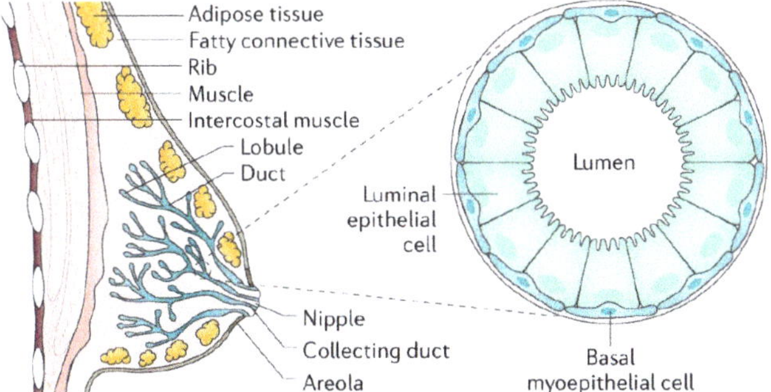

Figure 1. All breast cancers (BC) arise in the terminal duct lobular units (the functional unit of the breast) of the collecting duct. The histological and molecular characteristics have important implications for therapy. Several classifications based on molecular and histological characteristics have been developed. Reprinted by permission from Nature, Ref. [26], Copyright 2019.

3. BC Biomarkers

Due to progress in genomics, proteomics, and glycomics, various candidate biomarkers have been identified with a clinical potential for BC management [27]. A tumor marker was first discovered in 1847, and currently, there are more than 100 known different tumor markers [28]. Biomarkers have great potential for screening and diagnostics because they are present in blood and provide information about the health condition [7]. In healthy individuals, the tumor marker concentration is at a very low level or even in some cases absent, while increased values can reveal development and/or progression of a disease [29]. Serum biomarkers providing key information about the disease are important for the management of cancer patients, since blood aspiration is only a moderately invasive procedure.

The most relevant biomarkers that have occurred in BC include the presence of gene markers such as BReast CAncer Type (*BRCA1*, *BRCA2*), and protein-based biomarkers including cancer antigen CA 27.29, carcinoembryonic antigen (CEA), human epidermal growth factor receptor 2 (HER2), vascular endothelial growth factor (VEGF), polypeptide antigen (TPA), cytokeratin 19 fragment (CIFRA-21-1), platelet-derived growth factor (PDGF), and osteopontin (OPN). The basic characteristics of BC biomarkers are summarized in Table 1.

BC biomarkers (glycoproteins: mucin 1 (MUC1), HER2, carcinoembryonic antigen (CEA), epidermal growth factor receptor (EGFR), carbohydrate antigen 15-3 (CA15-3), CA 27-29, mammaglobin (MAM); DNA: *BRCA1*, *BRCA2*, proteins: Ki-67, OPN, microRNAs, and circulating tumor cells (CTC)) can be classified as diagnostic (healthy versus BC), prognostic (early BC versus advanced BC), predictive (provide information regarding whether a particular treatment will be beneficial for the BC patient) or therapeutic (a target biomolecule for therapeutics) based biomarkers [30,31].

Interestingly, although many studies have been published, only 9 biomarkers of cancer have been approved by the Food and Drug Administration (FDA) for clinical examinations so far. Since these biomarkers are all glycosylated proteins, changes in the glycan composition of these glycoproteins may serve as additional information for cancer diagnostics and/or prognosis. The following glycoprotein-based biomarkers have been published in the literature for BC management: HER2/NEU, CA15-3, CA27.29, MAM, galectin 3 binding protein, nectin 4, and fibronectin 1 with a typical concentration in human serum of 1–50 ng/mL [30]. The invasion, migration, and metastasis of cancer are caused by the deregulation of glycosylation related to an essential post-translational modification of proteins. In more than 90% of BC cases, there are observed changes in *O*-linked mucin-type glycosylation for example, expression of the Tn antigen, and the loss of core 2 *O*-glycans [32]. In BC, major glycan changes involve increased sialylation and fucosylation [33–36].

Table 1. Candidate breast cancer biomarkers (BRCA1, BRCA2, CA27.29, CA 15-3, CEA, HER-2, VEGF, tPA, CIFRA-21-1, PDGF, OPN).

Biomarker	Size/kDa	Incidence in Cancer	Level in Serum
BRCA1	207–220	breast, ovarian, prostate, pancreatic	ND
BRCA2	384	Fanconi anemia, breast, ovarian, lung, prostate, pancreatic	ND
CA27.29	250–1000	breast	≤37 U/mL
CA15-3	290–400	breast	3–30 U/mL
CEA	180–200	gastric, pancreatic, lung, breast, medullary thyroid	2–4 ng/mL
HER-2	185	breast, ovarian, gastric, prostate	15 ng/mL
VEGF	18–27	brain, lung, gastrointestinal, hepatobiliary, renal, breast, ovarian	~220 pg/mL
TPA	20–45	breast, lung, pancreatic	109 U/L
CIFRA-21-1	40	breast, lung, pancreatic	50 ng/mL
PDGF	35	glioblastoma, lung, colorectal, breast, liver and ovarian	(7.5 ± 3.1) ng/mL
OPN	41–75	breast, colon, liver, lung, ovarian, prostate	16 ng/mL

When new BC biomarkers are identified, it is of high importance to compare their clinical performance with already approved biomarkers in a form of AUC (area under curve) values in the receiver operating characteristic (ROC). The ROC curves calculated using the Youden index for the determination of CEA and CA15-3 showed AUC values of 0.616 or 0.678, respectively for CEA (cut-off value of 3.2 ng/mL) and CA15-3 (cut-off value of 13.3 ng/mL), when applied for disease-free survival examination [37]. We have found out clinical parameters for some of the novel BC biomarkers, showing significant advantage over already approved biomarkers. For example, Park et al. found out that the level of the human cytosolic thioredoxin correlated very well with the progress of BC [38]. At the cut-off value of 33.2 ng/mL, a sensitivity of 89.8%, a specificity of 78.0%, and an AUC value of 0.901 ± 0.025 were obtained by applying the ELISA format for thioredoxin analysis [38]. The ELISA method developed by Bernstein et al. resulted in an estimated AUC value of 0.892 using mammaglobin detection [39]. Thus, their method was able to distinguish healthy women from those having BC with high accuracy [39]. Yan et al. examined a clinical potential of the level of one type of fucosyltransferase (FUT4) determined in blood serum by ELISA for BC diagnostics [40]. AUC values of 0.784, 0.468, and 0.563 were determined for FUT4, CA15-3, and CEA, respectively. The results pointed out that FUT4 is in correlation with CA15-3 ($p < 0.05$). Moreover, FUT4 could be applied besides BC diagnostics also for BC prognosis [40].

4. Nanomaterials/Nanoparticles-Based Electrochemical Biosensors as Ultrasensitive Tools in Detection of BC Biomarkers

The speech of the physicist Richard Feynman entitled "There's plenty of room at the bottom", which took place at the Meeting of the American Physical Society in 1959 at CalTech, is considered to be the beginning of the nanotechnology era. Significant attention is currently being paid to nanomaterials. Nanomaterials are considered a pivotal tool for numerous applications in part due to their high surface area, compared to their respective bulk forms. Nanostructures with at least one dimension of size of 100 nm (1 nm = 1×10^{-9} m) or smaller are extremely useful in a number of areas, such as electronics, aerospace, military, pharmaceuticals, medicine, etc. Within last years, there has been an improvement in the synthesis and characterization of different nanomaterials, such as carbon-based nanomaterials, hydrogels, magnetic nanoparticles, metallic nanoparticles, polymer nanoparticles, and/or nanocomposites and two-dimensional nanomaterials [41,42].

One of the leading areas for practical application of the state-of-the-art nanoscience and nanotechnology is the development of various types of biosensors.

The application of nanomaterials to design biosensing platforms offers exceptional electronic, magnetic, mechanical, and optical properties for such devices. Nanomaterials can increase the surface of the transducing area of the sensors, which in turn provides enhanced catalytic activity. Electroactive properties of nanoparticles toward certain reactions have been widely exploited in biosensing applications. Nanometer-size structures have a large surface-to-volume ratio, controlled morphology, and structure that would scale down the characteristic size, which is a clear advantage when the sample volume is critical. The integration of advanced 2D nanomaterial MXene into biosensors architecture brings the advantage of hydrophilic character due to functional groups onto the nanoscale surface [43]. However, advances in nanomaterial biofunctionalization are crucial to achieve higher specificity in biosensing. To that end, nanomaterials can be "decorated" with different (bio)receptors offering specific recognition for biosensing [44–49]. There are basically two approaches applied to designing nanobiosensors: i.e., the application of nanoparticles for the modification of electrode surfaces (Approach 1, Figure 2), or the application of nanoparticles to make signal nanoprobes that enhance a generated signal (Approach 2, Figure 2). There are some nanobiosensors constructed using both amplification approaches (hybrid biosensing, i.e., Approach 3). In the forthcoming sections, when discussing particular nanobiosensors, amplification strategies are indicated as well.

Electrochemical biosensors (amperometric, potentiometric, conductometric, impedimetric, field-effect devices, etc.) are of particular interest for early-stage diagnostics of cancer diseases [15,50]. Electrochemical techniques such as cyclic voltammetry (CV), chronoamperometry (CA), differential pulse voltammetry (DPV), electrochemical impedance spectroscopy (EIS), and square wave voltammetry (SWV) offer an easy-to-use, affordable, highly sensitive, and reliable way for the ultrasensitive sensing of biomarkers related to such diseases [51,52]. Lab-on-chip biosensors presenting the compact and low-power portable miniaturized devices can be utilized in cancer biomarker discovery research, leading to potential clinical applications [53–57]. The surface architecture connecting the sensing element to the biological sample at the nanometer scale determines signal transduction and the general performance of electrochemical sensors. The eventual biosensor sensitivity is affected by the most common surface modification techniques with subsequent functionalization, various electrochemical transduction mechanisms, and by the choice of the recognition element (antibodies, nucleic acids, cells, micro-organisms, etc.). Electrochemical biosensors employing surface nanoarchitectures offer attractive features including robustness, easy miniaturization, excellent detection limits, as well as small analyte volumes and the ability to be used in turbid biofluids with optically absorbing and fluorescing compounds.

Nonetheless, there is still great room for improvement with regard to reproducibility, specificity, stability, and assay throughput of biosensing assay formats.

Regarding the sensitivity of detection by the biosensors, there is need to achieve a limit of detection (LOD) for the analytes that is at least comparable with ELISA assay format offering LODs of 0.75 ng/mL (HER2), 0.1 µg/L (kallikrein 5), and 0.17 ng/mL (thymidine kinase (TK1)). It is also important to outperform ELISA by the design of electrochemical biosensors offering to complete the whole assay procedure within 5 h and with a moderate throughput of analysis (up to 50 samples analyzed *per* run), which is typical for ELISA-based assay formats [58–60]. The novel generation of highly specific, sensitive, selective, and reliable micro (bio-)chemical sensors and sensor arrays can merge interdisciplinary knowledge in bio- and electrochemistry, solid-state chemistry, surface physics, bioengineering, integrated circuit silicon technology, and data processing. In the forthcoming sections, we discuss novel, nanoparticle-based approaches for the electrochemical detection of BC biomarkers.

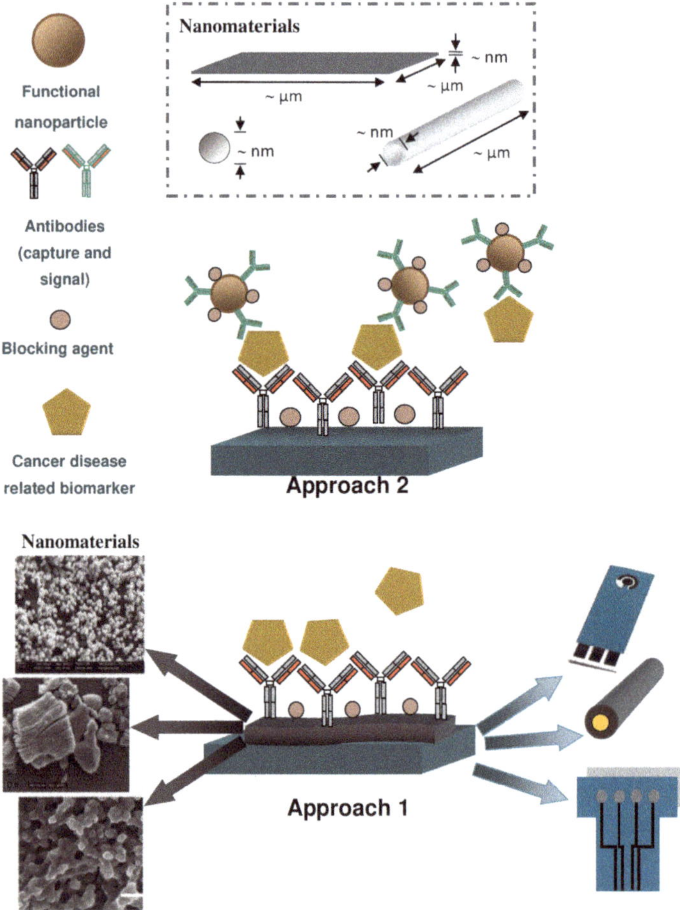

Figure 2. Schematic illustration of two different approaches applicable for the enhanced biosensing of cancer biomarkers using functional nanomaterials/nanoparticles either to enhance the electrode area, accessibility of analytes toward the interface, or interfacial properties with capture biorecognition elements (antibodies) immobilized (*Approach 1*) or for enhanced signal generation using a signal probe with biorecognition elements (antibodies) immobilized on the electrode (without being modified by nanoparticles) of a signal probe (*Approach 2*). Please note that we recognize *Approach 3* (not shown in the figure) applied to design biosensor devices by a combination of the nanomaterial/nanoparticle-modified electrode (*Approach 1*) with the use of a signal nanoprobe (*Approach 2*) within one biosensor device.

4.1. Detection of DNAs

The product of a BReast CAncer Type 1 (*BRCA1*) gene controls the cell cycle and ensures DNA repair. Mutation in the *BRCA1* gene leads to BC predisposition due to the loss of a gene function [61].

Benvidi and co-workers are topically focused on the development of DNA biosensors for the detection of *BRCA1* mutation at initial stages [62–65]. Benvidi and Jahanbani [63] applied a carbon paste electrode in combination with metallic nanocomposite, i.e., a magnetic bar carbon paste electrode decorated with magnetic iron oxide and silver nanoparticles by a physical method for label-free DNA detection. In the next step, the nanocomposite was modified with a self-assembled monolayer (SAM) of thiolated single-stranded DNA. The biosensor detected *BRCA1* 5382 mutation by the EIS method with an LOD of 3.0×10^{-17} M (a linear range from 1.0×10^{-16} M to 1.0×10^{-8} M) [63]

(Approach 1). Benvidi et al. [64] published results obtained with an improved strategy based on the application of glassy carbon electrode (GCE) modified by another type of carbon nanomaterial, e.g., reduced graphene oxide (RGO) or multi-walled carbon nanotubes (MWCNTs). 1-pyrenebutyric acid-*N*-hydroxysuccinimide ester was applied as a scaffold molecule for the immobilization of a *BRCA1* DNA probe for the detection of complementary DNA sequences. By applying this approach, the authors obtained a low LOD of 3.1×10^{-18} M and 3.5×10^{-19} M for a MWCNT-modified or RGO-modified device, respectively [64] (Approach 1). Other work from the same group of authors [65] was focused on an advanced coating of GCE with a dispersion of GO and a silk fibroin (SF) with subsequently electrochemically immobilized gold nanoparticles (AuNPs) for *BRCA1* 5382 mutation detection. In the next step, the analyte was incubated with the modified electrode and measured by CV and EIS techniques with complementary target DNA sequences. The impedimetric DNA sensor achieved an LOD of 3.3×10^{-17} M (a linear range from 1.0×10^{-16} M to 1.0×10^{-8} M) [65] (Approach 1).

A pre-treated GCE surface was coated with a hydrophilic material consisting of electrochemically deposited polydopamine, which was followed by the deposition of tannic acid assisted by Fe^{3+} ions [66]. In the next step, the branched structure of four-armed polyethyleneglycol was grafted onto a modified interface via a layer-by-layer technique. To enhance *BRCA1* gene detection, AuNPs with thiol-modified oligonucleotides were finally deposited onto the modified surface (Figure 3). An impedimetric biosensor detected *BRCA1* with an LOD of 0.05 fM in a linear range from 0.1 fM to 10 pM [66] (Approach 1). A carbon paste electrode (CPE) modified with electrospun ribbon conductive nanofibers of polyethersulfone and nanotubes were employed for *BRCA1* detection by Ehzari et al. [67]. DNA was detected with LOD of 2.4 pM with high selectivity, stability, reproducibility, and with a recovery index in the range from 101.5% to 105.2% [67] (Approach 1).

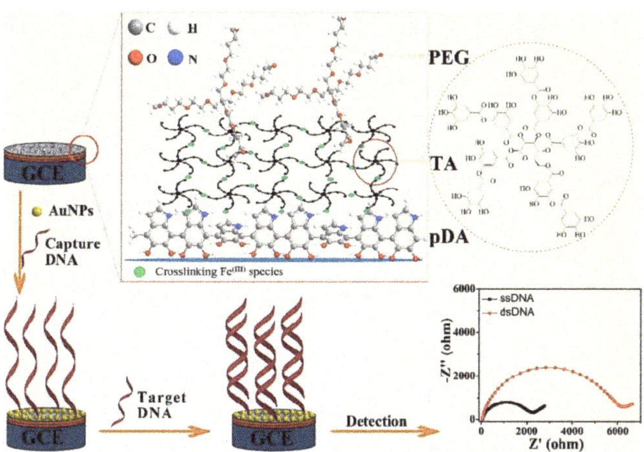

Figure 3. Fabrication process to design nanobiosensors for the detection of *BRCA1*. Reprinted from Ref. [66], Copyright (2017), with permission from Elsevier.

Graphene oxide (GO) was successfully applied as a promising nanomaterial with high surface area for the detection of the *BRCA1* gene. Kazerooni and Nassernejad developed a biosensor for detection of *BRCA1* with LOD of 2 pM by applying supramolecular ionic liquids grafted on nitrogen-doped graphene aerogel-modified GCEs by electrochemical reading [68] (Approach 1). The single-stranded DNA probe for the detection of *BRCA1* 5382 insC mutation was immobilized onto GCE electrochemically patterned with RGO and gold nanoparticles (AuNPs) [69]. The impedimetric biosensor was able to specifically recognize targets with LOD of 1.0×10^{-20} M [69] (Approach 1). RGO was also applied in combination with polypyrrole polymer by Shahrokhiana et al. for *BRCA1* detection [70]. A pyrrole-3-carboxylic acid monomer was electrochemically polymerized and applied for probe immobilization. *BRCA1* was determined with LOD of 3 fM in a linear range of 10 fM–0.1 μM by DPV and EIS [70] (Approach 1).

In addition to the development of a biosensor for the detection of the *BRCA1* gene, a DNA biosensor for the detection of the ERBB2c gene (producing HER2 protein) and CD24c was also prepared [71]. GCE was modified by GO, to which 4-aminothiophenol as a linker was covalently attached via amine coupling. The linker was in the subsequent step applied for the attachment of AuNPs. Then, a DNA capture probe was deposited on AuNPs via SAM formation. Then, DNA for the ERBB2c target was hybridized with a surface-confined capture DNA probe. Finally, the electrochemical signal was generated by hybridization with a conjugation DNA probe linked to horseradish peroxidase (HRP). The biosensor detected the ERBB2c gene down to 0.16 nM and CD24 down to 0.23 nM [71] (Approach 1).

The phosphatidylinositol-4,5-bisphosphate 3-kinase catalytic subunit alpha gene (*PIK3CA* gene) as a circulating tumor DNA was detected by a biosensor employing a nanocomposite of MoS_2 and poly(indole-6-carboxylic acid) as a surface-confined mediator, which was also applied for the covalent immobilization of –NH_2-modified ssDNA [72]. First, the surface of CPE was modified with exfoliated MoS_2 nanosheets, and then, it was incubated with a mediator monomer via π–π stacking with a subsequent potentiostatic polymerization of the monomer. Afterwards, ssDNA probes were covalently immobilized to such a modified electrode. The DNA biosensor could detect analytes down to 15 aM [72] (Approach 1).

4.2. Detection of MicroRNAs (MiRNAs)

The miRNAs are biomolecules consisting of 18–24 nucleotides that have a key role in biological processes such as cell proliferation, apoptosis, and tumorigenesis [73–75]. Abnormal expression has been observed in BC as well as in other cancer types [74,76].

GO was exploited as an effective part of several biosensors for miRNA detection. For example, the electrochemical nanobiosensor based on GCE that was step-by-step modified with GO and gold nanorods was fabricated for the detection of a serum miR-199a-5p level [77]. A thiolated oligonucleotide probe was immobilized on the modified electrode, and unspecific bindings were blocked by incubation with 6-mercapto-1-hexanol solution. The nanobiosensor exhibited LOD of 4.5 fM, which is a standard deviation of 2.9% for miR-199a-5p detection and a linear range from 15 fM to 148 pM [77] (Approach 1).

An impedimetric biosensor based on ZrO_2–RGO nanohybrids-modified GCE coupled with a catalytic hairpin assembly signal amplification strategy determined miRNA-21 in the range from 10 fM to 100 pM with LOD of 4.3 fM [78]. H1 modified with –NH_2 was covalently attached onto the ZrO_2–RGO-modified GCE surface via poly(acrylic acid) using amine coupling chemistry. In the absence of the analyte (miRNA-21), H1 and H2 did not hybridize. When the analyte was present, the hairpin of H2 hybridized with the analyte, which caused opening of the closed structure of H2. Subsequently, H1 hybridized with the unfolded H2. After this, target miRNA was released due to the DNA strand displacement reaction. At the end, H2 was attached to the electrode surface, and targeted miRNA started another cycle. This caused the amplification of the detected signal, since several H2 molecules per one analyte molecule were attached to the electrode surface (Figure 4) [78] (Approach 1).

The miRNA sensor using methylene blue as a redox mediator was fabricated by Rafiee-Pour et al. with a linear range from 0.1 to 500 pM with LOD of 84.3 fM [79]. In the experiment, the GCE electrode was modified with the dispersion of oxidized MWCNTs. Afterwards, 1.0 μM ss-DNA was immobilized, and half of the modified electrodes were incubated with target miRNA. The second half of the electrodes was used as a control. Non-hybridized miRNA was removed from the surface with saline sodium citrate. Both types of electrodes were immersed into 4.0 μM methylene blue, which was intercalated into a double-stranded helix, and DPV was applied to evaluate the change of the electrochemical signal [79] (Approach 1).

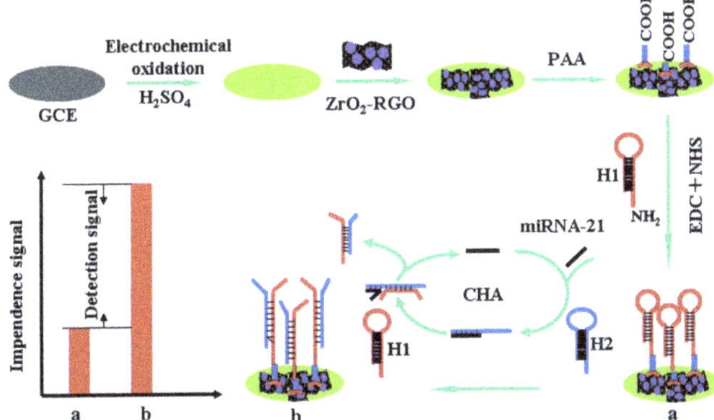

Figure 4. Immobilization of the capture probe H1 with subsequent miRNA-21 detection. For more information, please see the text. Published with permission by The Royal Society of Chemistry, Ref. [78].

Kilic et al. detected miRNA from cell lysates by using graphene-modified disposable pencil graphite electrodes [80]. The electrode was modified by an inosine substituted anti miRNA-2 probe. The analyte was detected with LOD of 2.1 μg/mL (EIS) or 5.8 μg/mL (DPV) [80] (Approach 1).

An enzyme-free biosensor based on a sandwich-type hybridization of two DNA probes with target miRNA was developed by Zouari et al. [81]. Thiol chemistry ensured the immobilization of a thiolated capture DNA onto the electrodes modified by a hybrid nanomaterial of RGO and AuNPs. Ferrocene-capped AuNPs were modified with streptavidin and conjugated with a biotinylated signal probe containing signal DNA. An enzymeless biosensor was able to determine the synthetic target miRNA with LOD of 5 fM (a linear range between 10 fM and 2 pM). Moreover, the biosensor was able to determine the target miRNA directly in diluted serum from BC patients. A 3-fold higher level of miRNA-21 was detected in serum samples of BC patients compared to a control [81] (Approach 3).

4.3. Detection of Mucins

Nowadays, there are more than 20 known types of mucins. They are encoded by MUC genes and represent high molecular weight glycoproteins expressed on epithelial cells. Aberrantly glycosylated mucins are expressed in cancer cells and serve as oncogenic molecules [82].

Nawaz et al. applied diazonium salt chemistry to modify single-walled carbon nanotubes (SWCNTs) for a biosensor development [83]. The MUC1 aptamer was immobilized onto modified SPCE via amine coupling. A DNA aptamer-based biosensor detected MUC1 with LOD of 0.02 U/mL with a linear range up to 2 U/mL [83] (Approach 1).

The MUC1 biosensor was also developed using GCE modified with core–shell nanofibers, MWCNTs, and AuNPs that were covalently modified with the anti MUC1-binding aptamer for the detection of MUC1 [84]. The impedimetric device using a soluble redox probe was able to detect MUC1 with LOD of 2.7 nM with a linear range up to 115 nM [84] (Approach 1).

Mouffouk together with colleagues applied bioconjugated self-assembled pH-responsive polymeric micelles loaded with ferrocene (Fc) and antiMUC1 antibodies as a signal probe [85]. The biosensor was able to detect MUC1 in a sample containing about 10 cells/mL [85] (Approach 2).

Nowadays, a novel 2D nanomaterial MXene (Ti$_3$C$_2$) due to its excellent electrical conductivity and large specific surface area with a large number of potential attachment binding sites is used as a conductive support for the immobilization of aptamer probes [53]. Wang et al. modified an electrode surface with MXene for the development of a MUC1 biosensor [86]. The Fc-labeled complementary DNA was bound onto MXene nanosheets to form a signal probe to amplify an electrochemical signal. GCE was modified by the electrodeposition of AuNPs with the MUC1 aptamer attached to the modified electrode via Au–S bonds. The modified electrode was blocked using bovine serum albumin (BSA) in order to resist non-specific interactions (Figure 5). Then, a signal probe was attached to the modified electrode via hybridization between complementary DNA and a MUC1 aptamer. Upon the interaction of MUC1 with such an electrode, the signal probe was detached from the working electrode, resulting in a decrease of an electrochemical signal (a signal-off response). This competitive aptasensor detected MUC1 with LOD of 0.33 pM with a linear range up to 10 mM [86] (Approach 3).

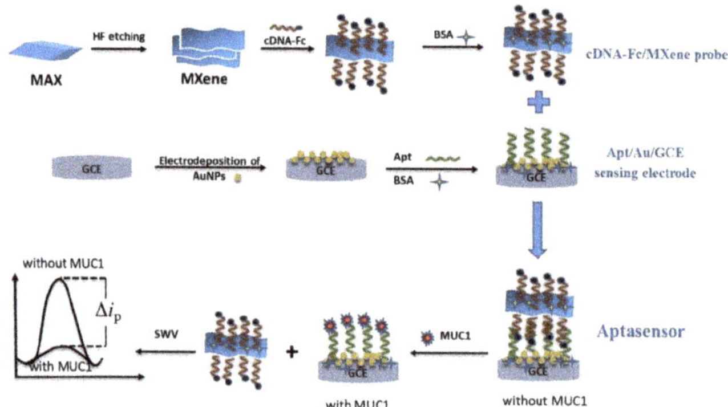

Figure 5. Procedure for fabrication of the competitive electrochemical aptasensor. Reprinted from Ref. [86], Copyright (2020), with permission from Elsevier.

CA15-3 (290–400 kDa) represents a soluble form of mucin 1 (MUC1): a transmembrane protein on the apical cellular surface. MUC-1 is a glycoprotein with three domains. The association between BC and elevated expression of CA15-3 has been experimentally confirmed [87].

Santos et al. used imprinting technology with a CA15-3 imprint within an electropolymerized layer of polypyrrole for CA15-3 detection [88]. Polypyrrole was deposited on a fluorine-doped tin oxide conductive glass support in the presence of the analyte. Then, the analyte was removed from the imprinted layer with ethanol, and the biomimetic material was then incorporated in a polyvinylchloride plasticized membrane acting as a potentiometric ionophore. The best results were obtained with electrodes covered by the imprinted polymer without any lipophilic additive with LOD of 1.07 U/mL and a linear response from 1.44 to 13.2 U/mL for CA15-3 [88] (Approach 1).

A CA15-3 immunosensor based on RGO and CuS NPs was fabricated using gold screen-printed electrode [89]. Firstly, anti CA15-3 antibodies were immobilized on the electrode. Once the analyte CA15-3 was bound to the surface of the electrode, the electrochemical response toward catechol was decreased. The sensor reached LOD of 0.3 U/mL, a sensitivity of 1.88 μA/(μM cm^2), and a linear response from 1.0 to 150 U/mL [89] (Approach 1).

Nakhjavani et al. prepared a sandwich-type of electrochemical immunosensor for the detection of CA15-3 [87]. Bare GE was incubated with streptavidin for 12 h with the subsequent immobilization of biotinylated anti-CA15-3 monoclonal antibodies. A considerable signal enhancement was reached due to the enhanced density of HRP delivered via streptavidin-coated magnetic beads (MBs) conjugated with biotinylated HRP and anti-CA15-3 antibodies. CA15-3 was detected employing the immunosensor in an electrolyte containing 0.1 M PBS pH 7.0 with a hydroquinone (HQ) as a redox mediator in the presence of H_2O_2 by CV and EIS techniques with LOD of 15×10^{-6} U/mL (a linear range from 50 to 15×10^{-6} U/mL). The lowest value of an electron-transfer resistance (R_{et}) at a bare electrode increased after the addition of streptavidin onto the surface, as well as after adding monoclonal antibodies and finally after CA15-3 addition. The R_{et} values decreased after the addition of a detection label, confirming attachment onto the electrode surface [87] (Approach 2).

The nanostructure-based immunosensor was developed by applying the non-covalent functionalization of GO with 1-pyrenecarboxylic acid as a modified electrode interface for the immobilization of a primary antibody (Ab_1) against the analyte [90]. Pre-treated GE were modified with a SAM of cysteamine, and the remaining empty places on the electrode were blocked with 2-mercaptoethanol. These electrodes were covalently patterned by GO already functionalized with 1-pyrenecarboxylic acid via amine coupling. Then, such modified electrodes were immobilized with monoclonal anti-CA15-3 Ab_1, blocked with BSA, incubated with CA15-3, and after immunoreaction took place, they were incubated with a signal probe (Figure 6). MWCNTs supporting a high density of ferritin molecules together with secondary antibody (Ab_2) against the analyte applied as a signal probe for the determination of CA15-3. MWCNTs were treated by a mixture of strong inorganic acids for the formation of carboxylic groups, for nanotube shortening, and for removing metallic and carbonaceous impurities. After the activation of MWCNTs, nanotubes were covalently modified by polyclonal anti-CA15-3 Ab_2 and ferritin. CA15-3 was detected through an enhanced bioelectrocatalytic reduction of H_2O_2 mediated by HQ at the immunosensor-offered LOD of 0.01 U/mL in human serum samples using DPV [90] (Approach 3).

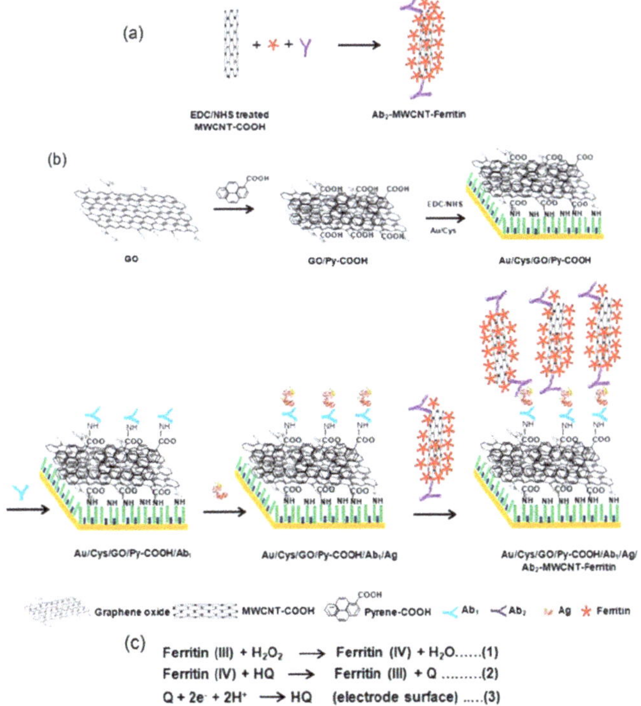

$$Ferritin\ (III) + H_2O_2 \longrightarrow Ferritin\ (IV) + H_2O \quad ......(1)$$
$$Ferritin\ (IV) + HQ \longrightarrow Ferritin\ (III) + Q \quad ....(2)$$
$$Q + 2e^- + 2H^+ \longrightarrow HQ \quad (electrode\ surface)\ .....(3)$$

Figure 6. (**a**) Preparation of the Ab_2 conjugate applied as a signal probe, (**b**) fabrication of the interfacial layer of the immunosensor device, and (**c**) mechanism behind the generation of an electrochemical signal. Reprinted from Ref. [90], Copyright (2016), with permission from Elsevier.

4.4. Detection of Human Epidermal Growth Factor Receptor-2 (HER2)

HER2 (185 kDa) i.e., human epidermal growth factor receptor-2, belongs to a family of receptor tyrosine kinases [91]. HER2 in BC is characterized by its high expression of growth factor receptor-related genes (*ERBB2*, *EGFR*, and/or *FGFR4*) and cell cycle-related genes [85].

There are several publications describing the development of biosensor platforms using various forms of graphene to enhance the selectivity and specificity of such devices. The in situ growth of 1D molybdenum trioxide anchored onto the 2D RGO via one-pot low-temperature hydrothermal synthesis and further functionalized using 3-aminopropyltriethoxysilane was fabricated as a suitable nanohybrid platform for HER2 detection [92]. In the following step, the surface conjugation of the monoclonal anti-HER2 antibodies onto the modified electrode was performed via amine coupling chemistry (Figure 7). The LOD of this nanohybrid-based immunosensor was 0.001 ng/mL, with a linear response in a concentration range of 0.001–500 ng/mL [92] (Approach 1).

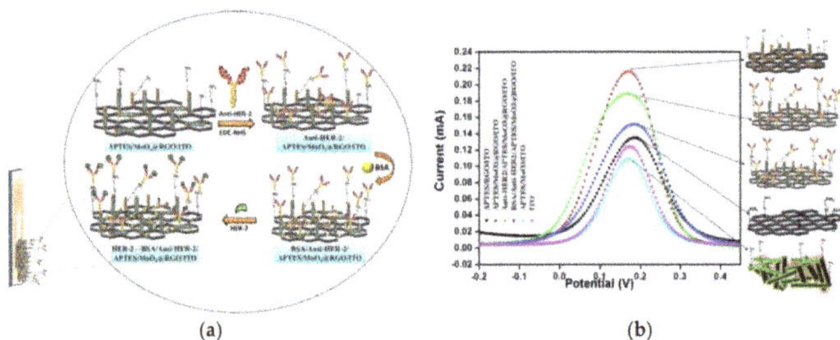

Figure 7. Development of an immunoelectrode for BC biomarker detection (**a**). Electrochemical peak response obtained via differential pulse voltammetry (DPV) at each step of electrode modification (**b**). Reprinted with permission from Ref. [92]. Copyright (2019) American Chemical Society.

An HER2 biosensor was prepared by the modification of GCE by a thin layer of RGO and SWCNTs to which a densely packed layer of AuNPs was deposited [93] (Approach 1). In the final step, the aptamer against HER2 was attached to the modified electrode and changes in the impedance were applied for the detection of HER2 with LOD of 50 fg/mL (a linear range from 0.1 pg/mL to 1 ng/mL). The recovery index of HER2 detection, when spiked into serum samples, was close to 100%, and the results of assaying HER2 levels in serum samples obtained by the biosensor device were in an excellent agreement with the ELISA method [93].

Arkan developed an impedimetric immunosensor using a hybrid nanomaterial modified electrode by the deposition of AuNPs and MWCNTs glued to the electrode by ionic liquid [94]. AuNPs were electrodeposited onto an electrode already patterned by MWCNTs and ionic liquid. Such an electrode was then immersed in an ethanol solution of 1,6-hexanedithiol. Then, another layer of AuNPs was deposited to which anti-HER2 antibodies were covalently grafted via amine coupling. It was found out that the charge transfer resistance increased linearly with increasing concentrations of HER2 antigen. The biosensor could detect HER2 in the linear range from 10 ng/mL to 110 ng/mL with LOD of 7.4 ng/mL. The results indicated the ability of HER2 detection in serum samples of BC patients, and such assays were in an excellent agreement with the results obtained by a commercial HER2 kit [94] (Approach 1).

An electrochemical molecularly imprinted polymer-based sensor (Figure 8) was developed for the detection of an extracellular domain of HER2 [95]. The sensor was prepared on a screen-printed gold electrode (AuSPE), where a molecularly imprinted layer was electropolymerized from a solution consisting of phenol and the analyte using the CV technique. The device exhibited a linear range for analyte detection from 10 to 70 ng/mL and LOD of 1.6 ng/mL, when DPV was applied as an electrochemical detection technique [95] (Approach 1).

Freitas et al. developed several biosensor devices for the detection of HER2 [96–99]. The first one was developed on SPCE modified either by AuNPs or combination of AuNPs with MWCNTs [96]. Such a modified electrode was then modified by primary anti-HER2 antibodies. Then, the biosensor was incubated with an analyte, and in the subsequent step, it was incubated with biotinylated secondary anti-HER2 antibodies. The electrochemical signal was generated by a final incubation of the biosensor by streptavidin-modified alkaline phosphatase, which catalytically reduced silver ions in the presence of 3-indoxyl phosphate. Under optimal conditions, the biosensors could detect HER2 in a concentration window of 7.5–50 ng/mL with LOD of 0.16 ng/mL (MWCNTs with AuNPs) or 8.5 ng/mL (AuNPs). The total assay time was 140 min, and the biosensor was applied for the analysis of HER2 spiked into serum samples [96] (Approach 1).

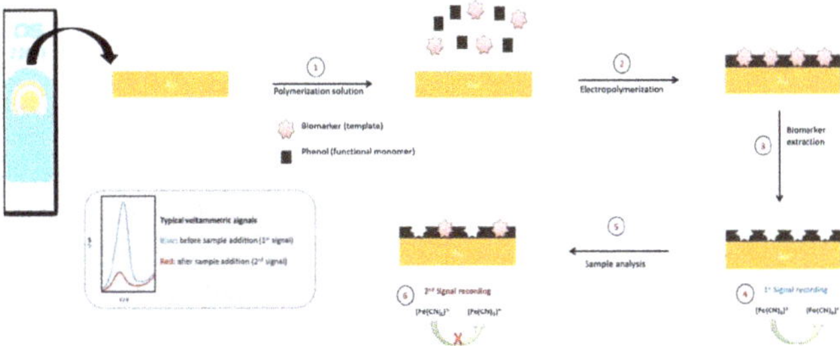

Figure 8. Fabrication and operation principles of the molecularly imprinted polymer-based sensor deposited on a gold screen-printed electrode (AuSPE). Reprinted from Ref. [95], Copyright (2018), with permission from Elsevier.

Malecka with colleagues constructed a cellulase-linked sandwich assay based on magnetic beads for HER2 detection [100]. The principle behind detection is the formation of an insulating layer consisting of nitrocellulose film on spectroscopic graphite electrode. HER2 interacts with primary aptamer/antibody-modified magnetic beads with the subsequent formation of a sandwich configuration on MBs by secondary aptamers/antibodies conjugated to cellulose. Once MBs are incubated with the electrode surface, nitrocellulose film is digested with the formation of holes within the film, resulting in a decrease of electrode capacitance (Figure 9). The chronocoulometry was measured for the determination of an electric charge, which was proportional to HER2 in the concentration window of 10^{-15}–10^{-10} M HER2 with LOD of 1 fM and with an overall assay time within 3 h. HER2 spiked into serum samples was detected with a recovery index of (109 ± 3)% [100] (Approach 1).

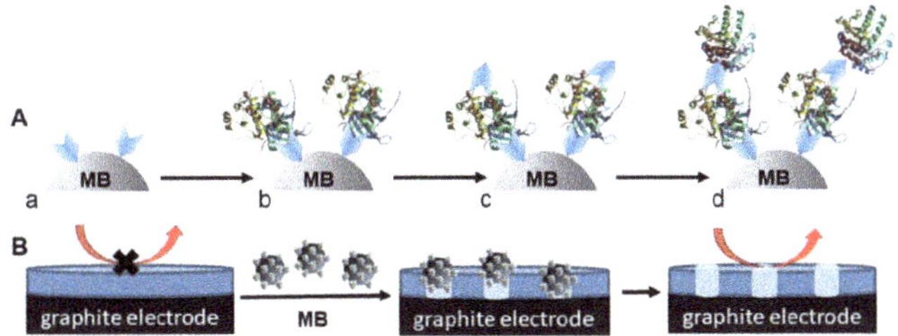

Figure 9. Schematic representation of (**A**) sandwich assembly on magnetic beads (MB) modified (**a**) with either an antibody or an aptamer, (**b**) binding of human epidermal growth factor receptor-2 (HER2)/neu via biorecognition to MB and the further reaction of MB (**c**) with a second antibody or an aptamer and (**d**) binding of the biotinylated cellulase label, through the biotin–streptavidin interaction. The protein structures' PDB (Protein Data Bank) IDs are: 3PP0 (HER-2/neu; DOI:10.2210/pdb3PP0/pdb) and 4IM4 (cellulase; DOI:10.2210/pdb4IM4/pdb). (**B**) Basic principle of the biosensor operation: electrochemically insulating nitrocellulose film on the surface of porous spectroscopic graphite is digested by MBs only when the analyte is present on the MB by cellulase, and that changes the electrochemical properties of the nitrocellulose-modified graphite surface. Reprinted from Ref. [100], Copyright (2019), with permission from Elsevier.

A DNA-based biosensor for the detection of HER2 was designed by the modification of GE with a DNA tetrahedron containing an aptamer against HER2 [101]. An electrochemical signal was generated by a signal probe consisting of gold nanorods with deposited PdNPs (5 nm), anti-HER2 aptamer, and HRP. Upon interaction of the modified electrode with HER2, a sandwich configuration was completed by a final incubation of the electrode with the signal probe. The biosensor could detect the analyte with LOD of 0.15 ng/mL and within a linear range from 10 to 200 ng/mL. Finally, the biosensor was applied for the analysis of HER2 spiked into serum samples [101] (Approach 2).

Lah and co-workers constructed a sandwich immunosensor for HER2 detection based on PbS quantum dots (QDs)-conjugated secondary anti-HER2 antibody as a signal probe [102]. Firstly, PbS QDs were synthesized, and anti-HER2 antibodies were attached to them. The application of QDs provided advantageous features such as a straightforward synthesis and well-defined electrochemical stripping signal of Pb(II) through acid dissolution. Primary anti-HER2 antibodies were immobilized onto pre-treated activated SPCE to capture the analyte. In the final incubation step, the signal probe formed a sandwich configuration. The biosensor could detect analyte down to 0.28 ng/mL with a linear calibration range up to 100 ng/mL. The biosensor was tested for the analysis of HER2 spiked into serum samples with a recovery index in the range from 91% to 104% [102] (Approach 2).

In the next work of Freitas et al., primary anti-HER2 antibodies were immobilized on MBs (Figure 10) [97] (Approach 2). The whole immunocomplex sandwich was formed directly in the solution phase, and then it was magnetically transferred to the electrode surface. The total assay time was 205 min with LOD down to 2.8 ng/mL. The biosensor was applied for the analysis of HER2 spiked into serum samples with a recovery index of 95%–99% [97].

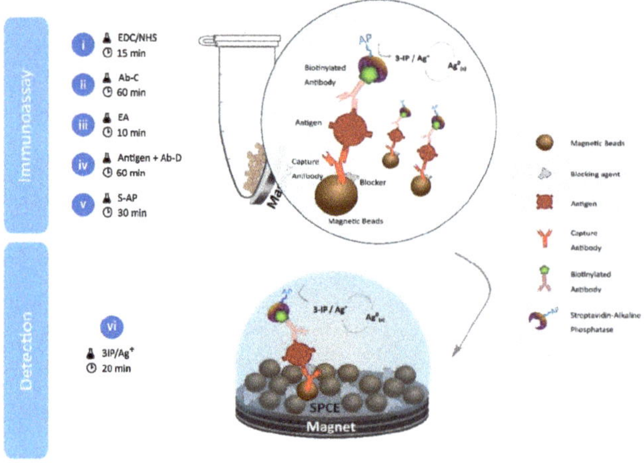

Figure 10. Graphical representation of operation of magnetic bead-based immunoassay. Reprinted from Ref. [97], Copyright (2020), with permission from Elsevier.

The advanced approach was achieved employing the core/shell CdSe@ZnS QDs as an electroactive detection probe for HER2 biosensing, requiring a total time assay of 2 h [98]. The sandwich configuration was formed on the SPCE involving primary and secondary anti-HER2 antibodies. The biosensor required only 40 µL of a sample volume with an LOD down to 2.1 ng/mL. The biosensor was applied for the analysis of HER2 spiked into serum samples with a recovery index between 104% and 106% [98] (Approach 2).

In next paper, the authors combined magnetic beads and core/shell streptavidin-modified CdSe@ZnS QDs as an electroactive detection probe for the affinity-based detection of HER2 with LOD of 0.29 ng/mL (a linear range of 0.50–50 ng/mL) [99]. The device was applied for the detection of HER2 spiked into serum samples with a recovery index of 100%–108%, assay time of 90 min, and with a good agreement with the reference ELISA method, which took 285 min to complete [99] (Approach 2).

Hartati et al. used a bioconjugate prepared by the covalent immobilization of anti-HER2 antibodies onto cerium oxide NPs previously modified by 3-aminopropyl trimethoxysilane (APTES) and polyethylene glycol-α-maleimide-ω-NHS (PEG–NHS–maleimide) [103]. Then, such a bioconjugate was covalently attached to SPCE modified by AuNPs. The interaction of HER2 with the modified electrode was analyzed by CV with a decrease of the peak current in the presence of the analyte (a signal-off approach). The biosensor could detect HER2 with an LOD of 34.9 pg/mL. The biosensor was finally used for the analysis of HER2 spiked into serum samples with a recovery index close to 100% [103] (Approach 3).

4.5. Detection of Carcinoembryonic Antigen (CEA)

CEA (180–200 kDa) is a glycoprotein that participated in cell adhesion. Normally, it is expressed by normal fetal intestinal tissue, and after birth, its expression is inhibited. The serum level can be increased in non-malignant diseases such as inflammatory bowel disease and also in many types of human cancers, such as gastric cancer, breast cancer, ovarian cancer, lung cancer, pancreatic cancer, and colorectal cancer [104].

Wang et al. developed a label-free aptasensor based on an electrochemiluminescent (ECL) strategy with ZnS–CdS NP-decorated molybdenum disulfide (MoS_2, a 2D nanomaterial [105]) nanocomposite for CEA detection [106]. The GCE was firstly modified with layered MoS_2 as an electrode matrix, and then ZnS–CdS NPs were electrodeposited directly onto MoS_2/GCE. In the next step, chitosan and glutaraldehyde covered the electrode for the immobilization of an anti-CEA aptamer. The aptasensor was completed by a final incubation with BSA to suppress non-specific interactions. The ECL aptasensor showed a linear range from 0.05 to 20 ng/mL with an LOD of 0.031 ng/mL. CEA spiked into human serum was analyzed with a recovery index in the range from 80% to 111%. The method was also applied for the determination of CEA in 8 human serum samples with an excellent agreement with a reference analytical method, showing the clinical application of the approach [106] (Approach 1).

Paimard with co-workers developed an immunosensor for CEA detection based on the CPE surface covered by the core–shell nanofibers prepared by electrospinning [107]. A nanofiber was made of honey (a core) electrospun with polyvinylalcohol (a shell) formed by a coaxial approach. Electrospun nanofibers were decorated with AuNPs and MWCNTs. Subsequently, anti-CEA antibodies were immobilized on the electrode surface. The impedimetric immunosensor exhibited high sensitivity toward the CEA biomarker with LOD of 0.09 ng/mL and with a linear range up to 125 ng/mL. The biosensor was applied for the analysis of CEA in human serum samples. Significantly higher levels of CEA were found in the serum samples of cancer patients compared to control, which was also verified using ELISA [107] (Approach 1).

Wang with colleagues employed flower-like Ag/MoS_2/RGO nanocomposites deposited onto GCE for CEA label-free detection with LOD of 1.6 fg/mL through the electrocatalytic H_2O_2 reduction [108]. Firstly AgNPs and GO were synthesized by a seed-mediated Lee–Meisel method and by an improved Hummer's method, respectively. Next, MoS_2/RGO was synthesized by applying $Na_2MoO_4\bullet 2H_2O$ and thiourea to obtain the final Ag/MoS_2/RGO nanocomposite. Anti-CEA antibodies were conjugated to the surface of AgNPs via amino groups for CEA determination in a wide concentration range from 0.01 pg/mL to 100 ng/mL. The analysis of CEA spiked into serum samples revealed a recovery index that was very close to 100% [108] (Approach 1). Another electrochemical platform for the detection of CEA using H_2O_2 reduction was developed by Su et al. [109]. Two-dimensional nanomaterial MoS_2 was modified by Prussian blue NPs, and such a hybrid nanomaterial was then deposited on GCE. The biosensor was finalized by the covalent immobilization of anti-CEA antibodies with subsequent surface blocking by BSA. CEA determination through the non-enzymatic detection of H_2O_2 offered LOD of 0.54 pg/mL (a linear range from 0.005 to 10 ng/mL). When the biosensor was applied for the detection of CEA spiked into human serum samples, a recovery index from 95% to 102% was obtained [109] (Approach 1).

Another sandwich-type electrochemical immunosensor for the determination of CEA was based on SPCE modified by AgNPs and RGO to which primary anti-CEA antibodies were immobilized [110]. After the electrode interface was incubated with an analyte, secondary anti-CEA antibodies labeled with HRP were added to complete a sandwich configuration, and a reduction of H_2O_2 was detected electrochemically. The modified SPCE-based biosensor detected CEA with LOD down to 0.035 μg/mL (a linear range of 0.05–0.50 μg/mL) [110] (Approach 1).

Rizwan et al. applied a layer-by-layer deposition of AuNPs, carbon nano-onions, SWCNTs, and chitosan layers onto GCE for the construction of a CEA immunosensor [111]. SWV was applied as an output signal in the presence of a soluble redox probe, and the device offered a linear range from 100 fg/mL to 400 ng/mL with LOD of 100 fg/mL for the detection of CEA. Only one serum sample spiked with three different CEA concentrations was applied for a clinical evaluation of the biosensor with recovery index in the range of 105%–110% [111] (Approach 1).

Wang and Hui [112] utilized the zwitterionic poly (carboxybetaine methacrylate) as a superhydrophilic matrix for the immobilization of anti-CEA antibodies and also as a layer resisting non-specific interactions. GCE was modified via electrodeposition by polyaniline nanowires, which were then activated to covalently graft zwitterionic monomers to the interfacial layer. In the subsequent step, a polymeric form of the zwitterions was prepared using UV irradiation. Finally, anti-CEA antibodies were immobilized via amine coupling. CEA concentration in the range from 1.0×10^{-14} g/mL to 1.0×10^{-10} g/mL with LOD of 3.05 fg/mL was determined by a DPV method. Four serum samples were analyzed by the biosensor, with the CEA values obtained being in excellent agreement with the reference ECL method, and when CEA was spiked in serum samples, a recovery index between 94% and 104% was obtained [112] (Approach 1).

Kumar et al. [113] functionalized ultrathin 2D nanomaterial Ti_3C_2 MXene nanosheets with aminosilane for the covalent immobilization of anti-CEA antibodies for ultrasensitive CEA detection with LOD down to 18 fg/mL (Figure 11). The label-free biosensor exhibited a linear detection range of 0.0001–2000 ng/mL using a soluble redox probe $[Ru(NH_3)_6]^{3+}$. The biosensor was applied for CEA detection when spiked into a human serum sample with a recovery index from 99% to 101% [113] (Approach 1).

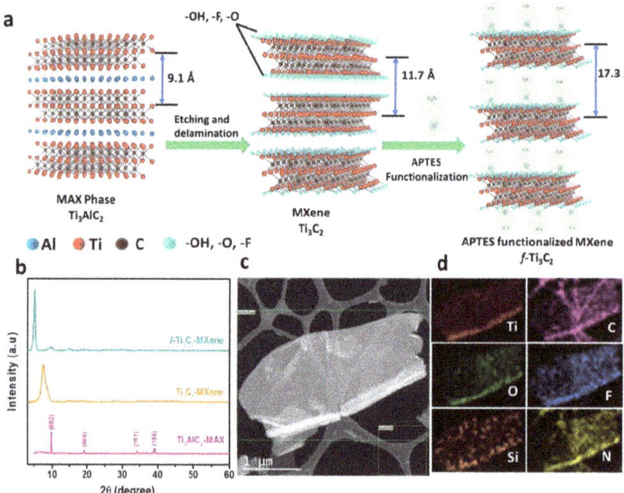

Figure 11. (**a**) Schematic illustration of MXene functionalization. Aluminum layered is etched from the Ti_3AlC_2-MAX phase, producing 2D nanosheets terminated with –OH, –O, and –F functional groups. Aminosilane (APTES) is then used to functionalize the MXene surface. (**b**) XRD pattern of the Ti_3AlC_2-MAX, MXene and functionalized MXene. (**c**,**d**) TEM image of a single MXene sheet and corresponding elemental map showing the uniform distribution of silicon (Si, brown), oxygen (O, green), and nitrogen (N, dark yellow), and revealing the homogenous functionalization of MXene with APTES. Reprinted from Ref. [113], Copyright 2018, with permission from Elsevier.

Yang and co-workers utilized a label-free amplification strategy based on an Au-Ag/RGO nanohybrid prepared using dopamine as a reducing agent, which was deposited on GCE [114]. Such a modified electrode was then used for the immobilization of anti-CEA antibodies. CEA was detected by the decrease of an electrochemical signal due to the oxidation of AgNPs present on a signal probe upon incubation with an analyte with LOD of 0.286 pg/mL (a linear range from 0.001 ng/mL to 80 ng/mL). A serum sample spiked with 3 different CEA concentrations was successfully analyzed by the biosensor with a recovery index of 96%–107% with excellent agreement with an ELISA method [114] (Approach 1).

Gu et al. [115] integrated ferrocene (Fc) derivative and AuNPs into their biosensor for CEA detection in order to increase the conductivity of the sensing surfaces and increase ferrocene loading. Firstly, AuNPs were reduced from chloroauric acid with trisodium citrate as a reducing agent, and subsequently, polyclonal secondary anti-CEA antibodies were immobilized onto their surface via physisorption. Further chemisorption of the electroactive ferrocene molecules in a form of thiolated ferrocene chains was accomplished on AuNPs. Finally, PEG8000 was applied to stabilize AuNPs, and repeated centrifugation was applied to remove excess antibodies and Fc, and such a bioconjugate was applied as a signal probe. Pre-treated GEs were first modified with lipoic acid *N*-hydroxysuccinimide ester to attach primary antibodies, and the surface was blocked with ethanolamine. After CEA was affinity captured on the modified electrode, the sandwich configuration was completed by incubation with a signal probe. The developed biosensor exhibited LOD of approximately 0.01 ng/mL (a linear range up to 20 ng/mL), when detecting CEA using a SWV method with a good performance after storage for 3 weeks (91.8% of the original response) [115] (Approach 2).

Wei et al. developed an electrochemical ratiometric method for CEA detection [116]. The method was based on an AuNPs functionalized Cu_2S-CuS/graphene composite as a SPCE-modifying nanomaterial to which primary anti-CEA antibodies were immobilized. A signal probe was developed using CeO_2 NPs modified by deposited AuNPs to which secondary anti-CEA antibodies and toluidine blue (TB) as a redox mediator were covalently immobilized. The adsorption capacity toward toluidine blue was improved with carboxymethyl chitosan (CMC)-doped ionic liquids containing active groups such as –OH, –COOH, and –NH$_2$. The change of dual signals "$\Delta I = \Delta I_{TB} + |\Delta I_{Cu2S\text{-}CuS}|$" ($\Delta I_{TB}$ and $|\Delta I_{Cu2S\text{-}CuS}|$ present the change values of the oxidation peak currents of toluidine blue and Cu_2S-CuS, respectively) was applied as the response signal for the quantitative determination of CEA with LOD of 0.78 pg/mL (a linear range of 0.001–100 ng/mL) (Figure 12). The biosensor was applied for the analysis of CEA in one serum sample, and the CEA level found out by the biosensor device was in an excellent agreement with an ELISA method [116] (Approach 3).

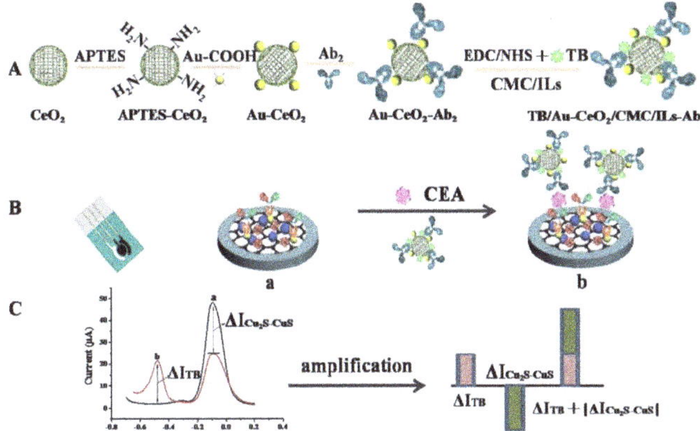

Figure 12. (**A**) Formation of a signal probe. (**B**) Schematic presentation of the immunosensor fabrication. (**C**) A dual-signaling amplification strategy. Reprinted from Ref. [116], Copyright (2018), with permission from Elsevier.

Another type of a label-based sandwich-type electrochemical immunosensor for CEA determination was developed by Li et al. [117], who used amino functionalized magnetic graphene sheets loaded with Au@Ag core–shell NPs to adsorb Ni^{2+} and secondary anti-CEA antibodies as a signal probe to reduce H_2O_2. AuNPs electrodeposited from $HAuCl_4$ solution onto GCE improved the immobilization of primary anti-CEA antibodies and the device exhibited an LOD of 0.07 pg/mL (a linear range from 0.1 pg/mL to 100 ng/mL). The biosensor offered a recovery index close to 100% for the determination of CEA spiked into serum samples [117] (Approach 3).

A similar strategy based on the application of multiple types of nanoparticles for the detection of CEA was also applied by Wu et al. [118]. GCE was patterned by aminated-graphene sheets to which primary anti-CEA antibodies were covalently immobilized using glutaraldehyde. A signal probe was made of magnetic NPs covered by a shell made of a MnO_2 layer with a deposition of PtNPs to which secondary anti-CEA antibodies were immobilized. CEA was detected with an LOD of 0.16 pg/mL in a linear range from 0.5 pg/mL to 20 ng/mL. Serum samples spiked with different CEA concentration provided reliable results with a recovery index from 95% to 106%, and the assay was validated using an ELISA method [118] (Approach 3).

4.6. Dual-Target Analysis

The dual-target detection of miRNA-21 and MUC1 based on a dual catalytic hairpin assembly was performed by Li and co-workers [119]. GCE was modified by Au nanoflowers to which hybridization probe 1 was immobilized to recognize miRNA-21. After incubation with miRNA-21, the electrode was incubated with a hybridization probe 2 conjugated with QDs, resulting in an increase of ECL signal (Cycle I, Figure 13). When such an electrode was incubated with anti-MUC1 aptamer and MUC1, both molecules were attached to the modified electrode surface. Incubation with a hybridization probe 3 conjugated with AuNPs in the subsequent step replaced the anti-MUC1 aptamer from the electrode surface and due to a fluorescence resonance energy transfer between QDs and AuNPs, a decrease of ECL signal was observed (Cycle II, Figure 13). The biosensor detected miRNA-21 with LOD of 11 aM and MUC1 with LOD of 0.40 fg/mL. When both analytes were spiked into human serum samples, a recovery index between 98% and 103% was obtained [119] (Approach 3).

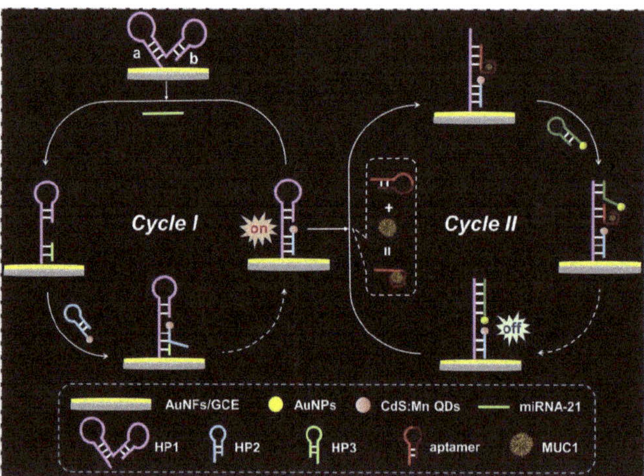

Figure 13. The principle of the fabricated biosensor for the sensitive detection of miRNA-21 and MUC1 based on dual catalytic hairpin assembly. Reprinted from Ref. [119], Copyright 2018, with permission from Elsevier.

4.7. Detection of Other Potential BC Biomarkers

In the next part, we will focus on an electrochemical performance for the detection of less known biomarkers present in the serum of BC patients.

Cancer antigen 27.29 (CA27.29, 250–1000 kDa) is a soluble form of glycoprotein MUC1. It is expressed mainly in BC, but CA 27.29 levels can also be elevated by colon, stomach, kidney, lung, ovary, pancreas, and liver cancers as well as other non-cancerous conditions such as benign breast disease, kidney, and liver diseases [120]. Alarfaj et al. constructed a label-free electrochemical immunosensor based on an Au/MoS$_2$/RGO nanocomposite system [121]. First, a hybrid Au/MoS$_2$/RGO nanocomposite was deposited on the GCE surface. Then, anti-CA 27-29 antibodies were immobilized on the modified electrode surface for selective capture of the analyte via affinity interactions. A signal amplification strategy was achieved by a synergy of all nanomaterial components of the nanocomposite to reduce H$_2$O$_2$. The biosensor could detect analyte down to an LOD of 0.08 U/mL. The device was finally applied for analysis of the analyte in 25 human serum samples with an excellent agreement with an ELISA method, and when CA27.29 was spiked into serum samples, an excellent recovery index of 96%–100% was obtained [121] (Approach 1).

Urokinase-type plasminogen activator receptor (uPa) belongs to cell membrane receptors with their expression increased in a number of different types of human cancers, including BC [122]. An immunosensor based on fluorine-doped tin oxide was modified with graphene nanosheets to enhance the loading of covalently immobilized antibodies [122]. The immunosensor could detect the analyte down to 4.8 fM using DPV assays in the presence of a soluble redox probe. The device offered a good stability (75% of an initial activity observed after 4 weeks) with the ability to detect an analyte spiked into serum samples [122] (Approach 1).

Tissue plasminogen activator (tPa, 20–45 kDa) belongs to serine proteases (enzymes ensuring cleaving peptide bonds in proteins). As a result of this fact, the protein is essential in the human body in relation to angiogenesis in cancer cells [123]. The protein was detected with LOD of 0.026 ng/mL in a linear range from 0.1 to 1.0 ng/mL [124]. A label-free biosensor was fabricated by the functionalization of SWCNTs with antibodies immobilized, and such a bionanoconjugate was subsequently immobilized onto a GCE surface (Figure 14) [124] (Approach 1).

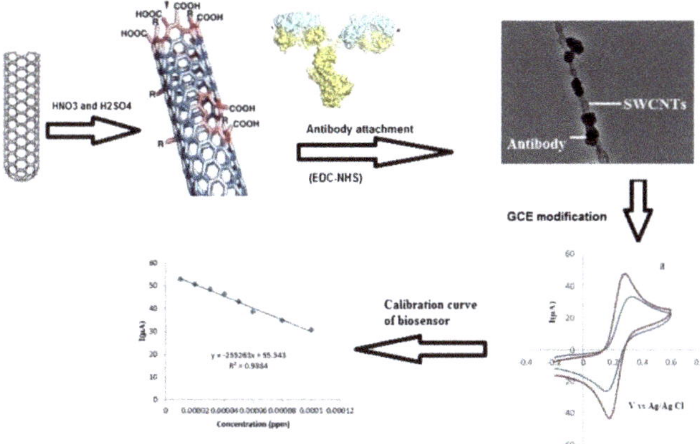

Figure 14. Fabrication of the biosensor for the determination of tissue plasminogen activator. Reprinted by permission from Springer, Ref. [124], Copyright 2018.

4.8. Detection of BC Cells

Circulatory tumor cells (CTC) are released from tumors and circulate in the bloodstream at a low concentration of up to 10 cells/mL, while the whole blood contains 10^9 erythrocytes and 10^6 leucocytes/mL [30]. This is why the detection of CTC is quite challenging.

The detection of CTCs, which are present in the blood at a very low level, is highly challenging and has not been done using affinity-based approaches. In the following text, we discuss some detection principles for the analysis of BC cells with some approaches potentially applicable for the analysis of CTCs. More details about the electrochemical detection of BC cells can be found elsewhere [125].

The Michigan Cancer Foundation-7 (MCF7) cell line is the most frequently studied BC cell line [126], since it is a suitable model for studying the development/progression of BC and anticancer drug therapies. The cells are non-invasive, expressing estrogen as well as progesterone receptors [127].

An interesting method for the electrochemical detection of CTCs within a microfluidic channel was proposed by Gurudatt et al. (Figure 15) [128]. Cells differing in their size, surface charge, and chemical state on the cell surface were effectively separated using such a device. In order to detect CTCs in an effective way, the surface of channels was chemically modified with an electrochemical polymerization of a monomer. In the subsequent step, a lipid layer by the deposition of phosphatidylserine was formed on the surface of the channels. In order to electrochemically detect cells, such cells were loaded with daunomycin prior to separation. Three different types of cancer cell lines were used for optimization of the assay, and optimized assay conditions allowed detecting single cells (approximately 7 cells/mL). Finally, the device was applied for the detection of CTCs from 37 cancer patients with (92.0 ± 0.5)% efficiency. The results showed differences in the retention time for different types of CTCs produced by different cancer types, suggesting differences in the size, surface charge, and chemical state on the cellular surface [128]. Another microfluidic electrochemical approach for the detection of CTCs was based on the measurement of changes in the impedance of the polydimethylsiloxane-based channel on a glass slide during the passage of CTCs [129]. A narrow constriction-based sensor was designed in a way allowing the passage of red and white blood cells without any restrictions, while much larger tumor cells needed to squeeze/deform in order to pass through the channel, causing changes in the impedance of the channel. As a result, only cancerous cells were able to generate an electrochemical signal in a label-free format, while smaller blood cells did not generate any measurable signal (Figure 16). The device was tested by an analysis of murine blood spiked with prostate or breast cancer cells with a throughput of 1 µL *per* min, but the throughput can be increased by analysis run in parallel. A signal processing of data generated was done automatically using MATLAB. The authors claim that false positive results can be obtained due to the presence of non-blood or non-cancer cells in blood such as epithelial cells, and this why the pre-enrichment of CTCs was suggested [129].

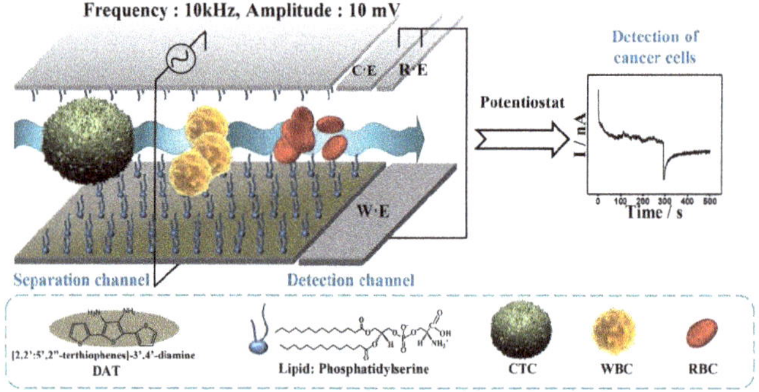

Figure 15. The schematic representation of the design and fabrication of the proposed microfluidic channel. Reprinted from Ref. [128], Copyright 2019, with permission from Elsevier.

Anti-MUC1 aptamers, hybrid AuNPs, and carbon dots (Au@CDs) modifying GE were applied for the label-free ECL detection of circulating MCF-7 cells (MCF-7 CTCs) [130]. The biosensor detected MCF-7 CTCs down to 34 cells/mL with a linear range up to 10,000 cells/mL. MCF-7 cells spiked into serum samples were in addition clinically tested with an obtained recovery index of 93–117% [130] (Approach 1). Tian et al. [131] investigated MCF-7 CTCs by using a supporting RGO/AuNPs composite deposited on GCE with a catalytic CuO nanozyme used as a signal probe (Approach 3). MCF-7 CTCs membranes contain specific a MUC1 protein, which was recognized by the MUC-1 aptamer. The reached LOD was as low as 27 cells/mL (a linear range from 50 to 7×10^3 cells/mL). MCF-7 cells were further successfully studied and determined by applying aptamer-based electrochemical biosensors [132]. A DNA aptamer was immobilized onto AuNPs supported by α-cyclodextrin on GE (Approach 1). The aptasensor determined MCF-7 cells in the range of 328–593 cells/mL with LOD of 328 cells/mL, when cells were lysed and an intracellular level of platelet-derived growth factor was electrochemically determined [132]. Yang with colleagues [133] utilized GCE modified by several nanomaterials using a layer-by-layer deposition process incorporating 3D graphene, Au nanocages, and MWCNTs to which primary antibodies were immobilized (Approach 1). Once the biosensor was incubated with MCF-7 cells, a sandwich configuration was established by incubation with secondary antibodies linked to DNA. In the next step, complementary DNA was applied, and to double-stranded DNA, methylene blue as a redox mediator was intercalated and detected using SWV. The biosensor detected BC cells with LOD of 80 cells/mL (a linear range of 1.0×10^2–1.0×10^6 cells/mL) and exhibited satisfactory stability [133]. Wang et al. [134] fabricated a sensitive sandwich-based aptamer biosensor for the label-free electrochemical detection of cells (Approach 2). The sensor was based on GE modified with polyadenine (polydA)-aptamer recognizing MUC1 on the surface of cells. A signal probe was designed by the immobilization of an aptamer recognizing MUC1 protein on an AuNP/GO hybrid nanomaterial. MCF-7 cells were recognized by polydA-aptamer, and then, a sandwich configuration was completed by incubation with a signal probe. BC cells were detected via a DPV method with LOD of 8 cells/mL and a linear range from 10 to 10^5 cells/mL using a soluble redox probe with satisfactory selectivity [134]. An interesting approach in order to differentiate between different BC cell lines was based on the detection of H_2O_2 produced by the cells [135]. A sandwich consisting of synthesized Bi_2Se_3 NPs as 3D topological insulators between the gold electrode and another Au-deposited thin layer was designed by Mohammadniaei et al. as a nanostructured working electrode. In order to detect H_2O_2 in an ultrasensitive fashion, the immobilization of double-stranded DNA loaded with Ag^+ ions was established through the Au–thiol interaction of thiolated DNA (Figure 17) (Approach 1). The developed biosensor showed LOD of 10×10^{-9} M for H_2O_2 with a dynamic range from 0.10×10^{-6} M to 27.3×10^{-6} M and a short response time of 1.6 s. The biosensor could distinguish the MCF-7 cell line from the MDA-MB-231 cell line based on the H_2O_2 produced [135].

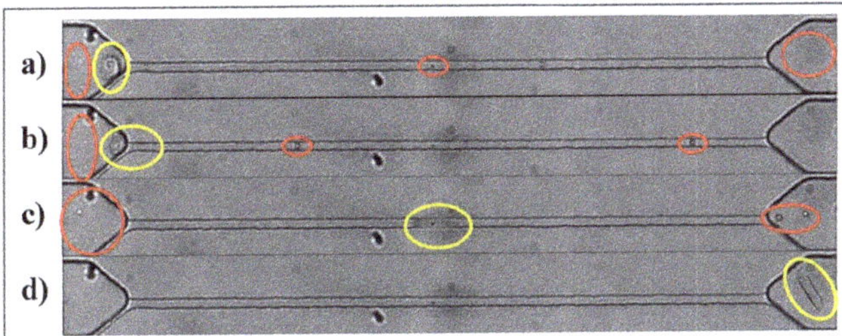

Figure 16. Transit of an MDA-MB-231 cell (circled in yellow) through the constriction region. (**a**) Cancer cell before deformation, (**b**) cell beginning to deform, (**c**) cell in constriction channel, and (**d**) cell after leaving the constriction channel. The surrounding white and red blood cells are indicated by the red circles. Reprinted from Ref. [129], Copyright 2020, with permission from Elsevier.

In the next study, a H_2O_2 sensor employing a trimetallic AuPtPd nanocomposite and RGO nanosheets deposited on GCE was applied for the electrocatalytic detection of H_2O_2 reduction with LOD of 2 nM (a linear range from 0.005 µM to 6.5 mM) [136]. The biosensor was applied for the detection of H_2O_2 released by two BC cell lines (MDA-MB-231 and T47D) [137] (Approach 1).

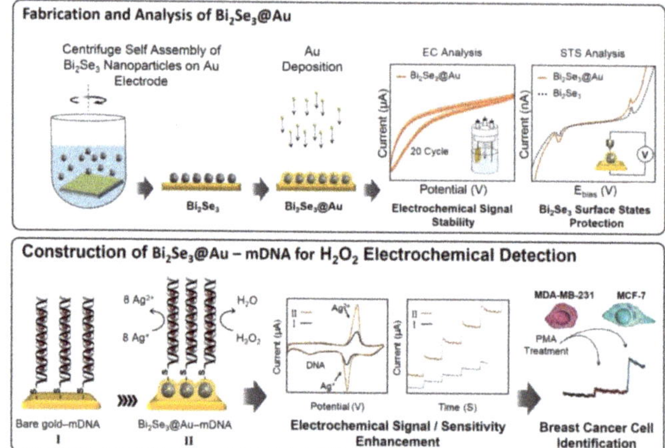

Figure 17. Fabrication process of a Bi$_2$Se$_3$@Au-DNA electrode, its electrical and electrochemical behavior, and the constitution for the electrochemical detection of H$_2$O$_2$. Reprinted from Ref. [135], Copyright 2018, with permission from John Wiley and sons.

Luo et al. applied hexagonal carbon nitride tubes as a photoactive material to determine the photocurrent in the presence of MCF-7 cells (Approach 1) [137]. The cells were detected down to 17 cells/mL (a linear range from 100 to 1×10^5 cells/mL). Glutaraldehyde was utilized as a cross-linker for the covalent immobilization of anti-MUC1 aptamers for the affinity capture of cells via surface-expressed MUC1. A clinical applicability of the biosensor was proved by the detection of cells spiked into blood samples at three different concentrations with a recovery index of 96%–104% [137].

Safavipour et al. developed an aptasensor using a hybrid nanomaterial composed of TiO$_2$ nanotubes attached to GO via UV irradiation [138]. GCE was modified by such a hybrid nanomaterial with the subsequent immobilization of anti-MUC1 aptamers for the affinity capture of MCF-7 cells via surface-expressed MUC1 proteins. An EIS-based device was able to ultrasensitively detect MCF-7 cells with LOD of 40 cells/mL within a linear concentration range from 10^3 to 10^7 cells/mL [138] (Approach 1).

GE modified by non-spherical AuNPs was made by electrodeposition in the presence of a shape-controlling agent for achieving an increased electrode active area [139]. A thiolated aptamer recognizing BC cells MDA-MB-231 was chemisorbed on the modified electrode surface, and the cells were electrochemically detected down to 2 cells/mL in an electrolyte using a soluble redox probe. The device was also applied for the analysis of cells spiked into blood serum samples with LOD of 5 cells/mL [139] (Approach 1).

Two immunomagnetic biosensors, which were described in Section 4.4. "Detection of HER2", were also applied for the detection of BC cells [97,99]. The first biosensor was applied for the determination of two BC cell lines: HER2$^+$ SK-BR-3 and HER2$^-$ MDA-MB-231 via surface-expressed HER2 proteins using Ag ions and 3-indoxyl phosphate for a signal generation [97] (Approach 2). The biosensor could detect cells in the linear range of 100–10,000 cells/mL with an LOD of 3 cells/mL [97]. Another immunomagnetic biosensor was applied for detection of the same BC cell lines as the first one [99] and an additional MCF-7 (a cell line with low HER2 expression) cell line via surface-expressed HER2 with a signal generated by a stripping voltammetry of Cd ions released from QDs (Approach 2). The selectivity toward SK-BR-3 cells was confirmed. A concentration-dependent signal that was 12.5× higher than the signal obtained for the HER2-negative cells (MDA-MB-231) and LOD of 2 cells/mL was obtained [99].

Cancer stem cells were discovered by Al-Hajj in 2003 [140]. Cancer stem cells were detected using a nanobiosensor with a thiolated aptamer against the CD44 surface protein immobilized on GE via chemisorption [141]. After stem-like cells were captured on the electrode surface, the electrode was modified by a self-assembled peptide-based multifunctional nanofiber containing CD44 binding protein and $-N_3$ groups, which were subsequently used for the clicking of AgNPs applied as a redox probe for an electrochemical signal generation. The LOD of the device was 6 cells/mL, and the device offered a wide linear range (from 10 cells/mL to 5×10^5 cells/mL). The selectivity of the device was successfully proved by the analysis of three different cancerous cell lines [141] (Approach 2).

4.9. Detection of Exosomes and Exosomal Content

Exosomes are characterized as endosome-derived vesicles involving the signal transduction within intercellular communication and in extracellular matrix remodeling. Exosomes are membrane-bound particles with a lipid bilayer structure carrying precious cargo: biomolecules that could be used as cancer biomarkers for more accurate cancer diagnostics in the future [51]. An increased number of exosomes circulating in body fluids was observed for cancer patients compared to healthy individuals, and a change in the exosome level can be applied as a diagnostic cancer biomarker on its own [51].

Kilic et al. fabricated a label-free electrochemical sensor to measure the increased release of nanoscale extracellular vesicles from the BC cell line, MCF-7, due to $CoCl_2$-induced hypoxia [142]. A pre-treated surface of AuSPE was modified with 11-mercaptoundenoic acid with the subsequent activation of –COOH groups for the attachment of neutravidin, which was applied for the immobilization of biotinylated anti-CD81 antibodies on the surface (Figure 18). Such a sensor was able to detect extracellular vesicles with LOD of 77 particles/mL or 379 particles/mL using EIS and DPV, respectively [142].

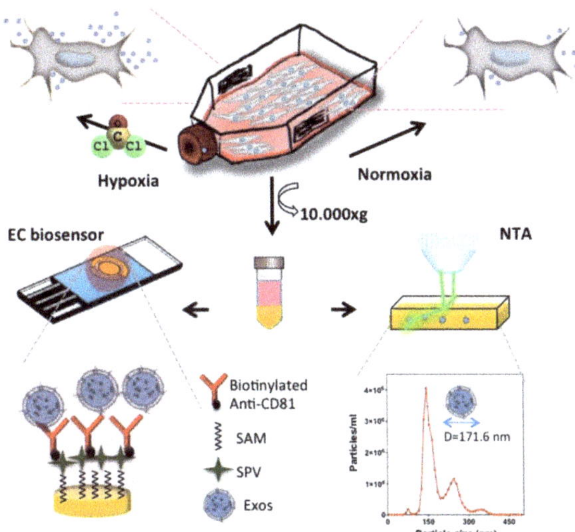

Figure 18. Experimental steps followed throughout the work. MCF-7 cells were exposed to either $CoCl_2$-induced hypoxic or normoxic conditions. The isolation of extracellular vesicles (EVs) was done via ultracentrifugation. The characterization and quantification of EVs were done via biosensors that are designed to capture exosomes on the electrode surface by antibodies raised against CD-81 protein expressed on the surface of exosomes. The gold electrode was patterned by thiolated SAM to which streptavidin was covalently immobilized, followed by the attachment of biotinylated anti-CD-81 antibodies. Reprinted with permission from Ref. [142]. Copyright 2018, Nature.

Exosomes released from 4 BC cell lines were detected using a magneto-mediated electrochemical sensor [143]. Magnetic beads were modified with anti-CD63 aptamer for the capture of exosomes.

The selective detection of 4 proteins (MUC1, HER2, EpCAM, and CEA) on the exosomal surface was achieved by using silica NPs modified with respective aptamers. Silica NPs were also functionalized using mercapto–ferrocene derivative. The sandwich structure magnetic beads–exosomes–SiNPs was separated using a magnet, and the ferrocene derivatives were released from the sandwich using dithiothreitol. Ferrocene derivatives released from the sandwich were electrochemically detected on SPCE modified by a GO layer. Using this approach, 4 different biomarkers on 4 different cell lines were sensitively detected (Figure 19) (Approach 2). The sensor was clinically tested by the analysis of expression profile of 4 proteins on exosomes isolated from one BC patient and one healthy individual, confirming statistically higher levels of all 4 exosomal proteins when using BC serum compared to the serum of a healthy individual [143].

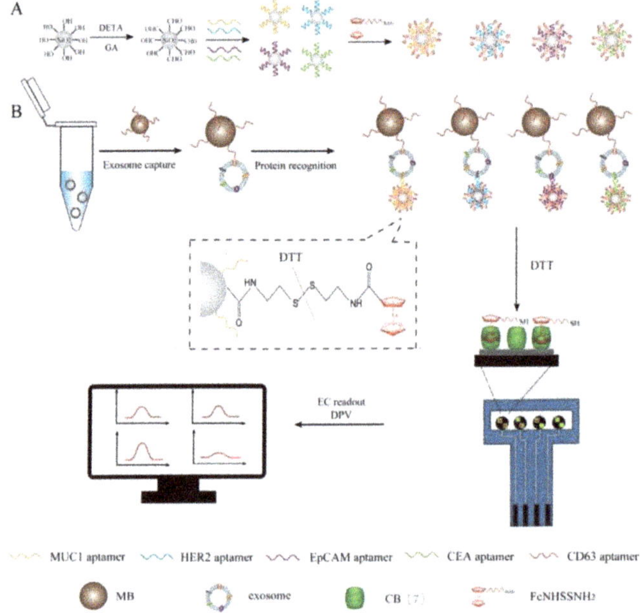

Figure 19. (**A**) Preparation of SiO2 NPs Probes; (**B**) Differential pulse voltammetry (DPV) responses of the magneto-mediated electrochemical sensor for MUC1, HER2, EpCAM, and carcinoembryonic antigen (CEA) markers. Magneto-mediated electrochemical sensor for exosomal proteins analysis based on host–guest recognition. Differential pulse voltammetry (DPV) responses of the magneto-mediated electrochemical sensor for MUC1, HER2, EpCAM, and carcinoembryonic antigen (CEA) markers for the MCF7, SKBR-3, MDA-MB-231, and BT474 cells-derived exosomes at a concentration of 1.2×10^6 particles/µL. Reprinted with permission from reference [143]. Copyright 2020, American Chemical Society.

Moura et al. prepared an electrochemical immunosensor for the detection of exosomes derived from three cell lines (MCF7, MDA-MB-231, and SKBR3) [144]. Exosomes were captured to MPs, which were modified with antibodies against general tetraspanins CD9, CD63, and CD81, as well as against specific receptors of cancer (CD24, CD44, CD54, CD326, and CD340) (Figure 20). Exosomes were immobilized on magnetic particles (MPs) in a direct and in an indirect format. The direct format was based on incubation of the exosomes–MP with the antiCD63-HRP antibodies with a final electrochemical signal readout. The indirect format was based on incubation of the exosomes–MPs with antiCDX mouse monoclonal antibodies (CDX being either CD9, CD24, CD44, CD54, CD63, CD81, CD326, or CD340 biomarkers) and the indirect labeling with antimouse–HRP antibodies. Hydroquinone was used as a mediator. The study also found out that there are differences in the size and amount of exosomes depending on the exosome origin. Moreover, the level and size distribution of exosomes purified from healthy individuals was strikingly different from the exosomes purified from BC patients. The approach offered LOD as low as 81 exosomes/µL. The method could be applied to distinguish exosomes from healthy donors and those isolated from BC patients [144] (Approach 2).

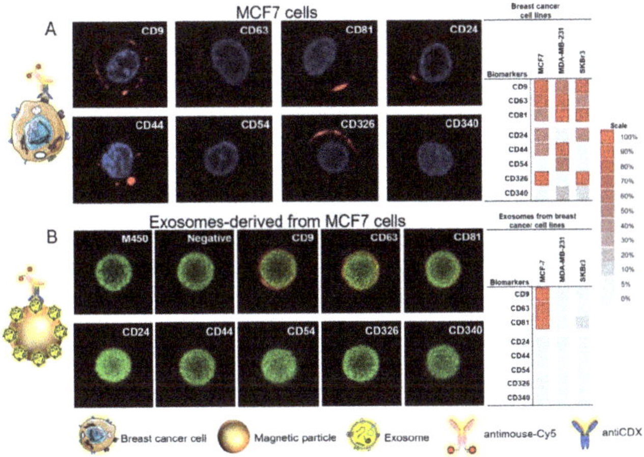

Figure 20. Confocal microscopy study for (**A**) MCF7 breast cancer cell lines and (**B**) their corresponding exosomes covalently immobilized on MPs (exosomes–MPs), followed by indirect labeling with mouse antiCDX antibodies (being CDX either CD9, CD24, CD44, CD54, CD63, CD81, CD326, and CD340 biomarkers) and antimouse-Cy5. The concentration of exosomes was set as 4×10^9 *per* assay. The scale indicates the percentage of positive entities (cells and exosomes-coated MPs in panels A and B, respectively). Reprinted from [144], Copyright 2020, with permission from Elsevier.

Luo et al. developed a ratiometric electrochemical DNA biosensor employing an immobilized locked nucleic acid (LNA)-modified in a form of a "Y" shape-like structure for the detection of miR-21 present in exosomes released by MCF-7 cell lines [145] (Approach 1). GCE was modified by a polylysine film to which a DNA probe 1 labeled with methylene blue was covalently attached followed by hybridization with a DNA probe 2 labeled with ferrocene. The DNA probe 2 was attached to the electrode in a way such that the ferrocene redox moiety was in proximity to the electrode surface, while methylene blue redox moiety was exposed to the solution phase (Figure 21). Upon binding of the analyte miRNA-21, a DNA probe 2 labeled with ferrocene was released from the electrode surface, leaving behind only a DNA probe 1 with a surface-confined methylene blue redox moiety. Thus, upon analyte binding, an increase ("signal-on" response) in the DPV signal for methylene blue was observed with a decrease ("signal-off" response) of the DPV peak attributed to ferrocene. Both single signal responses (i.e., "signal-on" response and "signal-off" response) exhibited significant signal variation, while a ratiometric signal was highly stable. When the biosensor response was expressed in a form of a ratiometric signal, the device offered LOD of 2.3 fM with a linear range from 10 to 70 fM. The biosensor exhibited high specificity with a negligible response obtained for single- and double-mismatched RNA sequences. The device showed high assay accuracy for the analysis of miRNA-21 released from exosomes produced by a MCF-7 cell line, which was confirmed by a reference analytical method [145].

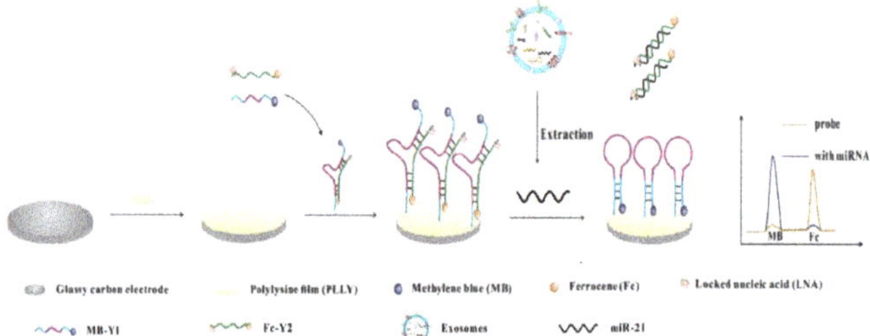

Figure 21. The ratiometric electrochemical biosensor for exosomal miR-21 detection. Reprinted from Ref. [145], Copyright 2020, with permission from Elsevier.

5. Conclusions and Perspectives

This review provides evidence that electrochemical detection principles can offer ultrasensitive detection platforms for the detection of various BC biomarkers with LODs in some cases down to the single molecule level thanks to the use of a wide range of nanoparticles (Table 3). Such biosensors discussed in this review were mainly based on modification of the electrodes by nanomaterials (i.e., Approach 1), followed by the design of signal probes based on nanomaterials (i.e., Approach 2) with the design of a hybrid approach using nanomaterials for the modification of electrodes and for the design of signal probes (i.e., Approach 3). In a significant number of studies, the clinical performance of the nanobiosensors was validated by using human serum samples. Unfortunately, only a minor fraction of papers was focused on the detection of true concentration of BC biomarkers in human serum, but rather analyte spiking into serum samples was applied to validate the clinical usefulness of the nanobiosensors developed. Furthermore, only a limited number of papers dealt with the validation of nanobiosensing by a standard reference method i.e., ELISA. Such comparison is really needed to verify the reliability of nanobiosensors for the analysis of cancer biomarkers in complex samples such as serum samples from BC patients.

In order to really verify the clinical usefulness of the nanobiosensing approach, a larger number of human serum samples divided into two cohorts—BC patients and healthy (non-cancerous) individuals—need to be analyzed for the level of BC biomarkers in order to see if a particular biomarker is present in serum samples of BC patients at a statistically higher level compared to serum samples of healthy individuals. At the same time, such a comparison needs to be evaluated in the form of an ROC (Receiver Operating Curve) with an AUC (Area Under Curve) determined, and only such information can be then applied for the direct comparison of a clinical performance of nanobiosensing with standard immunoassays. The other aspect worth investigating in the future is a multiplexed format of analysis, when several samples are run in a parallel or when several biomarkers are detected in a single sample in parallel. So far, there is only one study describing the simultaneous analysis of miRNA-21 and MUC1, and only one study that determined the level of 4 different proteins on an exosomal surface. Although the electrochemical sensing approach is an ideal tool for integration into lab-on-a-chip platforms, we have not identified any single study analyzing BC biosensors in such an advanced assay platform using electrochemical sensing.

It is also of the utmost importance to deal with the non-specific binding of proteins from complex samples especially in cases when an electrochemical signal readout is not done in a sandwich configuration and rather detects changes in the interfacial properties of the interfacial layer after incubation with a sample i.e., impedimetric signal reading.

Table 2. Key characteristics of electrochemical nanobiosensors for the detection of BC biomarkers.

Target Biomarker (Biomolecule)	Bare Electrode	Electrode Modification	Detection	LR	LOD	Refs.
BRCA 1	MBCPE	Fe_3O_4@Ag, DNA probe	EIS	100 aM–10 nM	30 aM	[63]
	GCE	RGO, MWCNTs, PANHS	CV, EIS	100 aM–10 nM	37 aM	[64]
PIK3CA gene	CPE	ssDNA/PIn6COOH/ MoS_2	CV, EIS	100 aM–10 pM	15 aM	[72]
	GCE	GO, GNR	EIS	15 fM–148 pM	4.5 fM	[77]
	GCE	ZrO_2-RGO	EIS	10 fM–100 pM	4.3 fM	[78]
MUC MUC	SPCE	CNTs	CV, EIS	0.1–2 U/mL	0.02 U/mL	[83]
	GCE	ferrocene-loaded polymeric micelle	CV	1–1000 cells/mL	10 cells/mL	[85]
	GCE	cDNA-Fc/MXene/Apt/Au	EIS, SWV	1.0 pM–10 mM	0.33 pM	[86]
CA15-3	GE	streptavidin-coated magnetic beads	CV, EIS	ND	15×10^{-12} U/mL	[87]
	GE	GO/Py-COOH, MWCNTs	DPV	0.1–20 U/mL	0.01 U/mL	[90]
HER2	ITO	APTES/MoO_3@RGO	CV, DPV, EIS	0.001–500 ng/mL	0.001 ng/mL (~5.41 fM)	[92]
	GCE	AuNP-ERGO-SWCNTs	EIS	0.1 pg/mL–1 ng/mL	50 fg/mL (~0.27 fM)	[93]
	SPGE	MIP	CV	10–70 ng/mL	1.6 ng/L (~8.65 fM)	[95]
	GE	GNR@Pd SSs—Apt—HRP	EIS	10–200 ng/mL	0.15 ng/mL (~0.81 pM)	[101]
	SPCE	MBs and CdSe@ZnS QDs	DPASV	0.50–50 ng/mL	0.29 ng/mL (~1.57 pM)	[99]
CEA	GCE	aptamer/GLD/CS/ZnS-CdS/MoS_2	CV	0.05–20 ng/mL	0.031 ng/mL (~0.16 pM)	[106]
	CPE	GNPs and MWCNTs.	CV, EIS	0.4–125 ng/mL	0.09 ng/mL (~0.45 pM)	[107]
	GCE	Au-AgNPs/RGO	CV	0.001–80 ng/mL	0.29 pg/mL (~1.45 fM)	[114]
miRNA-21 and MUC1	GCE	Au nanoflowers	ECL	20 aM–50 pM (miRNA-21) 1 fg mL^{-1}–10 ng mL^{-1} (MUC1)	11 aM (miRNA-21) 0.4 fg/mL (~7.27 aM) (MUC1)	[119]
CA 27-29	GCE	Au/MoS_2/RGO	CV	0.1–100 U/mL	0.08 U/mL	[121]
uPA	FTO	GNS	DPV, CV	1 fM–1 μM	4.8 fM	[122]

Table 3. Key characteristics of electrochemical nanobiosensors for the detection of BC biomarkers.

Target Biomarker (Biomolecule)	Bare Electrode	Electrode Modification	Detection	LR	LOD	Refs.
tPA	GCE	SWCNTs	CV, EIS	0.1–1.0 ng/mL	0.026 ng/mL (~0.37 pM)	[124]
MCF-7/CTC		RGO/AuNPs/CuO	CV, CA	50–7000 cells/mL	27 cells/mL	[131]
MCF-7	GCE	Au NCs/amino-functionalized MWCNT-NH$_2$	CV, EIS	100–1.0 × 10^6 cells/mL	80 cells/mL	[133]
	GE	Bi$_2$Se$_3$@Au-mDNA	CV, EIS	100 nM–27 μM	10 nM	[135]
	GCE	Hexagonal carbon nitride tubes	Photo-current	100–1 × 10^5 cells/mL	17 cells/mL	[137]
	GCE	TiO$_2$ nanotubes with graphene	EIS	1000–1 × 10^7 cells/mL	40 cells/mL	[138]
MDA-MB-231	GE	Non-spherical AuNPs	DPV	10–1 × 10^6 cells/mL	2 cells/mL	[139]
Cancer stem cells	GE	AgNPs	DPV	10–5 × 10^5 cells/mL	6 cells/mL	[141]
Exosomes Exosomal miRNA-21	Au SPE	11-MUA	EIS, DPV	10^2–10^9 particles/mL	77 particles/mL	[142]
	SPCE	MB SiO$_2$ NPs	DPC, EIS	1.2 × 10^3–1.2 × 10^7 particles/μL	1.0 × 10^7 particles/μL	[143]
	m-GEC magnetic	MPs	Ampero-metry	0–1 × 10^6 particles/μL	10^5 particles/μL	[144]
	GCE	Polylysine	DPV	10–70 fM	2.3 fM	[145]

Funding: The authors would like to acknowledge financial support from the Slovak Research and Development Agency APVV 17-0300. This publication was supported by Qatar University Grants IRCC-2020-004. The statements made herein are solely the responsibility of the authors.

Conflicts of Interest: The authors declare no conflict of interest.

Abbreviations

Apt	aptamer
APTES	3-aminopropyltriethoxysilane
Ar-CH$_2$-COOH	p-aminophenylacetic acid
AuNCs	gold nanocages
AuNPs	gold nanoparticles
Au-SPE	gold screen-printed electrode
BRCA1	BReast CAncer Type 1 gene
CA15-3	cancer antigen 15-3
CA 27-29	cancer antigen 27-29
CA	chronoamperometry
CEA	carcinoembryonic antigen
CdSe@ZnS QDs	core/shell quantum dots
CNTs	carbon nanotubes
CPE	carbon paste electrode
CV	cyclic voltammetry
CysA	cysteamine
CTES	carboxyethylsilanetriol
DPV	differential pulse voltammetry
DPASV	differential pulse anodic stripping voltammetry
EIS	electrochemical impedance spectroscopy
ERGO	electrochemically reduced graphene oxide
Fc	ferrocene
FTO	fluorine doped tin oxide
GE	gold electrode
GCE	glassy carbon electrode
GLD	glutaraldehyde
GNR	gold nanorods
GNS	graphene nanosheets
GO	graphene oxide
3-GOPE	3-glycidoxypropyl triethoxysilane
GSPE	Graphite-based screen-printed electrode
GQDs	graphene quantum dots
HER2	human epidermal growth factor receptor 2
HRP	horseradish peroxidase
ITO	indium tin oxide
mDNA	mediated double-stranded DNA
m-GEC	magnetic electrode—magneto-actuated graphite epoxy composite
MBs	magnetic beads
MAM	mammaglobin
MBCPE	magnetic bar carbon paste electrode
MPs	magnetic particles
11-MUA	11-mercaptoundenoic acid
MIP	molecularly imprinted polymer
Mt-HSA NCs	clay–protein based composite nanoparticles
MWCNTs	multi-walled carbon nanotubes
OM	one order of magnitude

OPN	osteopontin
PANHS	1-pyrenebutyric acid-*N*-hydroxysuccinimide ester
Pd SSs	Pd superstructures
PEG	polyethylene glycol
PGE	pencil graphite electrode
PIK3CA gene	phosphatidylinositol-4,5-bisphosphate 3-kinase catalytic subunit alpha gene
Ppy	polypyrrole polymer
Py-COOH	1-pyrenecarboxylic acid
RGO	reduced graphene oxide
SAM	self-assembled monolayer
SPCE	screen-printed carbon electrode
SPGE	screen-printed gold electrode
ssDNA	single-stranded deoxyribonucleic acid
SOX2	sex-determining region Y-box 2
SWCNTs	single-walled carbon nanotubes
SWV	square wave voltammetry
tPA	tissue plasminogen activator
uPA	urokinase plasminogen activator

References

1. Siegel, R.L.; Miller, K.D.; Jemal, A. Cancer statistics, 2019. *CA Cancer J. Clin.* **2019**, *69*, 7–34. [CrossRef] [PubMed]
2. Rahman, S.A.; Al-Marzouki, A.; Otim, M.; Khalil Khayat, N.E.H.; Yousuf, R.; Rahman, P. Awareness about Breast Cancer and Breast Self-Examination among Female Students at the University of Sharjah: A Cross-Sectional Study. *Asian Pac. J. Cancer Prev.* **2019**, *20*, 1901–1908. [CrossRef] [PubMed]
3. Dos Anjos Pultz, B.; da Luz, F.A.; de Faria, P.R.; Oliveira, A.P.; de Araujo, R.A.; Silva, M.J. Far beyond the usual biomarkers in breast cancer: A review. *J. Cancer* **2014**, *5*, 559–571. [CrossRef] [PubMed]
4. Khorrami, S.; Tavakoli, M.; Safari, E. Clinical Value of Serum S100A8/A9 and CA15-3 in the Diagnosis of Breast Cancer. *Iran. J. Pathol.* **2019**, *14*, 104–112. [CrossRef]
5. Begum, M.; Karim, S.; Malik, A.; Khurshid, R.; Asif, M.; Salim, A.; Nagra, S.A.; Zaheer, A.; Iqbal, Z.; Abuzenadah, A.M.; et al. CA 15-3 (Mucin-1) and physiological characteristics of breast cancer from Lahore, Pakistan. *Asian Pac. J. Cancer Prev.* **2012**, *13*, 5257–5261. [CrossRef]
6. Scott, D.A.; Drake, R.R. Glycosylation and its implications in breast cancer. *Expert Rev. Proteom.* **2019**, *16*, 665–680. [CrossRef]
7. Núñez, C. Blood-based protein biomarkers in breast cancer. *Clin. Chim. Acta* **2019**, *490*, 113–127. [CrossRef]
8. Torre, L.A.; Trabert, B.; DeSantis, C.E.; Miller, K.D.; Samimi, G.; Runowicz, C.D.; Gaudet, M.M.; Jemal, A.; Siegel, R.L. Ovarian Cancer statistics, 2018. *CA Cancer J. Clin.* **2018**, *68*, 284–296. [CrossRef]
9. Becker, S. A historic and scientific review of breast cancer: The next global healthcare challenge. *Int. J. Gynaecol. Obstet.* **2015**, *131* (Suppl. 1), S36–S39. [CrossRef]
10. Schopper, D.; de Wolf, C. How effective are breast cancer screening programmes by mammography? Review of the current evidence. *Eur. J. Cancer* **2009**, *45*, 1916–1923. [CrossRef]
11. Gebrehiwot, A.G.; Melka, D.S.; Kassaye, Y.M.; Gemechu, T.; Lako, W.; Hinou, H.; Nishimura, S.-I. Exploring serum and immunoglobulin G N-glycome as diagnostic biomarkers for early detection of breast cancer in Ethiopian women. *BMC Cancer* **2019**, *19*, 588. [CrossRef] [PubMed]
12. Sadighbayan, D.; Sadighbayan, K.; Khosroushahi, A.Y.; Hasanzadeh, M. Recent advances on the DNA-based electrochemical biosensing of cancer biomarkers: Analytical approach. *Trends Anal. Chem.* **2019**, *119*, 115609. [CrossRef]
13. Sadighbayan, D.; Sadighbayan, K.; Tohid-kia, M.R.; Khosroushahi, A.Y.; Hasanzadeh, M. Development of electrochemical biosensors for tumor marker determination towards cancer diagnosis: Recent progress. *Trends Anal. Chem.* **2019**, *118*, 73–88. [CrossRef]
14. Sharifi, M.; Avadi, M.R.; Attar, F.; Dashtestani, F.; Ghorchian, H.; Rezayat, S.M.; Saboury, A.A.; Falahati, M. Cancer diagnosis using nanomaterials based electrochemical nanobiosensors. *Biosens. Bioelectron.* **2019**, *126*, 773–784. [CrossRef] [PubMed]

15. Freitas, M.; Nouws, H.P.A.; Delerue-Matos, C. Electrochemical Biosensing in Cancer Diagnostics and Follow-up. *Electroanalysis* **2018**, *30*, 1584–1603. [CrossRef]
16. Sharifi, M.; Hasan, A.; Attar, F.; Taghizadeh, A.; Falahati, M. Development of point-of-care nanobiosensors for breast cancers diagnosis. *Talanta* **2020**, *217*, 121091. [CrossRef]
17. Campuzano, S.; Pedrero, M.; Pingarron, J.M. Non-invasive breast cancer diagnosis through electrochemical biosensing at different molecular levels. *Sensors* **2017**, *17*, 1993. [CrossRef]
18. Hasanzadeh, M.; Shadjou, N.; de la Guardia, M. Early stage screening of breast cancer using electrochemical biomarker detection. *Trends Anal. Chem.* **2017**, *91*, 67–76. [CrossRef]
19. Barber, M.D.; Thomas, J.S.J.; Dixon, J.M. *Breast Cancer: An Atlas of Investigation and Management*; Clinical Pub.: Oxford, UK, 2008; pp. 1–5.
20. Mills, D.; Gomberawalla, A.; Gordon, E.J.; Tondre, J.; Nejad, M.; Nguyen, T.; Pogoda, J.M.; Rao, J.; Chatterton, R.; Henning, S.; et al. Examination of Duct Physiology in the Human Mammary Gland. *PLoS ONE* **2016**, *11*, e0150653. [CrossRef]
21. Love, S.; Barsky, S. Anatomy of the nipple and breast ducts revisited. *Cancer* **2004**, *101*, 1947–1957. [CrossRef]
22. Bombonati, A.; Sgroi, D.C. The molecular pathology of breast cancer progression. *J. Pathol.* **2011**, *223*, 307–317. [CrossRef] [PubMed]
23. Rakha, E.A.; Putti, T.C.; Abd El-Rehim, D.M.; Paish, C.; Green, A.R.; Powe, D.G.; Lee, A.H.; Robertson, J.F.; Ellis, I.O. Morphological and immunophenotypic analysis of breast carcinomas with basal and myoepithelial differentiation. *J. Pathol.* **2006**, *208*, 495–506. [CrossRef] [PubMed]
24. Barroso-Sousa, R.; Metzger-Filho, O. Differences between invasive lobular and invasive ductal carcinoma of the breast: Results and therapeutic implications. *Ther. Adv. Med. Oncol.* **2016**, *8*, 261–266. [CrossRef] [PubMed]
25. Li, Q.; Li, G.; Zhou, Y.; Zhang, X.; Sun, M.; Jiang, H.; Yu, G. Comprehensive N-Glycome Profiling of Cells and Tissues for Breast Cancer Diagnosis. *J. Proteome Res.* **2019**, *18*, 2559–2570. [CrossRef]
26. Harbeck, N.; Penault-Llorca, F.; Cortes, J.; Gnant, M.; Houssami, N.; Poortmans, P.; Ruddy, K.; Tsang, J.; Cardoso, F. Breast cancer. *Nat. Rev. Dis. Primers* **2019**, *5*, 66. [CrossRef]
27. Ludwig, J.A.; Weinstein, J.N. Biomarkers in cancer staging, prognosis and treatment selection. *Nat. Rev. Cancer* **2005**, *5*, 845–856. [CrossRef]
28. Tzitzikos, G.; Saridi, M.; Filippopoulou, T.; Makri, A.; Goulioti, A.; Stavropoulos, T.; Stamatiou, K. Measurement of tumor markers in chronic hemodialysis patients. *Saudi J. Kidney Dis. Transplant.* **2010**, *21*, 50–53.
29. Estakhri, R.; Ghahramanzade, A.; Vahedi, A.; Nourazarian, A. Serum levels of CA15-3, AFP, CA19-9 and CEA tumor markers in cancer care and treatment of patients with impaired renal function on hemodialysis. *Asian Pac. J. Cancer Prev.* **2013**, *14*, 1597–1599. [CrossRef]
30. Mittal, S.; Kaur, H.; Gautam, N.; Mantha, A.K. Biosensors for breast cancer diagnosis: A review of bioreceptors, biotransducers and signal amplification strategies. *Biosens. Bioelectron.* **2017**, *88*, 217–231. [CrossRef]
31. Tkac, J.; Gajdosova, V.; Hroncekova, S.; Bertok, T.; Hires, M.; Jane, E.; Lorencova, L.; Kasak, P. Prostate-specific antigen glycoprofiling as diagnostic and prognostic biomarker of prostate cancer. *Interface Focus* **2019**, *9*, 20180077. [CrossRef]
32. Au, G.H.; Mejias, L.; Swami, V.K.; Brooks, A.D.; Shih, W.Y.; Shih, W.H. Quantitative assessment of Tn antigen in breast tissue micro-arrays using CdSe aqueous quantum dots. *Biomaterials* **2014**, *35*, 2971–2980. [CrossRef] [PubMed]
33. Christiansen, M.N.; Chik, J.; Lee, L.; Anugraham, M.; Abrahams, J.L.; Packer, N.H. Cell surface protein glycosylation in cancer. *Proteomics* **2014**, *14*, 525–546. [CrossRef] [PubMed]
34. Tousi, F.; Bones, J.; Hancock, W.S.; Hincapie, M. Differential Chemical Derivatization Integrated with Chromatographic Separation for Analysis of Isomeric Sialylated N-Glycans: A Nano-Hydrophilic Interaction Liquid Chromatography-MS Platform. *Anal. Chem.* **2013**, *85*, 8421–8428. [CrossRef] [PubMed]
35. Ju, L.L.; Wang, Y.P.; Xie, Q.; Xu, X.K.; Li, Y.; Chen, Z.J.; Li, Y.S. Elevated level of serum glycoprotein bifucosylation and prognostic value in Chinese breast cancer. *Glycobiology* **2016**, *26*, 460–471. [CrossRef] [PubMed]
36. Ideo, H.; Hinoda, Y.; Sakai, K.; Hoshi, I.; Yamamoto, S.; Oka, M.; Maeda, K.; Maeda, N.; Hazama, S.; Amano, J. Expression of mucin 1 possessing a 3'-sulfated core1 in recurrent and metastatic breast cancer. *Int. J. Cancer* **2015**, *137*, 1652–1660. [CrossRef] [PubMed]

37. Imamura, M.; Morimoto, T.; Nomura, T.; Michishita, S.; Nishimukai, A.; Higuchi, T.; Fujimoto, Y.; Miyagawa, Y.; Kira, A.; Murase, K.; et al. Independent prognostic impact of preoperative serum carcinoembryonic antigen and cancer antigen 15–3 levels for early breast cancer subtypes. *World J. Oncol.* **2018**, *16*, 26. [CrossRef]

38. Park, B.-J.; Cha, M.-K.; Kim, I.-H. Thioredoxin 1 as a serum marker for breast cancer and its use in combination with CEA or CA15-3 for improving the sensitivity of breast cancer diagnoses. *BMC Res. Notes* **2014**, *7*. [CrossRef]

39. Bernstein, J.L.; Godbold, J.H.; Raptis, G.; Watson, M.A.; Levinson, B.; Aaronson, S.A.; Fleming, T.P. Identification of mammaglobin as a novel serum marker for breast cancer. *Clin. Cancer Res.* **2005**, *11*, 6528–6535. [CrossRef]

40. Yan, X.; Lin, Y.; Liu, S.; Aziz, F.; Yan, Q. Fucosyltransferase IV (FUT4) as an effective biomarker for the diagnosis of breast cancer. *Biomed. Pharmacother.* **2015**, *70*, 299–304. [CrossRef]

41. Koo, J.H. (Ed.) An Overview of Nanomaterials. In *Fundamentals, Properties, and Applications of Polymer Nanocomposites*; Cambridge University Press: Cambridge, UK, 2016; pp. 22–108.

42. Pacheco-Torgal, F.; Jalali, S. Nanotechnology: Advantages and drawbacks in the field of construction and building materials. *Constr. Build. Mater.* **2011**, *25*, 582–590. [CrossRef]

43. Gajdosova, V.; Lorencova, L.; Prochazka, M.; Omastova, M.; Micusik, M.; Prochazkova, S.; Kveton, F.; Jerigova, M.; Velic, D.; Kasak, P.; et al. Remarkable differences in the voltammetric response towards hydrogen peroxide, oxygen and Ru(NH(3))(6)(3+) of electrode interfaces modified with HF or LiF-HCl etched Ti(3)C(2)T(x) MXene. *Mikrochim. Acta* **2019**, *187*, 52. [CrossRef] [PubMed]

44. Hughes, G.; Westmacott, K.; Honeychurch, K.C.; Crew, A.; Pemberton, R.M.; Hart, J.P. Recent Advances in the Fabrication and Application of Screen-Printed Electrochemical (Bio)Sensors Based on Carbon Materials for Biomedical, Agri-Food and Environmental Analyses. *Biosensors* **2016**, *6*, 50. [CrossRef]

45. Taleat, Z.; Khoshroo, A. Screen-printed electrodes for biosensing: A review (2008–2013). *Microchim. Acta* **2014**, *181*, 865–891. [CrossRef]

46. Arduini, F.; Micheli, L.; Moscone, D.; Palleschi, G.; Piermarini, S.; Ricci, F.; Volpe, G. Electrochemical biosensors based on nanomodified screen-printed electrodes: Recent applications in clinical analysis. *TrAC Trends Anal. Chem.* **2016**, *79*, 114–126. [CrossRef]

47. Renedo, O.D.; Alonso-Lomillo, M.A.; Martínez, M.J.A. Recent developments in the field of screen-printed electrodes and their related applications. *Talanta* **2007**, *73*, 202–219. [CrossRef] [PubMed]

48. Grieshaber, D.; MacKenzie, R.; Vörös, J.; Reimhult, E. Electrochemical Biosensors-Sensor Principles and Architectures. *Sensors* **2008**, *8*, 1400–1458. [CrossRef]

49. Tang, Y.; Ouyang, M. Tailoring properties and functionalities of metal nanoparticles through crystallinity engineering. *Nat. Mater.* **2007**, *6*, 754–759. [CrossRef]

50. Bertok, T.; Lorencova, L.; Chocholova, E.; Jane, E.; Vikartovska, A.; Kasak, P.; Tkac, J. Electrochemical Impedance Spectroscopy Based Biosensors: Mechanistic Principles, Analytical Examples and Challenges towards Commercialization for Assays of Protein Cancer Biomarkers. *ChemElectroChem* **2019**, *6*, 989–1003. [CrossRef]

51. Lorencova, L.; Bertok, T.; Bertokova, A.; Gajdosova, V.; Hroncekova, S.; Vikartovska, A.; Kasak, P.; Tkac, J. Exosomes as a Source of Cancer Biomarkers: Advances in Electrochemical Biosensing of Exosomes. *ChemElectroChem* **2020**, *7*, 1956–1973. [CrossRef]

52. Reddy, K.K.; Bandal, H.; Satyanarayana, M.; Goud, K.Y.; Gobi, K.V.; Jayaramudu, T.; Amalraj, J.; Kim, H. Recent Trends in Electrochemical Sensors for Vital Biomedical Markers Using Hybrid Nanostructured Materials. *Adv. Sci.* **2020**. [CrossRef]

53. Lorencova, L.; Gajdosova, V.; Hroncekova, S.; Bertok, T.; Blahutova, J.; Vikartovska, A.; Parrakova, L.; Gemeiner, P.; Kasak, P.; Tkac, J. 2D MXenes as Perspective Immobilization Platforms for Design of Electrochemical Nanobiosensors. *Electroanalysis* **2019**, *31*, 1833–1844. [CrossRef]

54. Maia, F.R.; Reis, R.L.; Oliveira, J.M.; Maia, F.R.; Reis, R.L.; Oliveira, J.M.; Maia, F.R.; Reis, R.L.; Oliveira, J.M. Nanoparticles and Microfluidic Devices in Cancer Research. *Adv. Exp. Med. Biol.* **2020**, *1230*, 161–171. [PubMed]

55. Meng, L.; Turner, A.P.F.; Mak, W.C. Soft and flexible material-based affinity sensors. *Biotechnol. Adv.* **2020**, *39*, 107398. [CrossRef]

56. Siemer, S.; Wunsch, D.; Khamis, A.; Lu, Q.; Hagemann, J.; Stauber, R.H.; Gribko, A.; Scherberich, A.; Filippi, M.; Krafft, M.P.; et al. Nano Meets Micro-Translational Nanotechnology in Medicine: Nano-Based Applications for Early Tumor Detection and Therapy. *Nanomaterials* **2020**, *10*, 383. [CrossRef]

57. Sierra, J.; Rodriguez-Trujillo, R.; Mir, M.; Samitier, J.; Sierra, J.; Marrugo-Ramirez, J.; Rodriguez-Trujillo, R.; Mir, M.; Samitier, J.; Mir, M.; et al. Sensor-Integrated Microfluidic Approaches for Liquid Biopsies Applications in Early Detection of Cancer. *Sensors* **2020**, *20*, 1317. [CrossRef]

58. Agnolon, V.; Contato, A.; Meneghello, A.; Tagliabue, E.; Toffoli, G.; Gion, M.; Polo, F.; Fabricio, A.S.C. ELISA assay employing epitope-specific monoclonal antibodies to quantify circulating HER2 with potential application in monitoring cancer patients undergoing therapy with trastuzumab. *Sci. Rep.* **2020**, *10*, 3016. [CrossRef] [PubMed]

59. Yousef, G.; Polymeris, M.-E.; Grass, L.; Soosaipillai, A.; Chan, P.-C.; Scorilas, A.; Borgoño, C.; Harbeck, N.; Schmalfeldt, B.; Dorn, J.; et al. Human Kallikrein 5 A Potential Novel Serum Biomarker for Breast and Ovarian Cancer. *Cancer Res.* **2003**, *63*, 3958–3965.

60. Kumar, J.K.; Aronsson, A.C.; Pilko, G.; Zupan, M.; Kumer, K.; Fabjan, T.; Osredkar, J.; Eriksson, S. A clinical evaluation of the TK 210 ELISA in sera from breast cancer patients demonstrates high sensitivity and specificity in all stages of disease. *Tumour Biol. J. Int. Soc. Oncodev. Biol. Med.* **2016**, *37*, 11937–11945. [CrossRef]

61. Li, D.; Harlan-Williams, L.M.; Kumaraswamy, E.; Jensen, R.A. BRCA1-No Matter How You Splice It. *Cancer Res.* **2019**, *79*, 2091–2098. [CrossRef]

62. Benvidi, A.; Dehghani Firouzabadi, A.; Dehghan Tezerjani, M.; Moshtaghiun, S.M.; Mazloum-Ardakani, M.; Ansarin, A. A highly sensitive and selective electrochemical DNA biosensor to diagnose breast cancer. *J. Electroanal. Chem.* **2015**, *750*, 57–64. [CrossRef]

63. Benvidi, A.; Jahanbani, S. Self-assembled monolayer of SH-DNA strand on a magnetic bar carbon paste electrode modified with Fe3O4@Ag nanoparticles for detection of breast cancer mutation. *J. Electroanal. Chem.* **2016**, *768*, 47–54. [CrossRef]

64. Benvidi, A.; Tezerjani, M.D.; Jahanbani, S.; Mazloum Ardakani, M.; Moshtaghioun, S.M. Comparison of impedimetric detection of DNA hybridization on the various biosensors based on modified glassy carbon electrodes with PANHS and nanomaterials of RGO and MWCNTs. *Talanta* **2016**, *147*, 621–627. [CrossRef] [PubMed]

65. Benvidi, A.; Abbasi, Z.; Tezerjani, M.; Banaei, M.; Zare, H.; Molahosseini, H.; Jahanbani, S. A Highly Selective DNA Sensor Based on Graphene Oxide-Silk Fibroin Composite and AuNPs as a Probe Oligonucleotide Immobilization Platform. *Acta Chim. Slov.* **2018**, *65*, 278–288. [CrossRef] [PubMed]

66. Chen, L.; Liu, X.; Chen, C. Impedimetric biosensor modified with hydrophilic material of tannic acid/polyethylene glycol and dopamine-assisted deposition for detection of breast cancer-related BRCA1 gene. *J. Electroanal. Chem.* **2017**, *791*, 204–210. [CrossRef]

67. Ehzari, H.; Safari, M.; Shahlaei, M. A simple and label-free genosensor for BRCA1 related sequence based on electrospinned ribbon conductive nanofibers. *Microchem. J.* **2018**, *143*, 118–126. [CrossRef]

68. Kazerooni, H.; Nassernejad, B. A novel biosensor nanomaterial for the ultraselective and ultrasensitive electrochemical diagnosis of the breast cancer-related BRCA1 gene. *Anal. Methods* **2016**, *8*, 3069–3074. [CrossRef]

69. Benvidi, A.; Firouzabadi, A.D.; Moshtaghiun, S.M.; Mazloum-Ardakani, M.; Tezerjani, M.D. Ultrasensitive DNA sensor based on gold nanoparticles/reduced graphene oxide/glassy carbon electrode. *Anal. Biochem.* **2015**, *484*, 24–30. [CrossRef]

70. Shahrokhian, S.; Salimian, R. Ultrasensitive detection of cancer biomarkers using conducting polymer/electrochemically reduced graphene oxide-based biosensor: Application toward BRCA1 sensing. *Sens. Actuators B Chem.* **2018**, *266*, 160–169. [CrossRef]

71. Saeed, A.A.; Sánchez, J.L.A.; O'Sullivan, C.K.; Abbas, M.N. DNA biosensors based on gold nanoparticles-modified graphene oxide for the detection of breast cancer biomarkers for early diagnosis. *Bioelectrochemistry* **2017**, *118*, 91–99. [CrossRef]

72. Yang, J.; Yin, X.; Zhang, W. Electrochemical determination of PIK3CA gene associated with breast cancer based on molybdenum disulfide nanosheet-supported poly(indole-6-carboxylic acid). *Anal. Methods* **2019**, *11*, 157–162. [CrossRef]

73. Babaei, K.; Shams, S.; Keymoradzadeh, A.; Vahidi, S.; Hamami, P.; Khaksar, R.; Norollahi, S.E.; Samadani, A.A. An insight of microRNAs performance in carcinogenesis and tumorigenesis; an overview of cancer therapy. *Life Sci.* **2020**, *240*, 117077. [CrossRef]

74. Kashyap, D.; Kaur, H. Cell-free miRNAs as non-invasive biomarkers in breast cancer: Significance in early diagnosis and metastasis prediction. *Life Sci.* **2020**, *246*, 117417. [CrossRef] [PubMed]

75. Valihrach, L.; Androvic, P.; Kubista, M. Circulating miRNA analysis for cancer diagnostics and therapy. *Mol. Asp. Med.* **2020**, *72*, 100825. [CrossRef] [PubMed]

76. Cardoso, A.R.; Moreira, F.T.C.; Fernandes, R.; Sales, M.G.F. Novel and simple electrochemical biosensor monitoring attomolar levels of miRNA-155 in breast cancer. *Biosens. Bioelectron.* **2016**, *80*, 621–630. [CrossRef] [PubMed]

77. Ebrahimi, A.; Nikokar, I.; Zokaei, M.; Bozorgzadeh, E. Design, development and evaluation of microRNA-199a-5p detecting electrochemical nanobiosensor with diagnostic application in Triple Negative Breast Cancer. *Talanta* **2018**, *189*, 592–598. [CrossRef] [PubMed]

78. Zhang, K.; Zhang, N.; Zhang, L.; Wang, H.; Shi, H.; Liu, Q. Label-free impedimetric sensing platform for microRNA-21 based on ZrO 2 -reduced graphene oxide nanohybrids coupled with catalytic hairpin assembly amplification. *RSC Adv.* **2018**, *8*, 16146–16151. [CrossRef]

79. Rafiee-Pour, H.A.; Behpour, M.; Keshavarz, M. A novel label-free electrochemical miRNA biosensor using methylene blue as redox indicator: Application to breast cancer biomarker miRNA-21. *Biosens. Bioelectron.* **2016**, *77*, 202–207. [CrossRef]

80. Kilic, T.; Erdem, A.; Erac, Y.; Seydibeyoglu, M.O.; Okur, S.; Ozsoz, M. Electrochemical Detection of a Cancer Biomarker mir-21 in Cell Lysates Using Graphene Modified Sensors. *Electroanalysis* **2015**, *27*, 317–326. [CrossRef]

81. Zouari, M.; Campuzano, S.; Pingarrón, J.M.; Raouafi, N. Femtomolar direct voltammetric determination of circulating miRNAs in sera of cancer patients using an enzymeless biosensor. *Anal. Chim. Acta* **2020**, *1104*, 188–198. [CrossRef]

82. Nath, S.; Mukherjee, P. MUC1: A multifaceted oncoprotein with a key role in cancer progression. *Trends Mol. Med.* **2014**, *20*, 332–342. [CrossRef]

83. Nawaz, M.A.H.; Rauf, S.; Catanante, G.; Nawaz, M.H.; Nunes, G.; Marty, J.L.; Hayat, A. One Step Assembly of Thin Films of Carbon Nanotubes on Screen Printed Interface for Electrochemical Aptasensing of Breast Cancer Biomarker. *Sensors* **2016**, *16*, 1651. [CrossRef] [PubMed]

84. Paimard, G.; Shahlaei, M.; Moradipour, P.; Karamali, V.; Arkan, E. Impedimetric aptamer based determination of the tumor marker MUC1 by using electrospun core-shell nanofibers. *Microchim. Acta* **2019**, *187*. [CrossRef] [PubMed]

85. Mouffouk, F.; Aouabdi, S.; Al-Hetlani, E.; Serrai, H.; Alrefae, T.; Leo Chen, L. New generation of electrochemical immunoassay based on polymeric nanoparticles for early detection of breast cancer. *Int. J. Nanomed.* **2017**, *12*, 3037–3047. [CrossRef] [PubMed]

86. Wang, H.; Sun, J.; Lu, L.; Yang, X.; Xia, J.; Zhang, F.; Wang, Z. Competitive electrochemical aptasensor based on a cDNA-ferrocene/MXene probe for detection of breast cancer marker Mucin1. *Anal. Chim. Acta* **2020**, *1094*, 18–25. [CrossRef]

87. Akbari Nakhjavani, S.; Khalilzadeh, B.; Samadi Pakchin, P.; Saber, R.; Ghahremani, M.H.; Omidi, Y. A highly sensitive and reliable detection of CA15-3 in patient plasma with electrochemical biosensor labeled with magnetic beads. *Biosens. Bioelectron.* **2018**, *122*, 8–15. [CrossRef] [PubMed]

88. Santos, A.R.T.; Moreira, F.T.C.; Helguero, L.A.; Sales, M.G.F. Antibody Biomimetic Material Made of Pyrrole for CA 15–3 and Its Application as Sensing Material in Ion-Selective Electrodes for Potentiometric Detection. *Biosensors* **2018**, *8*, 8. [CrossRef]

89. Amani, J.; Khoshroo, A.; Rahimi-Nasrabadi, M. Electrochemical immunosensor for the breast cancer marker CA 15–3 based on the catalytic activity of a CuS/reduced graphene oxide nanocomposite towards the electrooxidation of catechol. *Microchim. Acta* **2018**, *185*. [CrossRef]

90. Akter, R.; Jeong, B.; Choi, J.S.; Rahman, M.A. Ultrasensitive Nanoimmunosensor by coupling non-covalent functionalized graphene oxide platform and numerous ferritin labels on carbon nanotubes. *Biosens. Bioelectron.* **2016**, *80*, 123–130. [CrossRef]

91. Goutsouliak, K.; Veeraraghavan, J.; Sethunath, V.; De Angelis, C.; Osborne, C.K.; Rimawi, M.F.; Schiff, R. Towards personalized treatment for early stage HER2-positive breast cancer. *Nat. Rev. Clin. Oncol.* **2020**, *17*, 233–250. [CrossRef]

92. Augustine, S.; Kumar, P.; Malhotra, B.D. Amine-Functionalized MoO3@RGO Nanohybrid-Based Biosensor for Breast Cancer Detection. *ACS Appl. Bio Mater.* **2019**, *2*, 5366–5378. [CrossRef]

93. Rostamabadi, P.; Heydari-Bafrooei, E. Impedimetric aptasensing of the breast cancer biomarker HER2 using a glassy carbon electrode modified with gold nanoparticles in a composite consisting of electrochemically reduced graphene oxide and single-walled carbon nanotubes. *Microchim. Acta* **2019**, *186*. [CrossRef] [PubMed]

94. Arkan, E.; Saber, R.; Karimi, Z.; Shamsipur, M. A novel antibody-antigen based impedimetric immunosensor for low level detection of HER2 in serum samples of breast cancer patients via modification of a gold nanoparticles decorated multiwall carbon nanotube-ionic liquid electrode. *Anal. Chim. Acta* **2015**, *874*, 66–74. [CrossRef] [PubMed]

95. Pacheco, J.G.; Rebelo, P.; Freitas, M.; Nouws, H.P.A.; Delerue-Matos, C. Breast cancer biomarker (HER2-ECD) detection using a molecularly imprinted electrochemical sensor. *Sens. Actuators B Chem.* **2018**, *273*, 1008–1014. [CrossRef]

96. Freitas, M.; Nouws, H.P.A.; Delerue-Matos, C. Electrochemical Sensing Platforms for HER2-ECD Breast Cancer Biomarker Detection. *Electroanalysis* **2019**, *31*, 121–128. [CrossRef]

97. Freitas, M.; Nouws, H.P.A.; Keating, E.; Delerue-Matos, C. High-performance electrochemical immunomagnetic assay for breast cancer analysis. *Sens. Actuators B Chem.* **2020**, *308*, 127667. [CrossRef]

98. Freitas, M.; Neves, M.; Nouws, H.; Delerue-Matos, C. Quantum dots as nanolabels for breast cancer biomarker HER2-ECD analysis in human serum. *Talanta* **2019**, *208*, 120430. [CrossRef]

99. Freitas, M.; Nouws, H.P.A.; Keating, E.; Fernandes, V.C.; Delerue-Matos, C. Immunomagnetic bead-based bioassay for the voltammetric analysis of the breast cancer biomarker HER2-ECD and tumour cells using quantum dots as detection labels. *Microchim. Acta* **2020**, *187*. [CrossRef]

100. Malecka, K.; Pankratov, D.; Ferapontova, E.E. Femtomolar electroanalysis of a breast cancer biomarker HER-2/neu protein in human serum by the cellulase-linked sandwich assay on magnetic beads. *Anal. Chim. Acta* **2019**, *1077*, 140–149. [CrossRef]

101. Chen, D.; Wang, D.; Hu, X.; Long, G.; Zhang, Y.; Zhou, L. A DNA nanostructured biosensor for electrochemical analysis of HER2 using bioconjugate of GNR@Pd SSs—Apt—HRP. *Sens. Actuators B Chem.* **2019**, *296*, 126650. [CrossRef]

102. Lah, Z.M.A.N.H.; Ahmad, S.A.A.; Zaini, M.S.; Kamarudin, M.A. An Electrochemical Sandwich Immunosensor for the Detection of HER2 using Antibody-Conjugated PbS Quantum Dot as a label. *J. Pharm. Biomed. Anal.* **2019**, *174*, 608–617. [CrossRef]

103. Hartati, Y.W.; Letelay, L.K.; Gaffar, S.; Wyantuti, S.; Bahti, H.H. Cerium oxide-monoclonal antibody bioconjugate for electrochemical immunosensing of HER2 as a breast cancer biomarker. *Sens. BioSens. Res.* **2020**, *27*, 100316. [CrossRef]

104. Xiang, W.; Lv, Q.; Shi, H.; Xie, B.; Gao, L. Aptamer-based biosensor for detecting carcinoembryonic antigen. *Talanta* **2020**, *214*, 120716. [CrossRef] [PubMed]

105. Dalila, R.N.; Arshad, M.K.M.; Gopinath, S.C.B.; Norhaimi, W.M.W.; Fathil, M.F.M. Current and future envision on developing biosensors aided by 2D molybdenum disulfide (MoS2) productions. *Biosens. Bioelectron.* **2019**, *132*, 248–264. [CrossRef] [PubMed]

106. Wang, Y.-L.; Cao, J.-T.; Chen, Y.-H.; Liu, Y.-M. A label-free electrochemiluminescence aptasensor for carcinoembryonic antigen detection based on electrodeposited ZnS–CdS on MoS2 decorated electrode. *Anal. Methods* **2016**, *8*, 5242–5247. [CrossRef]

107. Paimard, G.; Shahlaei, M.; Moradipour, P.; Akbari, H.; Jafari, M.; Arkan, E. An Impedimetric Immunosensor modified with electrospun core-shell nanofibers for determination of the carcinoma embryonic antigen. *Sens. Actuators B Chem.* **2020**, *311*, 127928. [CrossRef]

108. Wang, Y.; Wang, Y.; Wu, D.; Ma, H.; Zhang, Y.; Fan, D.; Pang, X.; Du, B.; Wei, Q. Label-free electrochemical immunosensor based on flower-like Ag/MoS2/rGO nanocomposites for ultrasensitive detection of carcinoembryonic antigen. *Sens. Actuators B Chem.* **2018**, *255*, 125–132. [CrossRef]

109. Su, S.; Han, X.; Lu, Z.; Liu, W.; Zhu, D.; Chao, J.; Fan, C.; Wang, L.; Song, S.; Weng, L.; et al. Facile Synthesis of a MoS2–Prussian Blue Nanocube Nanohybrid-Based Electrochemical Sensing Platform for Hydrogen Peroxide and Carcinoembryonic Antigen Detection. *ACS Appl. Mater. Interfaces* **2017**, *9*, 12773–12781. [CrossRef]

110. Lee, S.X.; Lim, H.N.; Ibrahim, I.; Jamil, A.; Pandikumar, A.; Huang, N.M. Horseradish peroxidase-labeled silver/reduced graphene oxide thin film-modified screen-printed electrode for detection of carcinoembryonic antigen. *Biosens. Bioelectron.* **2017**, *89*, 673–680. [CrossRef]

111. Rizwan, M.; Elma, S.; Lim, S.A.; Ahmed, M.U. AuNPs/CNOs/SWCNTs/chitosan-nanocomposite modified electrochemical sensor for the label-free detection of carcinoembryonic antigen. *Biosens. Bioelectron.* **2018**, *107*, 211–217. [CrossRef]

112. Wang, J.; Hui, N. Zwitterionic poly(carboxybetaine) functionalized conducting polymer polyaniline nanowires for the electrochemical detection of carcinoembryonic antigen in undiluted blood serum. *Bioelectrochemistry* **2019**, *125*, 90–96. [CrossRef]

113. Kumar, S.; Lei, Y.; Alshareef, N.H.; Quevedo-Lopez, M.A.; Salama, K.N. Biofunctionalized two-dimensional Ti3C2 MXenes for ultrasensitive detection of cancer biomarker. *Biosens. Bioelectron.* **2018**, *121*, 243–249. [CrossRef] [PubMed]

114. Yang, Y.; Jiang, M.; Cao, K.; Wu, M.; Zhao, C.; Li, H.; Hong, C. An electrochemical immunosensor for CEA detection based on Au-Ag/rGO@ PDA nanocomposites as integrated double signal amplification strategy. *Microchem. J.* **2019**, *151*, 104223. [CrossRef]

115. Gu, X.; She, Z.; Ma, T.; Tian, S.; Kraatz, H.-B. Electrochemical detection of carcinoembryonic antigen. *Biosens. Bioelectron.* **2018**, *102*, 610–616. [CrossRef] [PubMed]

116. Wei, Y.; Ma, H.; Ren, X.; Ding, C.; Wang, H.; Sun, X.; Du, B.; Zhang, Y.; Wei, Q. A dual-signaling electrochemical ratiometric method for sensitive detection of carcinoembryonic antigen based on Au-Cu2S-CuS/graphene and Au-CeO2 supported toluidine blue complex. *Sens. Actuators B Chem.* **2018**, *256*, 504–511. [CrossRef]

117. Li, Y.; Zhang, Y.; Li, F.; Li, M.; Chen, L.; Dong, Y.; Wei, Q. Sandwich-type amperometric immunosensor using functionalized magnetic graphene loaded gold and silver core-shell nanocomposites for the detection of Carcinoembryonic antigen. *J. Electroanal. Chem.* **2017**, *795*, 1–9. [CrossRef]

118. Wu, D.; Ma, H.; Zhang, Y.; Jia, H.; Yan, T.; Wei, Q. Corallite-like Magnetic Fe3O4@MnO2@Pt Nanocomposites as Multiple Signal Amplifiers for the Detection of Carcinoembryonic Antigen. *ACS Appl. Mater. Interfaces* **2015**, *7*, 18786–18793. [CrossRef]

119. Li, J.; Liu, J.; Bi, Y.; Sun, M.; Bai, J.; Zhou, M. Ultrasensitive electrochemiluminescence biosensing platform for miRNA-21 and MUC1 detection based on dual catalytic hairpin assembly. *Anal. Chim. Acta* **2020**, *1105*, 87–94. [CrossRef]

120. Jeong, S.; Park, M.-J.; Song, W.; Kim, H.-S. Current immunoassay methods and their applications to clinically used biomarkers of breast cancer. *Clin. Biochem.* **2020**, *78*, 43–57. [CrossRef]

121. Alarfaj, N.A.; El-Tohamy, M.F.; Oraby, H. New label-free ultrasensitive electrochemical immunosensor-based Au/MoS2/rGO nanocomposites for CA 27–29 breast cancer antigen detection. *New J. Chem.* **2018**, *42*, 11046–11053. [CrossRef]

122. Roberts, A.; Tripathi, P.P.; Gandhi, S. Graphene nanosheets as an electric mediator for ultrafast sensing of urokinase plasminogen activator receptor-A biomarker of cancer. *Biosens. Bioelectron.* **2019**, *141*, 111398. [CrossRef]

123. Jilani, T.N.; Siddiqui, A.H. Tissue Plasminogen Activator. In *StatPearls*; StatPearls Publishing Copyright ©: Treasure Island, FL, USA, 2020; pp. 159–170.

124. Saify Nabiabad, H.; Piri, K.; Kafrashi, F.; Afkhami, A.; Madrakian, T. Fabrication of an immunosensor for early and ultrasensitive determination of human tissue plasminogen activator (tPA) in myocardial infraction and breast cancer patients. *Anal. Bioanal. Chem.* **2018**, *410*, 3683–3691. [CrossRef] [PubMed]

125. Vajhadin, F.; Ahadian, S.; Travas-Sejdic, J.; Lee, J.; Mazloum-Ardakani, M.; Salvador, J.; Aninwene, G.E., II; Bandaru, P.; Sun, W.; Khademhossieni, A. Electrochemical cytosensors for detection of breast cancer cells. *Biosens. Bioelectron.* **2019**, *151*, 111984. [CrossRef]

126. Lee, A.V.; Oesterreich, S.; Davidson, N.E. MCF-7 Cells—Changing the Course of Breast Cancer Research and Care for 45 Years. *J. Natl. Cancer Inst.* **2015**, *107*. [CrossRef] [PubMed]

127. Comşa, Ş.; Cîmpean, A.M.; Raica, M. The Story of MCF-7 Breast Cancer Cell Line: 40 years of Experience in Research. *Anticancer Res.* **2015**, *35*, 3147–3154. [PubMed]

128. Gurudatt, N.G.; Chung, S.; Kim, J.-M.; Kim, M.-H.; Jung, D.-K.; Han, J.-Y.; Shim, Y.-B. Separation detection of different circulating tumor cells in the blood using an electrochemical microfluidic channel modified with a lipid-bonded conducting polymer. *Biosens. Bioelectron.* **2019**, *146*, 111746. [CrossRef] [PubMed]

129. Ghassemi, P.; Ren, X.; Foster, B.M.; Kerr, B.A.; Agah, M. Post-enrichment circulating tumor cell detection and enumeration via deformability impedance cytometry. *Biosens. Bioelectron.* **2020**, *150*, 111868. [CrossRef] [PubMed]

130. Liu, P.; Wang, L.; Zhao, K.; Liu, Z.; Cao, H.; Ye, S.; Liang, G. High luminous efficiency Au@CDs for sensitive and label-free electrochemiluminescent detection of circulating tumor cells in serum. *Sens. Actuators B Chem.* **2020**, *316*, 128131. [CrossRef]

131. Tian, L.; Qi, J.; Qian, K.; Oderinde, O.; Liu, Q.; Yao, C.; Song, W.; Wang, Y. Copper (II) oxide nanozyme based electrochemical cytosensor for high sensitive detection of circulating tumor cells in breast cancer. *J. Electroanal. Chem.* **2018**, *812*, 1–9. [CrossRef]

132. Hasanzadeh, M.; Razmi, N.; Mokhtarzadeh, A.; Shadjou, N.; Mahboob, S. Aptamer based assay of plated-derived grow factor in unprocessed human plasma sample and MCF-7 breast cancer cell lysates using gold nanoparticle supported alpha-cyclodextrin. *Int. J. Biol. Macromol.* **2018**, *108*, 69–80. [CrossRef]

133. Yang, Y.; Fu, Y.; Su, H.; Mao, L.; Chen, M. Sensitive detection of MCF-7 human breast cancer cells by using a novel DNA-labeled sandwich electrochemical biosensor. *Biosens. Bioelectron.* **2018**, *122*, 175–182. [CrossRef]

134. Wang, K.; He, M.-Q.; Zhai, F.-H.; He, R.-H.; Yu, Y.-L. A novel electrochemical biosensor based on polyadenine modified aptamer for label-free and ultrasensitive detection of human breast cancer cells. *Talanta* **2017**, *166*, 87–92. [CrossRef] [PubMed]

135. Mohammadniaei, M.; Yoon, J.; Lee, T.; Bharate, B.G.; Jo, J.; Lee, D.; Choi, J.-W. Electrochemical Biosensor Composed of Silver Ion-Mediated dsDNA on Au-Encapsulated Bi2Se3 Nanoparticles for the Detection of H2O2 Released from Breast Cancer Cells. *Small* **2018**, *14*, 1703970. [CrossRef] [PubMed]

136. Dong, W.; Ren, Y.; Bai, Z.; Yang, Y.; Wang, Z.; Zhang, C.; Chen, Q. Trimetallic AuPtPd nanocomposites platform on graphene: Applied to electrochemical detection and breast cancer diagnosis. *Talanta* **2018**, *189*, 79–85. [CrossRef] [PubMed]

137. Luo, J.; Liang, D.; Li, X.; Deng, L.; Wang, Z.; Yang, M. Aptamer-based photoelectrochemical assay for the determination of MCF-7. *Microchim. Acta* **2020**, *187*. [CrossRef] [PubMed]

138. Safavipour, M.; Kharaziha, M.; Amjadi, E.; Karimzadeh, F.; Allafchian, A. TiO2 nanotubes/reduced GO nanoparticles for sensitive detection of breast cancer cells and photothermal performance. *Talanta* **2020**, *208*, 120369. [CrossRef]

139. Akhtartavan, S.; Karimi, M.; Sattarahmady, N.; Heli, H. An electrochemical signal-on apta-cyto-sensor for quantitation of circulating human MDA-MB-231 breast cancer cells by transduction of electro-deposited non-spherical nanoparticles of gold. *J. Pharm. Biomed. Anal.* **2020**, *178*, 112948. [CrossRef] [PubMed]

140. Al-Hajj, M.; Wicha, M.S.; Benito-Hernandez, A.; Morrison, S.J.; Clarke, M.F. Prospective identification of tumorigenic breast cancer cells. *Proc. Natl. Acad. Sci. USA* **2003**, *100*, 3983–3988. [CrossRef]

141. Tang, Y.; Dai, Y.; Huang, X.; Li, L.; Han, B.; Cao, Y.; Zhao, J. Self-assembling peptide-based multifunctional nanofibers for electrochemical identification of breast cancer stem-like cells. *Anal. Chem.* **2019**, *91*, 7531–7537. [CrossRef]

142. Kilic, T.; Valinhas, A.T.D.S.; Wall, I.; Renaud, P.; Carrara, S. Label-free detection of hypoxia-induced extracellular vesicle secretion from MCF-7 cells. *Sci. Rep.* **2018**, *8*, 9402. [CrossRef] [PubMed]

143. An, Y.; Li, R.; Zhang, F.; He, P. Magneto-Mediated Electrochemical Sensor for Simultaneous Analysis of Breast Cancer Exosomal Proteins. *Anal. Chem.* **2020**, *92*, 5404–5410. [CrossRef] [PubMed]

144. Moura, S.L.; Martín, C.G.; Martí, M.; Pividori, M.I. Electrochemical immunosensing of nanovesicles as biomarkers for breast cancer. *Biosens. Bioelectron.* **2020**, *150*, 111882. [CrossRef] [PubMed]

145. Luo, L.; Wang, L.; Zeng, L.; Wang, Y.; Weng, Y.; Liao, Y.; Chen, T.; Xia, Y.; Zhang, J.; Chen, J. A ratiometric electrochemical DNA biosensor for detection of exosomal MicroRNA. *Talanta* **2020**, *207*, 120298. [CrossRef] [PubMed]

Review

Molecularly Imprinted Polymers and Surface Imprinted Polymers Based Electrochemical Biosensor for Infectious Diseases

Feiyun Cui, Zhiru Zhou and H. Susan Zhou *

Department of Chemical Engineering, Worcester Polytechnic Institute, 100 Institute Road, Worcester, MA 01609, USA; fcui@wpi.edu (F.C.); zzhou4@wpi.edu (Z.Z.)
* Correspondence: szhou@wpi.edu

Received: 14 January 2020; Accepted: 11 February 2020; Published: 13 February 2020

Abstract: Owing to their merits of simple, fast, sensitive, and low cost, electrochemical biosensors have been widely used for the diagnosis of infectious diseases. As a critical element, the receptor determines the selectivity, stability, and accuracy of the electrochemical biosensors. Molecularly imprinted polymers (MIPs) and surface imprinted polymers (SIPs) have great potential to be robust artificial receptors. Therefore, extensive studies have been reported to develop MIPs/SIPs for the detection of infectious diseases with high selectivity and reliability. In this review, we discuss mechanisms of recognition events between imprinted polymers with different biomarkers, such as signaling molecules, microbial toxins, viruses, and bacterial and fungal cells. Then, various preparation methods of MIPs/SIPs for electrochemical biosensors are summarized. Especially, the methods of electropolymerization and micro-contact imprinting are emphasized. Furthermore, applications of MIPs/SIPs based electrochemical biosensors for infectious disease detection are highlighted. At last, challenges and perspectives are discussed.

Keywords: molecularly imprinted polymers (MIPs); surface imprinted polymers (SIPs); electrochemical biosensor; biomarkers for infectious diseases

---

1. Introduction

Infectious diseases can be disseminated widely in various ways. They are mainly caused by pathogenic microorganisms, such as viruses, bacteria, fungi, or parasites. Despite great achievements in diagnosis, treatment, and prevention, infectious diseases remain a serious global health threat [1,2]. The challenges of controlling infectious diseases include irrational use of antibiotics, an increase of multidrug-resistant pathogens, the emergence of new pathogenic microorganisms, and rapid spread owing to globalization and overpopulation [3]. Timely diagnosis and targeted antimicrobial treatment are important for the successful clinical control of infectious diseases. Current diagnostic methods for infectious diseases mainly rely on laboratory-based tests including culture, microscopy, enzyme-linked immunosorbent assay (ELISA), and polymerase chain reaction (PCR) [4]. These methods are time-consuming, expensive, and required to be operated by a specialist. Biosensors are ideal alternative methods for timely diagnosis of infectious diseases. They have many merits such as high sensitivity, quick read-out time, and are easier to be mass fabricated and miniaturized. They also can be used as point-of-care (POC) devices at a doctor's office or home because of their simplicity and affordability. Therefore, extensive research has been published to report ultrasensitive electrochemical biosensors for infectious disease detection with excellent performance.

Receptors and transducer are the two main components of biosensors. The receptor recognizes the analyte specifically and the transducer converts the binding activity into a measurable signal sensitively. Electrodes are used as a transducer in electrochemical biosensors [5]. Natural receptors (Figure 1) such as antibodies, DNA, aptamer, phage, lectin, and peptide are used as receptors. They have a high affinity to their targets, but there are also huge challenges in practical applications because of their poor durability and repeatability at high temperature, pressure, in organic solvents, and also low stability in low or high pH solutions. Alternatively, the molecular imprinting technique has been reported to overcome most of these drawbacks. Molecularly imprinted polymers (MIPs) [6] and surface imprinted polymers (SIPs) [7,8] have a great potential to be robust artificial receptors (also called plastic antibodies) [9]. Due to its chemical and physical stability, MIPs/SIPs have provided a new insight for creating receptors by forming specific cavities for binding analytes in the polymeric matrix. In contrast to natural receptors, MIPs/SIPs offer an inexpensive, rapid, sensitive, easy-to-use, and highly selective receptors for sensors, typically for the electrochemical biosensors. Hence, MIPs/SIPs based electrochemical biosensors have become very attractive for infectious diseases.

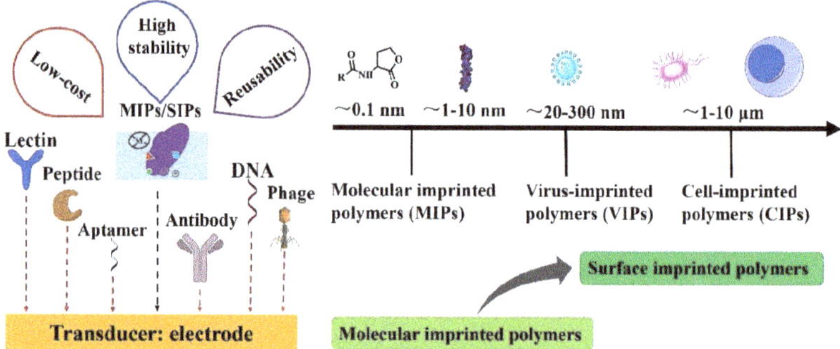

Figure 1. Various receptors for electrochemical biosensors applied in infectious diseases biomarker detection and size distribution covering all the analytes in this review, including small molecular, toxin protein, virus, bacteria, and fungal cells as plotted on a nanometer scale chart.

Several related reviews have been reported. Lahcen et al. [10] mainly presented the development of MIPs modified with nanomaterials for electrochemical biosensors. Good electrical catalytic properties and excellent conductivity of nanomaterials combined with the comparable selectivity of MIPs endow them with powerful performance for various kinds of biomarkers. The magnetic nanoparticles, carbon dots, multi/single-walled carbon nanotubes, and graphene oxides modified MIPs for electrochemical sensing were highlighted in their review paper. Origins, preparation methods, and applications of SIPs applied in larger biomarkers were reviewed by Eersels and coworkers [11]. They pointed out that the measurement of larger biomarkers such as viruses, bacteria, or cells met challenges when using the classical MIPs concept. SIPs can form binding cavities directly on the surface of cured polymers, thus making it easier to remove the templates and provide better use in larger biomarkers (Figure 1).

In this review, current trends in the development of MIPs/SIPs based electrochemical biosensors for rapid assessment of the infectious diseases, as well as future research directions are comprehensively summarized and discussed. Virus-imprinted polymers (VIPs) [12] for virus detection and cell-imprinted polymers (CIPs) [13] for bacteria detection are highlighted (Figure 1).

2. Recognition Mechanisms Between Imprinted Polymers with Biomarkers

The size and morphology of cavities are critical factors for specific recognition between MIPs/SIPs and biomarkers. Besides these, chemical recognition of the biomarkers is important. Three types of chemical recognition methods have been reported: non-covalent, semi-covalent, and covalent. Because of its excellent adaptability, the non-covalent recognition that includes hydrogen bonds, hydrophobic, and electrostatic interactions is the most widely applied for the fabrication of MIPs/SIPs [14,15]. Figure 2 presents various interactions of the template (analyte) and MIPs/SIPs.

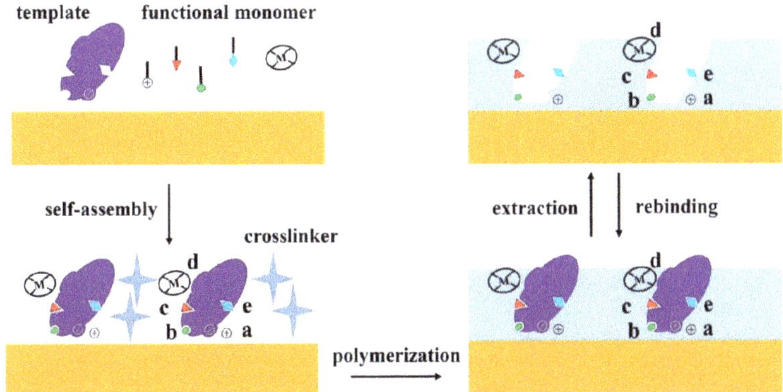

Figure 2. Preparation procedures of molecularly imprinted polymers (MIPs) and surface imprinted polymers (SIPs) on an electrode and various interactions of template (analyte) and MIPs/SIPs, (**a**) electrostatic interactions, (**b**) reversible covalent bonds, (**c**) van der Waals or hydrophobic interactions, (**d**) metal chelation, and (**e**) hydrogen bonds.

2.1. Small Molecular Biomarkers

Metabolites and small signaling molecules produced by microorganisms can be used as biomarkers of infectious diseases. For example, both L- and D-arabitol can be produced by human cells as natural metabolites. They are normally almost equal amounts in healthy humans' body fluids. However, only D-arabitol can be produced by fungi of the *Candida* family. Hence, excess of D-arabitol in body fluids can be used as a biomarker for the diagnosis of candidiasis [16,17]. Dabrowski et al. [18] developed electrochemical sensors based on MIPs for D-arabitol detection in urine samples of patients with candidiasis. They used 2,2′-bithiophene-5-boronic acid as a functional monomer because weak ester bonds can be formed by its boronic acid group and vicinal hydroxyl moieties of D -arabitol. The bithiophene group of 2,2′-bithiophene-5-boronic acid can be polymerized in position 5 of the thiophene ring (Figure 3A). The crosslinker 3,3′-bithiophene could be polymerized in its 2, 2′, 5, and 5′ four positions. The oxidation peak of the crosslinker and functional monomer was at ~1.45 V and ~1.10 V respectively with silver as a pseudo reference electrode. Hence, 0.50~1.20 V was used to induce the initiated polymerization to create a cation radical. Then the crosslinker 3,3′-bithiophene passively participated in the electropolymerization as an acceptor of the cation radical attack (Figure 3A).

N-acyl-homoserine-lactones (AHLs) are important signal molecules of gram-negative bacteria. They participate in the quorum sensing (QS) system to induce and regulate the expression of virulence [19,20]. Jiang et al. [21] used methacrylic acid (MAA) as a monomer and 2,5-dimethyl-4-hydroxy-3(2H)-furanone (DMHF) as an analog template to construct the magnetic molecularly imprinted polymers (MMIPs) which have the capability to selectively recognize AHLs. The hydrogen bond and the delicate binding microcavities are the main contributors to the specificity (Figure 3B).

2.2. Toxins and other Protein Biomarkers

Microbial toxins produced by microorganisms, including bacteria and fungi, are of high molecular weight and have antigenic properties. They can promote infectious diseases by directly damaging host tissues and disabling the immune system. Hence, the fast detection of microbial toxins is critical for the diagnosis of infectious diseases. Most of the microbial toxins are protein. For protein biomarkers, a simple way to improve the affinity of the target protein for its rebinding position is to locate specific charges at its specific rebinding site. A positively charged monomer, quaternary ammonium salt, holding a vinyl bond and an aromatic ring (VBTC), was used to assemble the MIPs for bovine serum albumin (BSA) which holds a negative charge under analytical conditions (pH 7.4, isoelectric point is 5.4). It promoted the ionic interaction between BSA and the MIPs [22]. π–π interaction was used to recognize toxic protein aflatoxin B1 by the p-aminothiophenol-based MIPs. The sensitivity of the imprinted sensor was 11 times greater than that of the non-imprinted sensor by applying the π-donor/π-acceptor interaction [23].

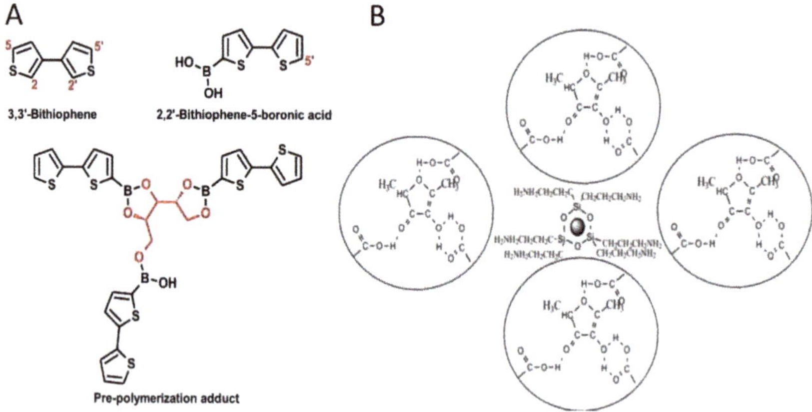

Figure 3. (**A**) Structural formulas of the D-arabitol template, 2,2′-bithiophene-5-boronic acid functional monomer, 3,3′-bithiophene crosslinker, and D-arabitol esterificated with three molecules of functional monomer. Reproduced from [18]—Published by the American Chemical Society. (**B**) Structural formulas of methacrylic acid (MAA) and 2,5-dimethyl-4-hydroxy-3(2H)-furanone (DMHF). Reproduced from [21]—Published by Elsevier B.V.

2.3. Virus

Creating virus-affinity MIPs/SIPs by imprinting techniques has great potential for the diagnosis of virus-related diseases. The most direct and simple method to prepare virus recognition sites is surface virus-imprinting. The viruses can be identified by their morphology and surface properties easily. Some molecules of the virus capsid are demonstrated to play a vital role in the chemical recognition between MIPs/SIPs and viruses [24]. Bai and Spivak developed a hydrogel-based SIP to recognize the Apple Stem Pitting Virus (ASPV). For preparing virus SIPs in the molecular-scale, the perhydro gel solution was fabricated by incubation of the impure ASPV extract with polymerizable ASPV-specific aptamers. Their results proved the need for aptamer preorganization by using the ASPV template, which illustrated the significance of the recognition mechanism for imprinting ASPV-specific sites (Figure 4A). Multivalent interactions of ASPV and aptamers-hydrogel based SIPs induce to evident visible volume-shrinking changes on the rebinding of the virus [25].

2.4. Bacterial and Fungal Cells

Direct bacteria imprinting and generation of recognition sites on polymeric matrices have demonstrated to be practicable [26]. It belongs to one type of cell-imprinted polymers. The recognition mechanism can give credit to the diversity of bacteria cells in shape (e.g., round-shaped *Staphylococcus aureus* and the rod-shaped *Escherichia coli* (*E. coli*)), the uniform size of the same bacteria and the relatively rigid cell wall, which enable size and shape-dependent physical space matching. More importantly, chemical recognition based on the multiple interactions between the cell surface and MIPs/SIPs is essential to recognizing bacteria cells. Ren and Zare [27] developed the bacteria cell-imprinted polydimethylsiloxane (PDMS) to investigate the role of chemical recognition (Figure 4B). The results showed that cell imprinted PDMS with methylsilane groups results in a cavity, thus losing much of its ability to capture the imprinted bacteria, although the shapes of the imprints were shown to be hardly affected which was proved by atomic force field microscopy. Hence, employing suitable functional groups or monomers to form efficient chemical interactions between MIPs/SIPs and the bacterial cell surface is a more important factor for cell-imprinting. Other studies also revealed that chemical recognition plays a dominant role in bacteria recognition. Phenylboronic acid (PBA) groups can significantly improve bacteria affinity of the MIPs with controllable bacteria recognition due to the reversibility between PBA and cis-diol groups of glycan chains presented on the bacterial surface [28]. Besides PBA groups, carbohydrate polymers like chitosan, which exhibits affinity for various bacteria strains, were also applied to create excellent MIPs matrices with high bacteria affinity [29].

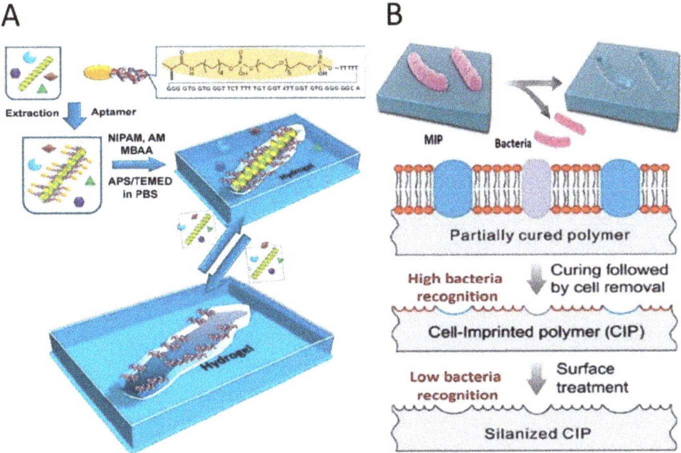

Figure 4. (**A**) Preparation process for virus sensitive super-aptamer hydrogels MIPs. Reproduced from [25]—Published by John Wiley and Sons. (**B**) Schematic diagram of cell-imprinted polymers for bacteria cells. Reproduced from [30]—Published by The Royal Society of Chemistry and [27]—Published by the American Chemical Society.

The factors influencing yeast cells (*Saccharomyces cerevisiae*) recognition by SIPs were studied by means of spectroscopic and microscopy techniques. The results indicated that cell imprinting creates selective binding sites on the surface of the SIPs layer in the form of binding cavities that match the cells in shape and size. Furthermore, it demonstrated that the incorporated phospholipids significantly enhance cell adhesion to the SIPs. The role of phospholipids in the SIP recognition mechanism is mediated by long-range hydrophobic forces [7].

3. Preparation of MIPs/SIPs for Electrochemical Biosensor

Various methods have been applied for the production of MIPs/SIPs on electrodes to prepare electrochemical biosensors. Generally, they can be synthesized by three main steps: (i) assembly of functional monomer and template, (ii) polymerization of monomer-template complex with cross-linkers, porogen, and initiators under photo-/thermal/electrical conditions, and (iii) template removal to reveal binding microcavities that are highly specific to the template [31]. Standard free radical polymerization and sol-gel process are usually used. Free radical polymerization can be further categorized into bulk, multi-step swelling, suspension, emulsion, seed, and precipitation polymerizations based on their synthesis methods [32–34]. As a result, the microcavities that resemble the original template molecules in terms of size, shape, and orientation are generated in the polymer matrix, like the "lock-and-key". Morphology of the polymer is determined by various factors, including polymer reaction time, the amount of pre-polymer, and porogenic solvent.

A broad range of markers associated with infectious diseases such as antibiotics [35], lipopolysaccharides [36], nucleotides [37], toxin proteins [38,39], virus [40,41], bacteria [42,43], and fungi [7] cells have been successfully used as templates in synthesizing MIPs/SIPs. Gast et al. [12]

highlighted synthesis strategies for virus imprinted polymers. Nowadays, double-templates [44,45] and multi-templates [46] methods have been developed, which makes MIPs/SIPs based-biosensors able to detect more target analytes in one complex sample.

The choice of a functional monomer is particularly essential to create highly specific microcavities for the templates. Interestingly, Su et al. [47] used computer-assisted molecular simulation calculations to select the suitable functional monomer and solvent for the template molecule. MAA is reported as the functional monomer which can form desirable pore shape and structure [48], meanwhile, it can be hydrogen bond based acceptor and donor [49]. Other monomers used in MIPs/SIPs synthesis include sulphonic acids (e.g., 2-acrylamido-2-methylpropane sulphonic acid), carboxylic acids (e.g., acrylic acid, vinylbenzoic acid), and heteroaromatic bases (e.g., vinylpyridine, vinylimidazole) were summarized by Choi and coworkers [33]. Typically, electropolymerizable monomers for the preparation of MIPs/SIPs were highlighted by Crapnell and coworkers [50]. MAA, polyvinylpyrrolidone (PVP), dimethylamino ethyl methacrylate (DMAEMA), and polyamine (PA) are usually used for bacteria imprinting to improve the recognition affinity for bacteria [30].

The crosslinker is another important component of MIPs/SIPs. It is responsible for the morphology and stability of imprinted binding sites. Ethylene glycoldimethacrylate (EGDMA), divinylbenzene (DVB), and trimethylolpropane trimethacrylate (TRIM) are the most reported cross-linkers [33]. The most common crosslinker for bacteria imprinting are polydimethylsiloxane (PDMS), polyacrylate, silica (SiO_2), and polyurethane (PU) [30].

Most recently, the combination of nanoparticles with MIPs/SIPs to enhance the performance of electrochemical biosensors is a popular topic. Noble metal nanoparticles (such as Au, Ag, Pt, Pd, etc.), metal oxide nanomaterials (such as TiO_2, Fe_2O_3, etc.), and carbon nanomaterials (such as carbon nanotubes, graphene, etc.) distinctly offer many unique advantages [51].

3.1. Deposition or Spin Coating on Electrodes

Deposition and spin coating are two simple methods for preparing MIPs/SIPs modified electrode. Tancharoen et al. [52] used spin coating method to prepare a SIPs for Zika virus (ZIKV) detection. In their procedures, a certain amount of the prepolymer−graphene oxide mixture was coated on a 1×1 cm^2 gold electrode before spinning at 1000 rpm for 10 s to remove excess prepolymer. Subsequently, the ZIKV template was dispersed on the composite film and exposed to UV light before keeping in an oven at 65 °C for 15 h to allow polymerization to occur. The proposed SIPs were obtained after removing the template from the composite polymer by washing in acetic acid and deionized water.

3.2. Assembly by Self-Assembled Monolayers

Self-assembled monolayers (SAMs) can be used to immobilize MIPs nanoparticles onto the gold surface. Unlike the in-situ synthesis of MIPs/SIPs on an electrode surface, the method dependent on SAMs includes two steps. Firstly, MIPs nanoparticles need to be prepared, then the MIPs nanoparticles can be fixed on a SAMs modified electrode by the covalent bond. The solid-phase synthesis method was used by Tothill's research group to fabricate the MIPs nanoparticles, then the amine coupling chemistry was used to fix nano MIPs receptors strongly to the gold chip. The principle of this method depends on the activation of carboxyl groups on the gold surface by an EDC/NHS mixture which forms reactive succinimide esters [53,54].

3.3. Electropolymerization or UV Light-Induced Polymerization

Electropolymerization is a simple and convenient deposition technique with a conductive polymer layer produced on an electrode surface combined with the template. The layer thickness can be controlled easily. The high-affinity binding sites can be formed by direct doping of templates into the polymer matrix [50]. Usually, the thickness of the film controlled by the electropolymerization conditions and can be characterized by electrical impedance spectroscopy (EIS) and cyclic voltammetry (CV). The charge-transfer resistance of the surface would be increased with the thicknesses added. It is mainly because the polymer holds a low-conductive nature. In order to fabricate a layer of effective MIPs/SIPs, it is critical to control the polymeric film so that it does not cover the whole template so that it can be removed easily and rebound later. If the MIPs/SIPs are too thin, there are no stable microcavities formed on the electrode. It also lowers sensitivity/affinity for the template, since a lower number of binding sites are available. In turn, if the MIPs/SIPs film is too thick, it may entrap the template within the polymeric matrix, hence make its removal/rebinding more difficult. Imprinted artificial capture antibodies (cAbs) for *Staphylococcus aureus (S. aureus)* were fabricated by electropolymerization [55]. By formation of a Schiff base linkage, *S. aureus* was fixed on the aldehyde functionalized ITO electrode surface first. Then, an in-situ electrochemically assisted polycondensation strategy was applied to deposit a silica film on the electrode surface around the *S. aureus*. Finally, a calcination treatment was used to achieve the cAbs. The cAbs with lots of regular cavities were observed after the removal of the template. The circular shape of the cavities was proven by images of higher magnification. The AFM image (Figure 5A) revealed that the depths of the cavities were ~160 nm. Apparently, the pathogen template was imprinted on the ITO surface successfully and the three-dimensional spheroidal architecture was observed. All the preparation process was also characterized by using electrochemical methods in the study.

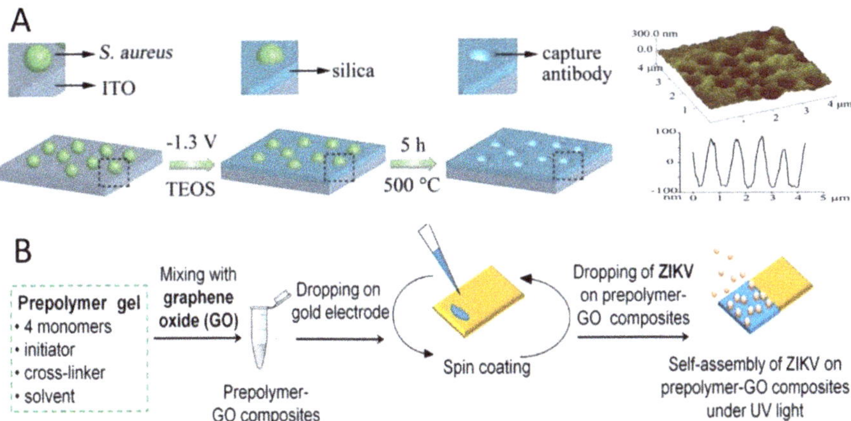

Figure 5. (**A**) Schematic preparation procedures for the artificial capture antibodies (cAbs), AFM images, and the corresponding height profiles of the cAbs. Reproduced from [55]—Published by The Royal Society of Chemistry. (**B**) Schematic preparation procedures for graphene oxide doped SIPs under UV light. Reproduced from [52]—Published by the American Chemical Society.

Tokonami et al. [56] applied a MIPs film consisting of overoxidized polypyrrole (OPPy) to recognize bacilliform bacteria specifically and rapidly. Polypyrrole (PPy) was synthesized using electrochemical polymerization combined with dielectrophoresis (DEP) technique. The DEP resulted in the *P. aeruginosa* being oriented in one direction, perpendicular to the film surface. The number of bacteria doped in the film was counted to be 1.8×10^9 cm^{-2}.

UV light-induced polymerization also can be used to prepare MIPs/SIPs on the electrode. It has been used to prepare SIPs-graphene oxide composites on the electrode for Zika virus (ZIKV) detection (Figure 5B) [52]. Idil et al. fabricated SIPs under UV-polymerization for *E. coli* detection [57].

3.4. Micro-Contact Imprinting

The micro-contact imprinting approach is a soft lithography method that involves the conformal stamping of a template-immobilized layer in a specific pattern on a polymer surface (e.g., PU, PDMS, or SiO$_2$), so that it is able to form shape-complementary recognition sites for relatively large templates on the surface. There are three main types of direct micro-contact imprinting methods: stamp imprinting, film imprinting, and sacrificial layer method imprinting (Figure 6A) [26]. It can also use an artificial template to generate the capturing of SIPs. This method is categorized as indirect micro-contact imprinting methods [58]. The preparation procedures are shown in Figure 6B. Stamp imprinting method was first used by Dickert et al. [59,60] to prepare SIPs to detect whole yeast cells. Recently, the micro-contact imprinting methods were considered as the most promising branch of MIPs/SIPs.

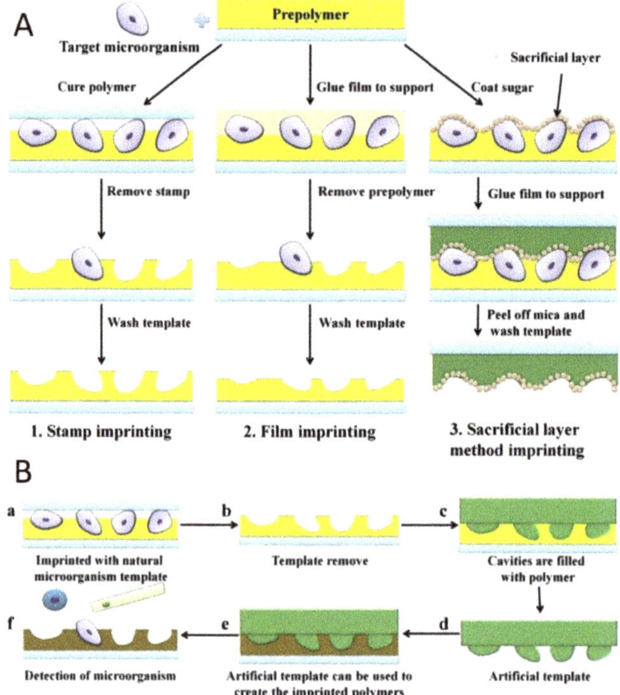

Figure 6. Schematic preparation procedures for three types of direct micro-contact imprinting (**A**) and indirect micro-contact imprinting (**B**). Reproduced from [26]—Published by Elsevier B.V.

4. Applications in Clinical Assays

4.1. Detection of Infectious Diseases Caused by Bacteria

Infectious diseases caused by bacteria are common in our life. MIPs/SIPs based electrochemical biosensors have been used as rapid diagnostic tools for these diseases. As a branch of MIPs/SIPs, CIPs are special for cell biomarkers. CIPs based electrochemical biosensor were reported for *Staphylococcus epidermidis* (*S. epidermidis*) detection [28]. 3-aminophenylboronic acid was used as a functional monomer for the electrochemical fabrication of the CIPs. EIS signal was shown to respond linearly to concentrations of *S. epidermidis* in the range of 10^3–10^7 cfu mL^{-1}. MIPs fabricated by polyphenol was used as an artificial receptor for the detection of flagellar filaments from *Proteus mirabilis* by Khan and coworkers [61]. EIS and square wave voltammetry (SWV) were applied to measure the interaction of flagellar filaments with the MIPs that was fixed on their home-made paper-printed electrodes. Their results showed that the limit of detection (LOD) for the flagellar filaments was as low as 0.6 ng/mL.

4.2. Detection of Infectious Diseases Caused by Viruses

MIPs/SIPs based biosensors have wide applications for the detection of virus in medical diagnostics. Malik and coworkers [62] summarized the state-of-the-art application of MIPs for virus detection.

The detection performance for influenza, Dengue virus, Japanese encephalitis virus (JEV), human immunodeficiency virus (HIV), hepatitis A virus, hepatitis B virus, adenovirus, and picornaviruses were discussed. However, the studies cited in their review paper mainly used quartz crystal microbalance (QCM), surface plasmon resonance (SPR), fluorescence resonance energy transfer (FRET), and resonance light scattering (RLS) as transducers. In this section, the MIPs/SIPs based electrochemical biosensors for virus detection are emphasized.

Human papillomavirus (HPV) is a group of more than 200 related viruses, some of which are spread through anal or vaginal sex. Long-lasting or chronic infections caused by HPV can induce cancer. Cai and coworkers [63] presented a MIPs based nano-sensor to detect human papillomavirus derived E7 protein. Analysis of EIS data revealed that the detection of E7 protein can be as low as sub pg L-1 levels. Notably, the human papillomavirus E6 protein (type-16) was not recognized by the E7 imprinted polymers. It shows outstanding specificity.

As a member of the Flaviviridae virus family, Zika virus usually infects human beings and typically causes a skin rash, conjunctivitis, red eyes, malaise, muscle and joint pain, headache, or mild fever. Recently, Tancharoen et al. [52] developed an electrochemical sensor based on SIPs and graphene oxide composite for Zika virus detection. The sensor was applied to detect virus in both PBS solutions and serum. In the PBS solution, LOD was found to be 2×10^{-2} PFU/mL in the presence of the dengue virus. For serum samples, dilution steps were added to reduce the background signal. The LOD found to be 2×10^{-3} in 10% serum samples and 5×10^{-2} PFU/mL (10~250 RNA copies/mL) in 1% serum samples. Generally, the lowest LOD in real samples should be 6000 (~10^4) particles (or ~10^{-3} PFU) per mL. This performance is sufficient for Zika virus detection in practical applications.

Acquired immune deficiency syndrome (AIDS) is a severe infectious disease caused by HIV. HIV is a member of retroviruses, it is disseminated mainly by contaminated blood transfusions, unprotected sex, and others. Ma et al. [64] developed an electrochemical biosensor based on multi-walled carbon nanotubes modified MIPs for the detection of HIV-p24. They proved that MIPs have a specific recognition capacity for HIV-p24. The linear range was found to be from 1.0×10^{-4} ng cm^{-3} to 2.0 ng/cm^{-3}. The LOD was tested to be 0.083 pg/cm^3. The reported biosensor showed excellent selectivity and stability. It was successfully used for the detection of HIV-p24 in a human serum sample.

5. Conclusion and Look into the Future

Molecular imprinting is an attractive technology used to create selective recognition sites within a polymer network. MIPs/SIPs as tailor-made biomimetic materials have the obvious priority over other recognition elements. The major advantages are their robustness, long-term stability, and cost-effectiveness, which cannot be obtained by fragile biomolecules. In this review, applications of MIPs and SIPs based electrochemical biosensors are focused on, especially in the detection of infectious diseases. Recognition mechanisms, preparation methods, and application performance of MIPs/SIPs were discussed. Although tremendous progress has been achieved, there still exist several challenges. The most important one is that the sensitivity (Table 1) and selectivity need further improvement since MIPs/SIPs do not always possess properties comparable to antibodies. In this case, more functional monomers are worth exploring to promote chemical recognition. Another strategy is using nanopatterned electrodes as the transducer. The design and application of nanopatterned electrodes could promote MIPs/SIPs to generate more effective cavities with excellent spatial matching effect. Moreover, in the era of artificial intelligence, using machine learning to design MIPs/SIPs and improve the recognizing ability of MIPs/SIPs based electrochemical biosensors is very promising.

Table 1. Analytical performance of MIPs/SIPs based electrochemical biosensors for infectious diseases.

Analytes	Preparation Methods of MIPs/SIPs	Device/Indicator	Label/Label Free	Method	LOD	LR	Ref.
N-acyl-homoserine-lactones (AHLs)	MMIPs: Fe3O4@ SiO2-MIP	MGCE/ $[Fe(CN)_6]^{3-/4-}$	Label free	DPV	10^{-10} M	2.5×10^{-9}–10^{-7} M	[21]
Bacterial surface proteins	3-aminophenol electropolymerization	SPEs-SWCNTs/ $[Fe(CN)_6]^{3-/4-}$	Label free	EIS	0.60 nM	NR	[65]
Bacterial flagellar filaments	Phenol electropolymerization	PPE/$[Fe(CN)_6]^{3-/4-}$	Label free	SWV	0.6 ng mL^{-1}	0.01–100 µg mL^{-1}	[61]
Staphylococcus epidermidis	3-APBA electropolymerization	GE//$[Fe(CN)_6]^{3-/4-}$	Label free	EIS	NR	10^3–10^7 CFU mL^{-1}	[28]
E. coli O157:H7	PDA-SIPs	N-GQDs	Label	ECL	8 CFU mL^{-1}	10–10^7 CFU mL^{-1}	[66]
E. coli	UV-polymerization	NR	Label free	Capacitance	70 CFU mL^{-1}	1.0×10^2–1.0×10^7 CFU mL^{-1}	[57]
Bacillus cereus spores	Pyrrole electropolymerization	CPE/$[Fe(CN)_6]^{3-/4-}$	Label free	CV	10^2 CFU mL^{-1}	10^2–10^5 CFU mL^{-1}	[67]
Zika virus	Prepolymer-GO composites under UV light	SPGE//$[Fe(CN)_6]^{3-/4-}$	Label free	CV/EIS	~10^{-3} PFU	10^{-3}–10^2 PFU mL^{-1}	[52]
HIV-1 gene	Directly electropolymerization of phenylenediamine	ITO electrode/EsNCs	Label	ECL	0.3 fM	3.0 fM–0.3 nM	[37]
HIV-p24	polymerization using AAM as functional monomer, MBA as crosslinking agent and APS as initiator.	GCE		DPV	0.083 pg mL^{-1}	1.0×10^{-4}–2 ng mL^{-1}	[64]
Aflatoxin B1	PATP-AuNPs electropolymerization	GE/$[Fe(CN)_6]^{3-/4-}$	Label free	LSV	3 fM	3.2 fM–3.2 µM	[23]

3-APBA: 3-aminophenylboronic acid. AAM: acrylamide. APS: ammonium persulfate. CPE: carbon paste electrode. CV: cyclic voltammetry. ECL: electrochemiluminescence. EsNCs: Europium sulfide nanocrystals. GCE: glassy carbon electrode. GE: gold electrode. LOD: limit of detection. LR: linear range. LSV: linear sweep voltammetry. MBA: N,N′-methylenebisacrylamide. N-GQDs: nitrogen-doped graphene quantum dots (N-GQDs). NR: not reported. PDA: polydopamine. PPE: paper-printed electrodes. SIPs: surface imprinted polymers. SPGE: screen-printed gold electrode.

Until now, few studies explored the recognition mechanism of MIPs/SIPs and larger bioparticles (viruses and bacteria). Research on the exact mechanisms behind target recognition should be emphasized because that can lead to an in-depth understanding, which will eventually help in designing MIPs/SIPs and electrochemical biosensors with even higher selectivity, sensitivity, and accuracy.

Author Contributions: All authors have read and agreed to the published version of the manuscript.

Funding: This research was funded by the National Science Foundation, grant number CBET-1805514.

Conflicts of Interest: The authors declare no conflict of interest.

References

1. Fauci, A.S.; Morens, D.M. The Perpetual Challenge of Infectious Diseases. *N. Engl. J. Med.* **2012**, *366*, 454–461. [CrossRef]
2. França, R.; Da Silva, C.; De Paula, S. Recent advances in molecular medicine techniques for the diagnosis, prevention, and control of infectious diseases. *Eur. J. Clin. Microbiol. Infect. Dis.* **2013**, *32*, 723–728. [CrossRef] [PubMed]
3. Mabey, D.; Peeling, R.W.; Ustianowski, A.; Perkins, M.D. Diagnostics for the developing world. *Nat. Rev. Microbiol.* **2004**, *2*, 231. [CrossRef] [PubMed]
4. Sin, M.L.; Mach, K.E.; Wong, P.K.; Liao, J.C. Advances and challenges in biosensor-based diagnosis of infectious diseases. *Expert Rev. Mol. Diagn.* **2014**, *14*, 225–244. [CrossRef] [PubMed]
5. Cui, F.; Zhou, Z.; Zhou, H.S. Review—Measurement and Analysis of Cancer Biomarkers Based on Electrochemical Biosensors. *J. Electrochem. Soc.* **2020**, *167*, 037525. [CrossRef]
6. Chen, L.; Xu, S.; Li, J. Recent advances in molecular imprinting technology: Current status, challenges and highlighted applications. *Chem. Soc. Rev.* **2011**, *40*, 2922–2942. [CrossRef]
7. Yongabi, D.; Khorshid, M.; Losada-Pérez, P.; Eersels, K.; Deschaume, O.; D'Haen, J.; Bartic, C.; Hooyberghs, J.; Thoelen, R.; Wübbenhorst, M.; et al. Cell detection by surface imprinted polymers SIPs: A study to unravel the recognition mechanisms. *Sens. Actuators B Chem.* **2018**, *255*, 907–917. [CrossRef]
8. Givanoudi, S.; Gennaro, A.; Yongabi, D.; Cornelis, P.; Wackers, G.; Lavatelli, A.; Robbens, J.; Heyndrickx, M.; Wübbenhorst, M.; Wagner, P. An imaging study and spectroscopic curing analysis on polymers for synthetic whole-cell receptors for bacterial detection. *Jap. J. Appl. Phys.* **2020**, *59*, SD0802. [CrossRef]
9. Cieplak, M.; Kutner, W. Artificial Biosensors: How Can Molecular Imprinting Mimic Biorecognition? *Trend Biotechnol.* **2016**, *34*, 922–941. [CrossRef]
10. Lahcen, A.A.; Amine, A. Recent Advances in Electrochemical Sensors Based on Molecularly Imprinted Polymers and Nanomaterials. *Electroanalysis* **2019**, *31*, 188–201. [CrossRef]
11. Eersels, K.; Lieberzeit, P.; Wagner, P. A Review on Synthetic Receptors for Bioparticle Detection Created by Surface-Imprinting Techniques From Principles to Applications. *ACS Sens.* **2016**, *1*, 1171–1187. [CrossRef]
12. Gast, M.; Sobek, H.; Mizaikoff, B. Advances in imprinting strategies for selective virus recognition a review. *TrAC Trend Anal. Chem.* **2019**, *114*, 218–232. [CrossRef]
13. Piletsky, S.; Canfarotta, F.; Poma, A.; Bossi, A.M.; Piletsky, S. Molecularly Imprinted Polymers for Cell Recognition. *Trends Biotechnol.* **2019**. [CrossRef] [PubMed]
14. Saylan, Y.; Akgönüllü, S.; Yavuz, H.; Ünal, S.; Denizli, A. Molecularly imprinted polymer based sensors for medical applications. *Sensors* **2019**, *19*, 1279. [CrossRef] [PubMed]
15. Sharma, P.S.; Iskierko, Z.; Pietrzyk-Le, A.; D'Souza, F.; Kutner, W. Bioinspired intelligent molecularly imprinted polymers for chemosensing: A mini review. *Electrochem. Commun.* **2015**, *50*, 81–87. [CrossRef]
16. Yeo, S.F.; Huie, S.; Sofair, A.N.; Campbell, S.; Durante, A.; Wong, B. Measurement of serum D-arabinitol/creatinine ratios for initial diagnosis and for predicting outcome in an unselected, population-based sample of patients with Candida fungemia. *J. Clin. Microbiol.* **2006**, *44*, 3894–3899. [CrossRef] [PubMed]
17. Arendrup, M.; Bergmann, O.; Larsson, L.; Nielsen, H.; Jarløv, J.O.; Christensson, B. Detection of candidaemia in patients with and without underlying haematological disease. *Clin. Microbiol. Infect.* **2010**, *16*, 855–862. [CrossRef]

18. Dabrowski, M.; Sharma, P.S.; Iskierko, Z.; Noworyta, K.; Cieplak, M.; Lisowski, W.; Oborska, S.; Kuhn, A.; Kutner, W. Early diagnosis of fungal infections using piezomicrogravimetric and electric chemosensors based on polymers molecularly imprinted with D-arabitol. *Biosens. Bioelectron.* **2016**, *79*, 627–635. [CrossRef]

19. Rumbaugh, K.P. Convergence of hormones and autoinducers at the host/pathogen interface. *Anal. Bioanal. Chem.* **2007**, *387*, 425–435. [CrossRef]

20. Shiner, E.K.; Rumbaugh, K.P.; Williams, S.C. Interkingdom signaling: Deciphering the language of acyl homoserine lactones. *FEMS Microbiol. Rev.* **2005**, *29*, 935–947. [CrossRef]

21. Jiang, H.; Jiang, D.; Shao, J.; Sun, X. Magnetic molecularly imprinted polymer nanoparticles based electrochemical sensor for the measurement of Gram-negative bacterial quorum signaling molecules (N-acyl-homoserine-lactones). *Biosens. Bioelectron.* **2016**, *75*, 411–419. [CrossRef] [PubMed]

22. Ferreira, N.S.; Moreira, A.P.; de Sá, M.; Sales, M.G.F. New electrochemically-derived plastic antibody on a simple conductive paper support for protein detection: Application to BSA. *Sens. Actuators B Chem.* **2017**, *243*, 1127–1136. [CrossRef]

23. Jiang, M.; Braiek, M.; Florea, A.; Chrouda, A.; Farre, C.; Bonhomme, A.; Bessueille, F.; Vocanson, F.; Zhang, A.; Jaffrezic-Renault, N. Aflatoxin B1 Detection Using a Highly-Sensitive Molecularly-Imprinted Electrochemical Sensor Based on an Electropolymerized Metal Organic Framework. *Toxins* **2015**, *7*, 3540–3553. [CrossRef] [PubMed]

24. Wangchareansak, T.; Thitithanyanont, A.; Chuakheaw, D.; Gleeson, M.P.; Lieberzeit, P.A.; Sangma, C. A novel approach to identify molecular binding to the influenza virus H5N1: Screening using molecularly imprinted polymers (MIPs). *Med. Chem. Commun.* **2014**, *5*, 617–621. [CrossRef]

25. Bai, W.; Spivak, D.A. A Double-Imprinted Diffraction-Grating Sensor Based on a Virus-Responsive Super-Aptamer Hydrogel Derived from an Impure Extract. *Angew. Chem. Int. Ed.* **2014**, *53*, 2095–2098. [CrossRef]

26. Jia, M.; Zhang, Z.; Li, J.; Ma, X.; Chen, L.; Yang, X. Molecular imprinting technology for microorganism analysis. *TrAC Trend Anal. Chem.* **2018**, *106*, 190–201. [CrossRef]

27. Ren, K.; Zare, R.N. Chemical Recognition in Cell-Imprinted Polymers. *ACS Nano* **2012**, *6*, 4314–4318. [CrossRef]

28. Golabi, M.; Kuralay, F.; Jager, E.W.H.; Beni, V.; Turner, A.P.F. Electrochemical bacterial detection using poly(3-aminophenylboronic acid)-based imprinted polymer. *Biosens. Bioelectron.* **2017**, *93*, 87–93. [CrossRef]

29. Shen, X.; Svensson Bonde, J.; Kamra, T.; Bülow, L.; Leo, J.C.; Linke, D.; Ye, L. Bacterial Imprinting at Pickering Emulsion Interfaces. *Angew. Chem. Int. Ed.* **2014**, *53*, 10687–10690. [CrossRef]

30. Pan, J.; Chen, W.; Ma, Y.; Pan, G. Molecularly imprinted polymers as receptor mimics for selective cell recognition. *Chem. Soc. Rev.* **2018**, *47*, 5574–5587. [CrossRef]

31. Wackerlig, J.; Lieberzeit, P.A. Molecularly imprinted polymer nanoparticles in chemical sensing–Synthesis, characterisation and application. *Sens. Actuators B Chem.* **2015**, *207*, 144–157. [CrossRef]

32. Mattiasson, B.; Ye, L. *Molecularly Imprinted Polymers in Biotechnology*; Springer: Berlin, Germany, 2015; Volume 150.

33. Choi, J.R.; Yong, K.W.; Choi, J.Y.; Cowie, A.C. Progress in molecularly imprinted polymers for biomedical applications. *Comb. Chem. High Throughput Screen.* **2019**, *22*, 78–88. [CrossRef] [PubMed]

34. Chen, L.; Wang, X.; Lu, W.; Wu, X.; Li, J. Molecular imprinting: Perspectives and applications. *Chem. Soc. Rev.* **2016**, *45*, 2137–2211. [CrossRef] [PubMed]

35. Devkota, L.; Nguyen, L.T.; Vu, T.T.; Piro, B. Electrochemical determination of tetracycline using AuNP-coated molecularly imprinted overoxidized polypyrrole sensing interface. *Electrochem. Acta* **2018**, *270*, 535–542. [CrossRef]

36. Long, Y.; Li, Z.; Bi, Q.; Deng, C.; Chen, Z.; Bhattachayya, S.; Li, C. Novel polymeric nanoparticles targeting the lipopolysaccharides of Pseudomonas aeruginosa. *Int. J. Pharm.* **2016**, *502*, 232–241. [CrossRef]

37. Babamiri, B.; Salimi, A.; Hallaj, R. A molecularly imprinted electrochemiluminescence sensor for ultrasensitive HIV-1 gene detection using EuS nanocrystals as luminophore. *Biosens. Bioelectron.* **2018**, *117*, 332–339. [CrossRef]

38. Abdin, M.; Altintas, Z.; Tothill, I. In silico designed nanoMIP based optical sensor for endotoxins monitoring. *Biosens. Bioelectron.* **2015**, *67*, 177–183. [CrossRef]

39. Altintas, Z.; Abdin, M.J.; Tothill, A.M.; Karim, K.; Tothill, I.E. Ultrasensitive detection of endotoxins using computationally designed nanoMIPs. *Anal. Chem. Acta* **2016**, *935*, 239–248. [CrossRef]

40. Yang, B.; Gong, H.; Chen, C.; Chen, X.; Cai, C. A virus resonance light scattering sensor based on mussel-inspired molecularly imprinted polymers for high sensitive and high selective detection of Hepatitis A Virus. *Biosens. Bioelectron.* **2017**, *87*, 679–685. [CrossRef]

41. He, K.; Chen, C.; Liang, C.; Liu, C.; Yang, B.; Chen, X.; Cai, C. Highly selective recognition and fluorescent detection of JEV via virus-imprinted magnetic silicon microspheres. *Sens. Actuators B Chem.* **2016**, *233*, 607–614. [CrossRef]

42. Van Grinsven, B.; Eersels, K.; Akkermans, O.; Ellermann, S.; Kordek, A.; Peeters, M.; Deschaume, O.; Bartic, C.; Diliën, H.; Steen Redeker, E. Label-free detection of Escherichia coli based on thermal transport through surface imprinted polymers. *ACS Sens.* **2016**, *1*, 1140–1147. [CrossRef]

43. Erdem, Ö.; Saylan, Y.; Cihangir, N.; Denizli, A. Molecularly imprinted nanoparticles based plasmonic sensors for real-time Enterococcus faecalis detection. *Biosens. Bioelectron.* **2019**, *126*, 608–614. [CrossRef] [PubMed]

44. Liu, J.; Song, H.; Liu, J.; Liu, Y.; Li, L.; Tang, H.; Li, Y. Preparation of molecularly imprinted polymer with double templates for rapid simultaneous determination of melamine and dicyandiamide in dairy products. *Talanta* **2015**, *134*, 761–767. [CrossRef] [PubMed]

45. Tang, W.; Li, G.; Row, K.H.; Zhu, T. Preparation of hybrid molecularly imprinted polymer with double-templates for rapid simultaneous purification of theophylline and chlorogenic acid in green tea. *Talanta* **2016**, *152*, 1–8. [CrossRef]

46. Xie, X.; Ma, X.; Guo, L.; Fan, Y.; Zeng, G.; Zhang, M.; Li, J. Novel magnetic multi-templates molecularly imprinted polymer for selective and rapid removal and detection of alkylphenols in water. *Chem. Eng. J.* **2019**, *357*, 56–65. [CrossRef]

47. Su, C.; Li, Z.; Zhang, D.; Wang, Z.; Zhou, X.; Liao, L.; Xiao, X. A highly sensitive sensor based on a computer-designed magnetic molecularly imprinted membrane for the determination of acetaminophen. *Biosens. Bioelectron.* **2020**, *148*, 111819. [CrossRef]

48. Golker, K.; Olsson, G.D.; Nicholls, I.A. The influence of a methyl substituent on molecularly imprinted polymer morphology and recognition–Acrylic acid versus methacrylic acid. *Eur. Polym. J.* **2017**, *92*, 137–149. [CrossRef]

49. Lay, S.; Ni, X.; Yu, H.; Shen, S. State-of-the-art applications of cyclodextrins as functional monomers in molecular imprinting techniques: A review. *J. Sep. Sci.* **2016**, *39*, 2321–2331. [CrossRef]

50. Crapnell, R.D.; Hudson, A.; Foster, C.W.; Eersels, K.; Grinsven, B.V.; Cleij, T.J.; Banks, C.E.; Peeters, M. Recent Advances in Electrosynthesized Molecularly Imprinted Polymer Sensing Platforms for Bioanalyte Detection. *Sensors* **2019**, *19*, 1204. [CrossRef]

51. Maduraiveeran, G.; Sasidharan, M.; Ganesan, V. Electrochemical sensor and biosensor platforms based on advanced nanomaterials for biological and biomedical applications. *Biosens. Bioelectron.* **2018**, *103*, 113–129. [CrossRef]

52. Tancharoen, C.; Sukjee, W.; Thepparit, C.; Jaimipuk, T.; Auewarakul, P.; Thitithanyanont, A.; Sangma, C. Electrochemical biosensor based on surface imprinting for zika virus detection in serum. *ACS Sens.* **2018**, *4*, 69–75. [CrossRef] [PubMed]

53. Altintas, Z.; Pocock, J.; Thompson, K.-A.; Tothill, I.E. Comparative investigations for adenovirus recognition and quantification: Plastic or natural antibodies? *Biosens. Bioelectron.* **2015**, *74*, 996–1004. [CrossRef] [PubMed]

54. Altintas, Z.; Gittens, M.; Guerreiro, A.; Thompson, K.-A.; Walker, J.; Piletsky, S.; Tothill, I.E. Detection of waterborne viruses using high affinity molecularly imprinted polymers. *Anal. Chem.* **2015**, *87*, 6801–6807. [CrossRef] [PubMed]

55. Zhang, Z.; Guan, Y.; Li, M.; Zhao, A.; Ren, J.; Qu, X. Highly stable and reusable imprinted artificial antibody used for in situ detection and disinfection of pathogens. *Chem. Sci.* **2015**, *6*, 2822–2826. [CrossRef] [PubMed]

56. Tokonami, S.; Nakadoi, Y.; Takahashi, M.; Ikemizu, M.; Kadoma, T.; Saimatsu, K.; Dung, L.Q.; Shiigi, H.; Nagaoka, T. Label-Free and Selective Bacteria Detection Using a Film with Transferred Bacterial Configuration. *Anal. Chem.* **2013**, *85*, 4925–4929. [CrossRef] [PubMed]

57. Idil, N.; Hedström, M.; Denizli, A.; Mattiasson, B. Whole cell based microcontact imprinted capacitive biosensor for the detection of Escherichia coli. *Biosens. Bioelectron.* **2017**, *87*, 807–815. [CrossRef]

58. Schirhagl, R.; Ren, K.; Zare, R.N. Surface-imprinted polymers in microfluidic devices. *Sci. China Chem.* **2012**, *55*, 469–483. [CrossRef]

59. Hayden, O.; Dickert, F.L. Selective Microorganism Detection with Cell Surface Imprinted Polymers. *Adv. Mater.* **2001**, *13*, 1480–1483. [CrossRef]

60. Dickert, F.L.; Hayden, O. Bioimprinting of Polymers and Sol–Gel Phases. Selective Detection of Yeasts with Imprinted Polymers. *Anal. Chem.* **2002**, *74*, 1302–1306. [CrossRef]

61. Khan, M.A.R.; Cardoso, A.R.A.; Sales, M.G.F.; Merino, S.; Tomas, J.M.; Rius, F.X.; Riu, J. Artificial receptors for the electrochemical detection of bacterial flagellar filaments from Proteus mirabilis. *Sens. Actuators B Chem.* **2017**, *244*, 732–741. [CrossRef]

62. Malik, A.A.; Nantasenamat, C.; Piacham, T. Molecularly imprinted polymer for human viral pathogen detection. *Mater. Sci. Eng. C* **2017**, *77*, 1341–1348. [CrossRef] [PubMed]

63. Cai, D.; Ren, L.; Zhao, H.; Xu, C.; Zhang, L.; Yu, Y.; Wang, H.; Lan, Y.; Roberts, M.F.; Chuang, J.H.; et al. A molecular-imprint nanosensor for ultrasensitive detection of proteins. *Nat. Nanotechnol.* **2010**, *5*, 597–601. [CrossRef] [PubMed]

64. Ma, Y.; Shen, X.-L.; Zeng, Q.; Wang, H.-S.; Wang, L.-S. A multi-walled carbon nanotubes based molecularly imprinted polymers electrochemical sensor for the sensitive determination of HIV-p24. *Talanta* **2017**, *164*, 121–127. [CrossRef] [PubMed]

65. KKhan, M.A.R.; Moreira, F.T.; Riu, J.; Sales, M.G.F. Plastic antibody for the electrochemical detection of bacterial surface proteins. *Sens. Actuators B Chem.* **2016**, *233*, 697–704. [CrossRef]

66. Chen, S.; Chen, X.; Zhang, L.; Gao, J.; Ma, Q. Electrochemiluminescence Detection of Escherichia coli O157:H7 Based on a Novel Polydopamine Surface Imprinted Polymer Biosensor. *ACS Appl. Mater. Inter.* **2017**, *9*, 5430–5436. [CrossRef] [PubMed]

67. Ait Lahcen, A.; Arduini, F.; Lista, F.; Amine, A. Label-free electrochemical sensor based on spore-imprinted polymer for Bacillus cereus spore detection. *Sens. Actuators B Chem.* **2018**, *276*, 114–120. [CrossRef]

Article

Simple and Cost-Effective Electrochemical Method for Norepinephrine Determination Based on Carbon Dots and Tyrosinase

Sylwia Baluta [1], Anna Lesiak [1,2] and Joanna Cabaj [1,*]

[1] Faculty of Chemistry, Wrocław University of Science and Technology, Wybrzeże Wyspiańskiego 27, 50-370 Wrocław, Poland; sylwia.baluta@pwr.edu.pl (S.B.); anna.lesiak@pwr.edu.pl (A.L.)

[2] Faculty of Fundamental Problems of Technology, Wrocław University of Science and Technology, Wybrzeże Wyspiańskiego 27, 50-370 Wrocław, Poland

* Correspondence: joanna.cabaj@pwr.edu.pl; Tel.: +48-71-320-4641

Received: 6 July 2020; Accepted: 10 August 2020; Published: 14 August 2020

Abstract: Although neurotransmitters are present in human serum at the nM level, any dysfunction of the catecholamines concentration may lead to numerous serious health problems. Due to this fact, rapid and sensitive catecholamines detection is extremely important in modern medicine. However, there is no device that would measure the concentration of these compounds in body fluids. The main goal of the present study is to design a simple as possible, cost-effective new biosensor-based system for the detection of neurotransmitters, using nontoxic reagents. The miniature Au-E biosensor was designed and constructed through the immobilization of tyrosinase on an electroactive layer of cysteamine and carbon nanoparticles covering the gold electrode. This sensing arrangement utilized the catalytic oxidation of norepinephrine (NE) to NE quinone, measured with voltammetric techniques: cyclic voltammetry and differential pulse voltammetry. The prepared bio-system exhibited good parameters: a broad linear range (1–200 μM), limit of detection equal to 196 nM, limit of quantification equal to 312 nM, and high selectivity and sensitivity. It is noteworthy that described method was successfully applied for NE determination in real samples.

Keywords: biosensor; carbon dots; norepinephrine; tyrosinase; voltammetry

1. Introduction

One of the primary goals of worldwide scientific endeavors is to improve quality of life. Achieving this is directly related to the rapid analysis of common disorders, quality control in the food industry and environment monitoring. Constant, fast and sensitive in situ monitoring is a priority in diagnostic control, and most of all, in medical diagnostics. The devices that meet these requirements are biosensors. According to Cammann, biosensors are analytical devices allowing for conversion of a biological signal to a measurable signal, like for instance amperometric response in the case of electrochemical sensors [1].

In electrochemical biosensors, differential pulse voltammetry (DPV) is frequently used. DPV applies a linear sweep voltammetry with a series of regular voltage pulses superimposed on the linear potential sweep [2]. Due to this, the current can be estimated instantly before every change of potential. In consequence, the influence of the charging current is minimized, reaching a better sensitivity [3]. DPV is frequently used in voltammetry-based techniques, not only due to its good sensitivity, but also because of resolving power.

Norepinehrine (NE) is a monoamine neurotransmitter engaged in a broad range of physiological actions. The main function of NE is to help the organism adapt to internal and external environmental changes. Noradrenergic neurons influence many effects in the human body, such as processing sensory information, regulating sleep, and mediating attentional functions [4–6]. Any changes in NE level may directly affect the mental state, neurodegenerative disorders or cardiovascular function [7]. Therefore, NE monitoring is crucial from a medical point of view.

Many of the electrochemical biosensors for neurotransmitter detection, e.g., dopamine (DA), epinephrine (EP), and NE utilize oxidoreductases as a recognition element [8–12]. For instance, tyrosinase and laccase are multifunctional copper containing enzymes with two types of catalytic activity in the presence of oxygen: hydroxylation of monophenols to *o*-diphenols by cresolase activity, and oxidation of diphenols to *o*-quinones by catecholase activity [13]. The resulting quinones can be further reduced electrochemically on the electrode without any mediator, reforming the original *o*-diphenol. This reaction constitutes the basis of amperometric detection at negative potentials and quantification of phenolic compounds. By the same principle, catechol-like phenolic compounds (e.g., NE) can be detected by the amperometric method with an electrode modified with oxidoreductases [14]. There are also systems without any biologically active compound for NE detection, such as the most recent work presented by D. Ji et al., who demonstrated an electrochemical, smartphone-based system for NE detection [15]. Screen-printed graphene electrodes were modified, e.g., with GOx, and used as working electrodes. NE has been determined with square wave voltammetry (SWV) in the concentration range 1–30 μM. The detection limit obtained for such system was equal to 265 nM. The main advantage of the described system is the facility of miniaturization and the possibility of constant measurement, as a possible wearable device. It also presents the ability to work over a longer period of time (in comparison with biological systems).

Carbon dots (CDs) are an interesting platform for sensors because of their characteristic properties, such as water solubility, low toxicity, high emission intensity or chemical stability in time [16]. CDs' structure consists of a carbon core and the surface passivation layer with functional groups (mainly hydroxyl-, carboxyl- or amine groups), which allow a conjugation with other molecules (like proteins) [17].

Data from the literature show the use of the CDs matrix to determine various substances, with the focus on their fluorescent properties, e.g., hydrazine in water [18]. The use of CDs in electrochemical sensors is slowly gaining popularity, among others, for the determination of ascorbic acid [19] or riboflavin [20].

In addition, an increase in interest in electrochemical methods based on nanoparticles for neurotransmitter determination has been observed [21–25]. Samdani et al. presented an electrochemical method for NE determination using a glassy carbon electrode (GCE), modified with nanorods based on $FeMoO_4$. Synthesized nanorods were used as an active electrode material for the oxidation of NE by cyclic voltammetry (CV) and DPV techniques. The amperometric response of NE to the $GCE/FeMoO_4$ nanorods showed a linear increase in the current between 5.0×10^{-8} M and 2.0×10^{-4} M with a detection limit of 3.7×10^{-9} M [26]. Another detection method for NE was investigated by Mohammadi et al. A glassy carbon electrode was modified with single-walled carbon nanotubes (SWCNTs) and immobilized tyrosinase. The DPV technique for NE determination was applied. The detection limit of the modified electrode towards NE was found at 0.1 μM, and the calibration curves were linear over the concentration range of 1.0–21 μM [27].

In this paper, we present a CDs-based electrochemical biosensor for NE determination, using a tyrosinase-dependent redox system. The described novel technique has many advantages, such as: the application of stable materials, synthesis of nanomaterials with nontoxic and inexpensive materials, simple and quick analysis, high sensitivity, good selectivity and measurements in a broad linear range. The designed biosensor is as simple as possible to reduce costs and to avoid using toxic species, which is essential for the medical or diagnostic industry, such as point-of-care (POC) testing. The gold electrode modified with cysteamine, CDs and tyrosinase (Tyr) presents excellent electrochemical behavior and provides a facile, selective and sensitive method, indicating that the biosensor is a good candidate for catecholamines detection.

2. Materials and Methods

2.1. Reagents and Materials

Tyrosinase (from *Agaricus bisporus*, EC 1.14.18.1, ≥1000 U/mg) as well as norepinephrine hydrochloride (NE), epinephrine hydrochloride (EP), dopamine hydrochloride (DA), cysteamine (CA), glutaraldehyde (GA), uric acid (UA), ascorbic acid (AA), L-cysteine (CYS) and 4-*tert*-Butylcatechol (4*t*BC) were purchased from Sigma-Aldrich Co (Poznań, Poland). Citric acid, NaOH, NaH_2PO_4, KH_2PO_4, Tris, HCl, CH_3COONa, CH_3COOH, Na_2HPO_4, and K_2HPO_4 were purchased from POCH (Part of Avantor, Performance Materials, Gliwice, Poland). All chemicals were of analytical grade and were not further purified before use. All buffers (phosphate buffer, acetate buffer and Tris-HCl) were prepared according to commonly known, obligatory standards. In Section 2.2.2. the specific concentrations and pH values of prepared buffer solutions are listed. Synthetic urine CLEANU® was produced by CleanU (Poznań, Poland), CU-25 mL, and it contained creatinine, uric acid, urea, mineral salts, dyes and water (ingredients reserved—patent).

2.2. Apparatus and Procedures

2.2.1. Synthesis and Characterization of CDs

The preparation process for CDs was carried out following a modified procedure of Sahu et al. [28]. Briefly, 10 mL of 100% orange juice was added to 20 mL of ethanol and the mixture was vigorously stirred for 60 min (at 75 °C). Afterwards, the mixture was cooled down to room temperature and 10 mL of dichloromethane was added. Subsequently, the mixture was centrifuged 3 times at 6000 rpm for 10 min to dispose of unreacted organic compounds. From the resulting biphasic solution, the aqueous layer was collected and 10 mL of acetone was added. The mixture was again placed in the centrifuge for 10 min at high speed. The supernatant solution was decanted and the precipitate was dried. Deionized water was added and mixed to completely dissolve the precipitate. The emission (excitation wavelength 405 nM) and absorbance (Spectrophotometer UV/VIS/NIR V-570 JASCO) spectra of the resulting solution were measured.

The presence of homogenous CDs was confirmed by transmission electron microscopy (TEM) picture (FEI Tecnai G2 X-TWIN, Dawson Creek Drive, Hillsboro, OR, USA). The Fourier Transform Infrared Spectroscopy (FT-IR) spectrum, using the Nicolet iS10 spectrometer (ThermoFisher, Waltham, MA, USA), was executed for observation of bonding between CDs/cysteamine (CA) and the enzyme.

In the present study, CDs are used as a semi-conducting material, which, thanks to the ability to improve the electron transport between the active site of the enzyme and the electrode surface, allows the sensor to present a short response time and high sensitivity. In addition, this material acts as a platform for protein immobilization. The solution containing synthesized CDs was stable for at least 6 months and stored at 4 °C when not used. The stability of CDs was confirmed by the emission measurements.

2.2.2. Modification of Electrodes

The gold electrode (Au-E, Au electrode, diameter 2.5 mM, produced by BASi, MF-2014) was polished before the experiment with 3 μM fine diamond polish and rinsed thoroughly with double distilled water. The electrode was modified with a thin layer of CA, CDs and tyrosinase. CA solution (0.1 M) was applied onto the gold electrode for 24 h to link with the electrode surface by creating thiol bonds. CA molecule has an amine group, which stays exposed for the next step of modification. CDs have carboxylic groups with negative charge, which not only ensure good stability, but also enabled interaction with amine functional groups of CA and CDs. CA was physically adsorbed for 24 h to link with CA/CDs and thus created CDs-modified electrode (Figure 1). CDs, like most nanoparticles, possess semi-conducting properties which could make the surface of the electrode a more conductive and monodispersed surface as a platform for protein binding. The immobilization process was executed by physical adsorption of Tyr (2 mg/mL) in Phosphate Buffered Saline (PBS) buffer (0.1 M; pH 7.0) at room temperature onto the surface of Au electrode modified with CA/CDs for 2 h, and then crosslinked with 10% glutaraldehyde (GA) (10 min). Additional use of the GA cross-linking allows a relatively stable and active immobilization of the tyrosinase onto the CA/CDs surface, because of the creation of covalent bonds [29].

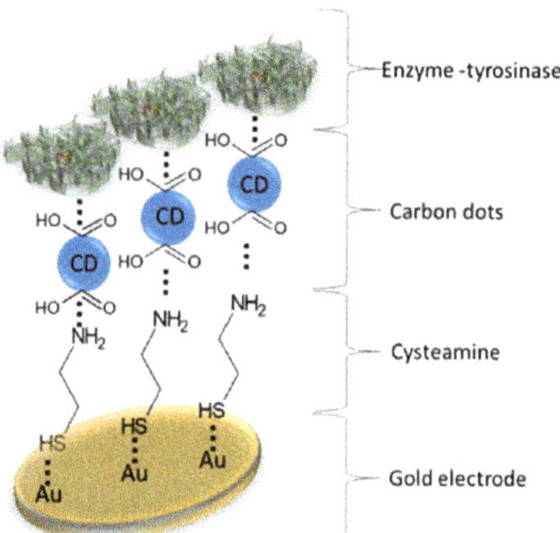

Figure 1. Scheme of measuring system of norepinephrine based on Gold-Electrode/Cysteamine/Carbon Dots/Tyrosinase.

The excess of unbounded proteins was washed with phosphate (0.1 M; pH 7.0), acetate (0.1 M; pH 5.2) and Tris-HCl (0.25 M; pH 7.2) buffers.

An enzyme immobilized by physical adsorption with cross-linker does not require any other activation. Modified Au-E/CA/CDs/Tyr electrode was obtained and was stored at 4 °C when not used. The described modified electrode was active for c.a. 80 cycles.

2.2.3. Electrochemical Measurements

All electrochemical experiments for NE determination were conducted using DPV and CV methods with a potentiostat/galvanostat AUTOLAB PGSTAT128N (serial nr. AUT84866; Utrecht, The Netherlands) with GPES software (version 4.9). A conventional three-electrode system was used for all electrochemical measurements in 8-mL cell. Gold electrode (unmodified or modified with CA/CDs/Tyr) was used as a working electrode, together with a coiled platinum wire as an auxiliary electrode and a silver-silver chloride reference electrode (Ag/AgCl). CV measurements were carried out by repeated potential scanning in range -0.2–1.2 V. DPV analysis was conducted with potential -0.2–0.6 V. All electrochemical measurements were performed at room temperature and in open-air conditions.

2.2.4. Electrochemical Determination of Norepinephrine

NE detection was determined using DPV technique in 8-mL cell. NE tests solutions were prepared by dissolving NE in 0.1 M PBS buffer (pH 7.0). The measurements were conduct in 1–200 µM range and the current response was proportional to the proper concentration.

In addition, CV technique was employed for showing the whole enzyme-based redox reaction during NE determination. Such test was provided in potential range -0.2 V–1.2 V, with scan rate 50 mV/s, for 10 cycles each.

2.2.5. Influence of Interfering Substances

Interfering species (ascorbic acid (AA), uric acid (UA), L-cysteine (CYS), dopamine (DA), epinephrine (EP) and 4-tert-Butylcatechol (4tBC)) were added to each NE standard solution in a concentration of 50 μM to investigate the selectivity of the proposed method. Listed compounds were mixed each time with NE solutions in the volume ratio 1:1.

3. Results and Discussion

3.1. Characterization of CDs

Figure 2A shows collected absorbance and photoluminescence spectra of CDs after synthesis. As can be seen, CDs have a maximum of emission at 503 nm, which corresponds to a green color. In order to confirm the presence of CDs, TEM measurements were performed (Figure 2B). The morphological display of the CDs shows that they were nearly homogeneous and monodisperse, with particle size of approximately 3 nm.

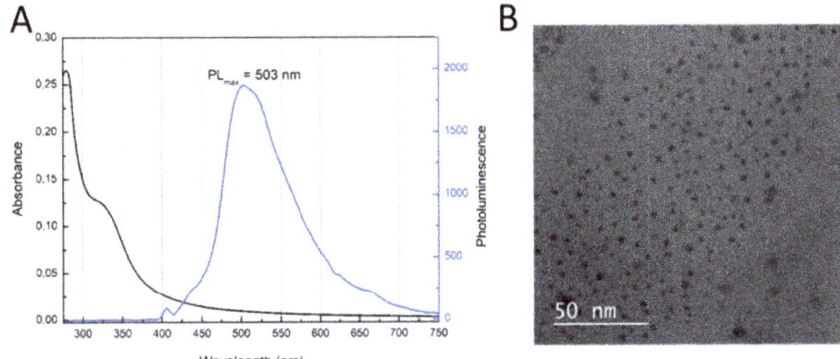

Figure 2. Characteristic information about CDs: (**A**): Absorbance and photoluminescence spectra, (**B**): Picture from TEM.

CDs are significant as a matrix for enzyme anchor. Synthesized CDs with carboxyl groups on their surface represent a suitable platform for amine groups present in the enzyme. Proteins can be successfully attached to CDs to initiate biorecognition element for NE detection.

Formation of specific functional groups in CDs (needed for anchoring the protein onto the surface of CDs), as well as formation of the bond between CDs and cysteamine and formation of the system (CA-CDs-Tyr) was confirmed by FT-IR spectroscopic analysis, as shown in Figure 3A,B. Because of the presence of aqueous media, samples were prepared on glass plates (background) using a layer-by-layer method.

The FT-IR spectrum of CA shows the bands at 3090 cm^{-1}, which are attributed to N-H_2 in the amine group. The same characteristic peaks were shifted to 2919 cm^{-1} on the CA-CDs spectrum, which shows the creation of amide group between CA and CDs [30]. Peaks from thiol group of CA (2516 cm^{-1}) disappeared on the CA-CDs spectrum, which proves an interaction between the gold electrode and CA-CDs.

The peak at 3328 cm^{-1} on the CDs is characteristic for the hydroxyl group (−OH), and the peak at 1036 cm^{-1} corresponds to C-O interactions from the carboxylic group on the CDs surface. Additionally, peaks at 1569 and 1457 cm^{-1} come from an O-H interaction (Figure 3A) [31].

The disappearance of O-H peaks in the region near 1500 cm^{-1} on the CA-CDs-Tyr can suggest, that protein is immobilized on CDs-based platform successfully [32]. It can be assumed that there is a thorough electrode coverage. Although the enzyme influences the peak intensity, characteristic chemical bands of CA-CDs were preserved (Figure 3B).

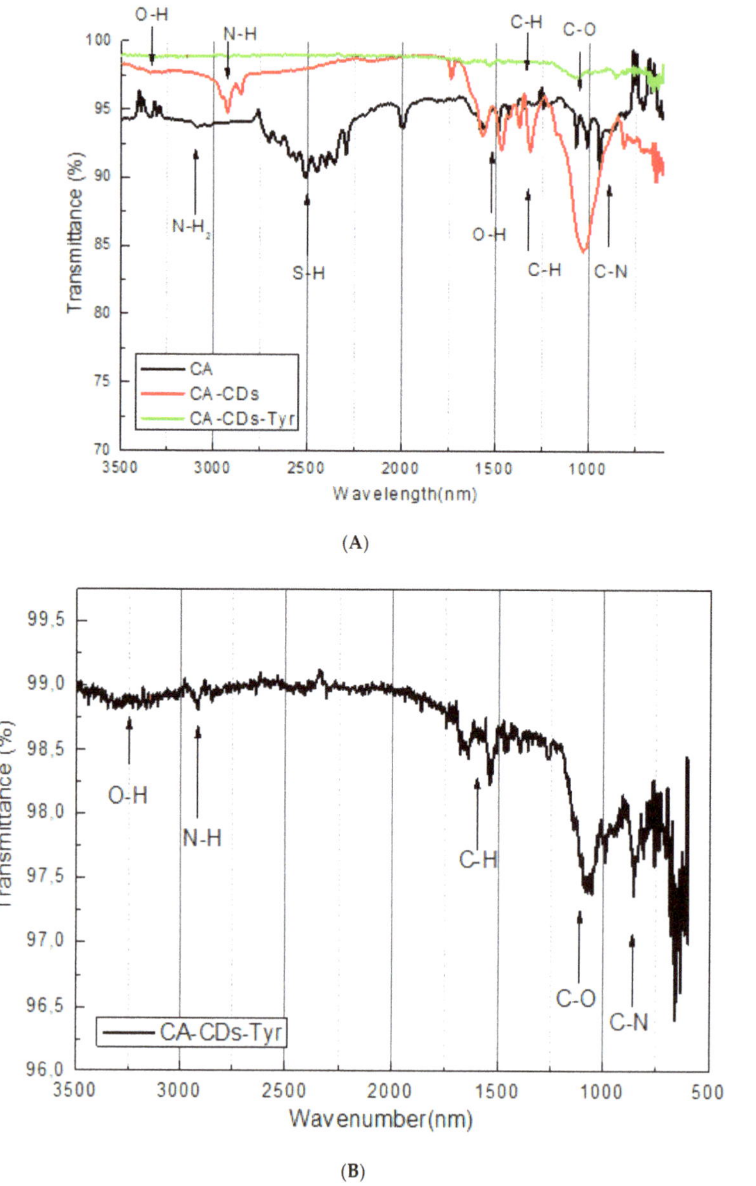

Figure 3. FT-IR spectrum of (**A**): cysteamine (CA), Carbon dots attached to cysteamine (CA-CDs), system with enzyme (CA-CDs-Tyr) and (**B**): enlargement of the system.

3.2. Cyclic Voltammetric Behavior of NE

The electrochemical signals of NE at Au-E, Au-E/CA/CDs and Au-E/CA/CDs/Tyr were measured using CV in 200 µM NE solution in PBS buffer (pH 7.0). PBS buffer is the optimal buffer for tyrosinase activity. As shown in Figure 4A, at a bare Au electrode a pair of NE redox peaks occur, however low signals are obtained. After coating the Au electrode surface with CA and CDs, the redox peaks increased slightly. Nevertheless, for Au-E/CA/CDs/Tyr, a significant signal growth was observed, which was due to the unique properties (e.g., stability in time) of CA/CDs; the CA together with CDs and the catalyst, tyrosinase, showed the highest oxidation and reduction signals of NE. Therefore, such bio-platform was used for the determination of sensitivity towards NE. Negatively charged carboxyl groups make CDs a suitable matrix for protein immobilization. Application in the receptor part layer of a conductive nanomaterial allows improvement of the parameters of a biosensor, like a sensor sensitivity, increasing the limit of detection and lengthening the life of the sensor. The nanomaterial also provides a place for anchoring the protein while maintaining its catalytic activity. Tyrosinase belongs to electron transfer proteins; immobilization of Tyrosinase on the surface of CA/CDs can promote electron transfer between semi-conductive CA/CDs-film and the electrode, with simultaneous oxidation of NE. Additionally, to register the enzymatic activity of tyrosinase immobilized on modified electrode, supplementary measurements for detection system (Au-E/CA/CDs/Tyr) without NE in the investigated sample were conducted (Figure 4B).

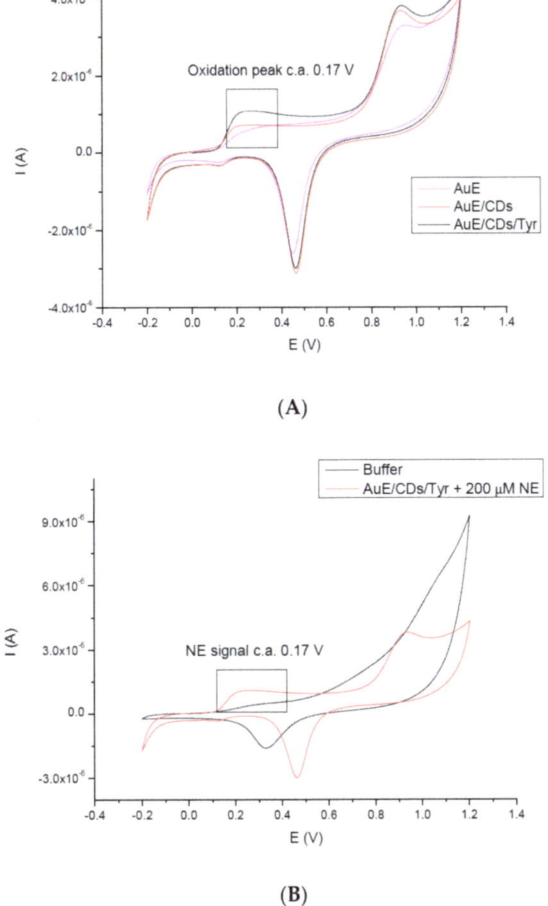

Figure 4. (**A**): Representative CV-scans of the bare Au electrode, Au electrode modified with cysteamine and carbon dots, and Au-E/CA/CDs/Tyr system in the presence of norepinephrine (200 µM) under applied potential in range −0.2–1.2 V, scan rate 50 mV/s vs. Ag/AgCl (0.1 M); (**B**): Representative CVs of the Au-E/CA/CDs/Tyr in the absence and in the presence of 200 µM NE in buffer solution (pH 7.0, scan rate 50 mV/s, applied potential: −0.2–1.2 V vs. Ag/AgCl 0.1 M).

3.3. Calibration and Limit of Detection of NE

Electrochemical signals of NE were investigated in a wide range of concentration (1–200 µM), employing the more sensitive DPV technique in oxygen-saturated conditions. Open-air conditions are in most cases crucial in the construction of such bio-tools, as these devices should work continuously in an oxygen-saturated state. According to Solomon et al. [13], the resting state is mainly in the oxidized form [Tyr–Cu(II)] which can interact with phenol derivatives, such as NE. The products of this catalytic reaction are the oxidized form of NE–*o*-quinone, and the reduced form of the enzyme [Tyr–Cu(I)], as shown in Figure 5A. In Figure 5B the produced *o*-quinone is present, which can be again changed into NE [33].

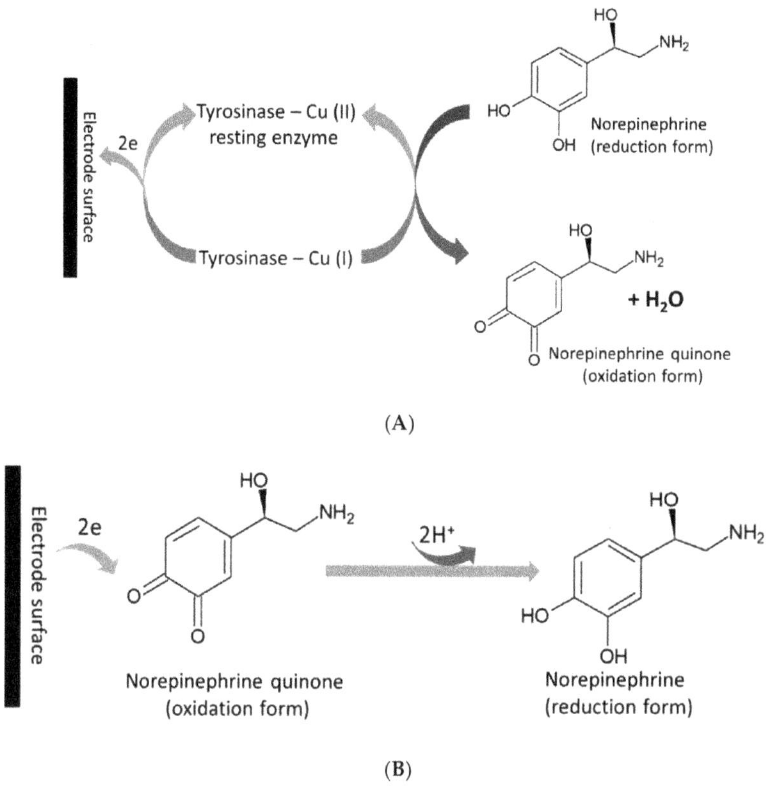

(A)

(B)

Figure 5. (**A,B**). Schematic representation of electron transfer between Tyr and Au-modified electrode (**A**): anodic reaction and (**B**): cathodic reaction.

The main problem during phenol-derivatives (e.g., NE) electrochemical oxidation is the electrode surface deactivation. It is caused by the formation of a passivating-polymeric film produced by the coupling of electrogenerated phenoxy radical [34]. While providing electrochemical measurements (e.g., using CV technique), such deactivation may be visible as a decrease in the oxidation current and an increase in the oxidation potential, when consecutive cycles are performed on the same electrode [2]. In consequence, the sensor loses reproducibility of the measurements, which is one of the main parameters characterizing such tools. Due to this, to characterize the work of the biosensor, the linear range and detection limit were examined using DPV method. Figure 6A shows the oxidation current peaks (I_{pa}) of NE increasing with its concentration (ranging from 1 µM to 200 µM). These signals precisely respond to the given NE concentration. Figure 6B presents a linear response to NE in the studied concentration range; good linear coefficient (R^2 = 0.998) was observed. The biosensor represents an excellent linear response in a broad range of concentrations.

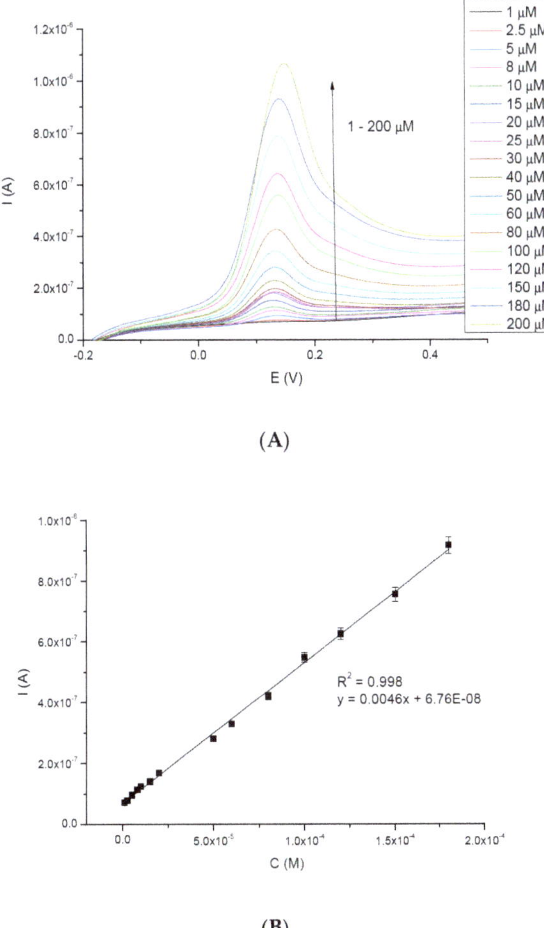

Figure 6. (**A**): Differential Pulse Voltammetry-scans for different concentrations of NE in range 1–200 µM vs. Ag/AgCl (0.1 M) electrode and (**B**): Linear relationship between current and NE concentration (1–200 µM).

Limit of detection for described method (LOD) was calculated as (1) [35]:

$$LOD = 3.29\ \sigma_B/b \tag{1}$$

where σ_B is the standard deviation of the population of blank responses and b is the slope of the regression line.

LOD calculated this way was found at 196 nM. There are only a few reports in the literature describing nanomaterial-based biosensing electrochemical platforms for NE determination. However, this LOD value is comparable to or lower than published results [36–39] which illustrates that Au-E/CA/CDs/Tyr has good sensitivity in wide linear range (Table 1).

Table 1. Comparison of biosensors and sensors for NE determination.

	Biosensor/Sensor	Technique	Linear Range	LOD	Ref.
1	GCE/ECR *	CV, DPV	2–50 μM	1.5 μM	[36]
2	CPE/BH/TiO$_2$ **	DPV	4–1100 μM	0.5 μM	[37]
3	GCE/DNA/AuNPs ***	DPV	0.5–80 μM	5 nM	[38]
4	CPE/NMM ****	CV, DPV	0.07–2000 μM	0.04 μM	[39]
5	Au-E/Cys/CDs/Tyr	DPV	1–200 μM	196 nM	This work

* GCE—glassy carbon electrode, ECR—Eriochrome Cyanine R. ** CPE—carbon paste electrode, BH—2,2′-[1,2 buthanediyl bis(nitriloethylidyne)]-bis-hydroquinone. *** AuNPs—gold nanoparticles. **** NMM—nanostructured mesoporous material.

Limit of quantification (LOQ) was also determined (calculated using Equation (2)) and it equals 312 nM.

$$LOQ = 5 \; \sigma_B/b \tag{2}$$

where σ_B is the standard deviation of the population of blank responses and b is the slope of the regression line [35].

Furthermore, sensitivity of proposed biosensor was found at 4.2 μA mM^{-1}cm^{-2}.

3.4. Selectivity

The selectivity is an extremely important parameter in designing bio-devices for diagnostic purposes. The tyrosinase-based biosensor presented here was developed for quantitative NE monitoring in human samples using DPV technique. In that case, the biosensor has to be selective only for NE detection, with minimal or no influence from other species present in the sample. Human body fluids contain a number of interfering substances, such as the most common ascorbic acid and uric acid. Figure 7 represents a wide range of species which may disturb NE signals, including ascorbic acid (AA), uric acid (UA), cysteine (CYS), epinephrine (EP), dopamine (DA) and 4-tert-Butylcatechol (4tBC). These compounds were added (50 μM) to every investigated NE sample (concentrations of NE 1, 50, 100 μM) to check an impact, while carrying out measurements into their high excess, equilibrium, and deficiency. L-DOPA, as a neurotransmitter and precursor of NE, has been selected as a possible interfering compound. What is more, its chemical structure is similar to NE, so it was necessary to check for possible effects on NE determination using the described procedure. The highest impact on the NE measurements had other neurotransmitters: EP and DA (21.25% and 18.5%, respectively), due to their similar structure to NE. It is known that DPV is an adequate technique for the analysis of the mixture of electrochemically active compounds, because a relatively small difference in their potential peak is needed. Despite the influence of EP and DA equal to c.a. 20%, the distinction of NE from other neurotransmitters and sensitive detection of NE was possible. Other species present negligible effects (<8.7%) on the current peak of the samples, compared to the blank (AA: 10.6%, UA: 5.55%, CYS: 4.59%, 4tBC: 14.15%). The presented results (Figure 7) confirm an insignificant impact on the selectivity of fabricated bio-tool, and prove that existing interferences do not interrupt the prominence of proposed NE test.

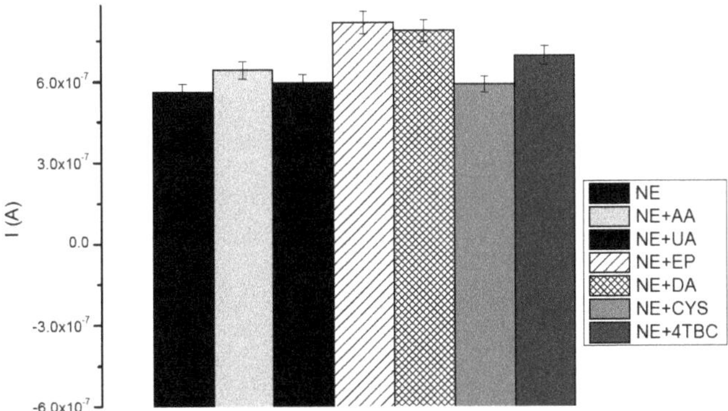

Figure 7. Effect of interfering substances (50 µM) on NE determination.

3.5. Real Application

The practicability of the introduced method was tested by the detection of NE dissolved in synthetic urine CLEANU®, containing, among others, creatinine, uric acid, mineral salts, dyes, urea and water (exact information reserved according to patent). Executed determination of the NE concentration in synthetic urine based on DPV method for three measurements showed an exquisite recovery value (Table 2). These results (calculated as ratio of detected concentration to the real concentration of norepinephrine in the synthetic urine (%)) demonstrate that the proposed strategy is selective, sensitive and suitable for real detection of NE.

Table 2. Results obtained for NE determination based on proposed method.

Concentration of NE in Real Sample [µM]	$C_{detected}$ [µM]	Recovery [%] (Average)	RSD (Calculated for 20 Repetitions)
100.00	98.44		
100.00	99.11	98.70	±0.73
100.00	98.57		

4. Conclusions

In the present study, a new NE biosensing assay was developed and characterized. NE biosensor based on the Au-E/CA/CDs/Tyr bioplatform with high sensitivity, selectivity and practicability. The described method shows an exquisite electrocatalytic activity over a broad linear concentration range ($1–200 \times 10^{-6}$ M) with a detection limit of 196 nM, quantification limit of 312 nM and sensitivity of 4.2 µA mM^{-1}cm^{-2}. It is noteworthy that the modified electrode exhibited good selectivity when tested in a wide range of interfering compounds (AA, UA, CYS, 4tBC, EP, DA). In addition, the obtained biosensor successfully validates the proposed strategy, with adequate recovery, for NE detection in biological samples. All characteristics made for NE determination with Au-E/CA/CDs/Tyr establish a convenient, stable, simple and long-term technique and it can be recommended as an excellent analytical bio-device.

Author Contributions: Conceptualization, S.B. and J.C.; Methodology, S.B.; Synthesis and characterization of CDs A.L., Calculations, S.B.; Writing—Original Draft Preparation, S.B. and A.L.; Writing—Review & Editing, J.C.; Supervision, J.C. All authors have read and agreed to the published version of the manuscript.

Funding: This research received no external funding.

Acknowledgments: The authors gratefully acknowledge for the financial support to Wrocław University of Science and Technology.

Conflicts of Interest: The authors declare no conflict of interest.

References

1. Cammann, K. Bio-sensors based on ion-selective electrodes. *Fresenius' Z. Anal. Chem.* **1977**, *287*, 1–9. [CrossRef]
2. Scott, K. Electrochemical principles and characterization of bioelectrochemical systems. *Micro. Electrochem. Fuel Cells* **2016**, *88*, 29–66.
3. Schröder, U.; Nießen, J.; Scholz, F. A Generation of Microbial Fuel Cells with Current Outputs Boosted by More Than One Order of Magnitude. *Angew. Chem. Int. Ed.* **2003**, *42*, 2880–2883. [CrossRef] [PubMed]
4. Berridge, C.W.; Waterhouse, B.D. The locus coeruleus-noradrenergic system: Modulation of behavioral state and state-dependent cognitive processes. *Brain Res. Rev.* **2003**, *42*, 33–84. [CrossRef]
5. Berridge, C.W.; Schmeichel, B.E.; España, R.A. Noradrenergic modulation of wakefulness/arousal. *Sleep Med. Rev.* **2012**, *16*, 187–197. [CrossRef]
6. McBurney-Lin, J.; Lu, J.; Zuo, Y.; Yang, H. Locus Coeruleus-Norepinephrine Modulation of Sensory Processing and Perception: A Focused Review. *Neurosci. Biobehav. Rev.* **2019**, *105*, 190–199. [CrossRef]
7. O'Donnell, J.; Zeppenfeld, D.; McConnell, E.; Pena, S.; Nedergaard, M. Norepinephrine: A Neuromodulator That Boosts the Function of Multiple Cell Types to Optimize CNS Performance. *Neurochem. Res.* **2012**, *37*, 2496–2512. [CrossRef]
8. Brondani, D.; Weber, S.C.; Dupont, J.; Cruz, V.I. Biosensor based on platinum nanoparticles dispersed in ionic liquid and laccase for determination of adrenaline. *Sens. Actuators B* **2009**, *140*, 252–259. [CrossRef]
9. Florescu, M.; David, M. Tyrosinase-Based Biosensors for Selective Dopamine Detection. *Sensors* **2017**, *17*, 1314. [CrossRef]
10. Arslan, F.; Durmus, S.; Colak, O.; Arslan, H. A New Laccase-Based Biosensor for Epinephrine Determination. *Gazi Univ. J. Sci.* **2015**, *28*, 1–9.
11. Baluta, S.; Zając, D.; Szyszka, A.; Malecha, K.; Cabaj, J. Enzymatic Platforms for Sensitive Neurotransmitter Detection. *Sensors* **2020**, *20*, 423. [CrossRef] [PubMed]
12. Apetrei, I.M.; Apetrei, C. Amperometric Tyrosinase based Biosensors for Serotonin Detection. *Rom. Biotechnol. Lett.* **2013**, *18*, 8253–8262.
13. Solomon, E.I.; Sundaram, U.M.; Machonkin, T.E. Multicopper Oxidases and Oxygenases. *Chem. Rev.* **1996**, *96*, 2563–2605. [CrossRef] [PubMed]
14. Jackowska, K.; Krysiński, P. New trends in the electrochemical sensing of dopamine. *Anal. Bioanal. Chem.* **2013**, *405*, 3753–3771. [CrossRef]
15. Ji, D.; Shi, Z.; Liu, Z.; Low, S.S.; Zhu, J.; Zhang, T.; Chen, Z.; Yu, X.; Lu, Y.; Lu, D.; et al. Smartphone-based square wave voltammetry system with screen-printed graphene electrodes for norepinephrine detection. *Smart Mater. Med.* **2020**, *1*, 1–9. [CrossRef]
16. Bonet-San-Emeterio, M.; Algarra, M.; Petković, M.; del Valle, M. Modification of electrodes with N-and S-doped carbon dots. Evaluation of the electrochemical response. *Talanta* **2020**, *212*, 1–8. [CrossRef]
17. Ding, X.; Niu, Y.; Zhang, G.; Xu, Y.; Li, J. Electrochemistry in Carbon-based Quantum Dots. *Chem. Asian J.* **2020**, *15*, 1–12. [CrossRef]
18. Hiremath, S.D.; Priyadarshi, B.; Banerjee, M.; Chatterjee, A. Carbon dots-MnO$_2$ based turn-on fluorescent probe for rapid and sensitive detection of hydrazine in water. *J. Photochem. Photobiol. A* **2020**, *389*, 1–8. [CrossRef]
19. Zhou, X.; Qu, Q.; Wang, L.; Li, L.; Li, S.; Xia, K. Nitrogen dozen carbon quantum dots as one dual function sensing platform for electrochemical and fluorescent detecting ascorbic acid. *J. Nanopart. Res.* **2020**, *22*, 1–13. [CrossRef]
20. Priyadarshini, E.; Rawat, K.; Bohidar, H.B. Multimode sensing of riboflavin via Ag@ carbon dot conjugates. *Appl. Nanosci.* **2020**, *10*, 281–291. [CrossRef]
21. Yu, H.W.; Jiang, J.H.; Zhang, Z.; Wan, G.C.; Liu, Z.Y.; Chang, D.; Pan, H.Z. Preparation of quantum dots CdTe decorated graphene composite for sensitive detection of uric acid and dopamine. *Anal. Biochem.* **2017**, *519*, 92–99. [CrossRef] [PubMed]
22. Retna, R.C.; Okajima, T.; Ohsaka, T. Gold nanoparticle arrays for the voltammetric sensing of dopamine. *J. Electroanal. Chem.* **2003**, *543*, 127–133.
23. Yang, Z.; Hu, G.; Chen, X.; Zhao, J.; Zhao, G. The nano-Au self-assembled glassy carbon electrode for selective determination of epinephrine in the presence of ascorbic acid. *Colloids Surf. B* **2007**, *54*, 230–235. [CrossRef]

24. Thiagarajan, S.; Chen, S.M. Applications of nanostructured Pt-Au hybrid film for the simultaneous determination of catecholamines in the presence of ascorbic acid. *J. Solid State Electrochem.* **2009**, *13*, 445–453. [CrossRef]

25. Mphuthi, N.G.; Adekunle, A.S.; Ebenso, E.E. Electrocatalytic oxidation of Epinephrine and Norepinephrine at metal oxide doped phthalocyanine/MWCNT composite sensor. *Sci. Rep.* **2016**, *26938*, 1–20. [CrossRef] [PubMed]

26. Samdani, K.J.; Samdani, J.S.; Kim, N.; Lee, J. FeMoO$_4$ based, enzyme free Electrochemical biosensor for ultrasensitive detection of norepinephrine. *Biosens. Bioelectron.* **2016**, *81*, 445–453. [CrossRef]

27. Mohammadi, A.; Moghaddam, A.B.; Hosseini, S.; Kazemzad, M.; Dinarvand, R. A norepinephrine biosensor based on a glassy carbon electrode modified with carbon nanotubes. *Anal. Methods* **2011**, *3*, 2406–2411. [CrossRef]

28. Sahu, S.; Behera, B.; Maiti, T.K.; Mohapatra, S. Simple one-step synthesis of highly luminescent carbon dots from orange juice: Application as excellent bio-imaging agents. *Chem. Commun.* **2012**, *48*, 8835–8837. [CrossRef]

29. Sun, J.; Yang, L.; Jiang, M.; Shi, Y.; Xu, B.; Ma, H.L. Stability and activity of immobilized trypsin on carboxymethyl chitosan-functionalized magnetic nanoparticles cross-linked with carbodiimide and glutaraldehyde. *J. Chromatogr. B* **2017**, *1054*, 57–63. [CrossRef]

30. Pandey, S.; Mewada, A.; Thakur, M.; Tank, A.; Sharon, M. Cysteamine hydrochloride protected carbon dots as a vehicle for the efficient release of the anti-schizophrenic drug haloperidol. *RSC Adv.* **2013**, *3*, 26290–26296. [CrossRef]

31. Arroyave, M.; Springer, V.; Centurión, M.E. Novel Synthesis Without Separation and Purification Processes of Carbon Dots and Silver/Carbon Hybrid Nanoparticles. *J. Inorg. Organomet. Polym.* **2020**, *30*, 1352–1359. [CrossRef]

32. Baluta, S.; Lesiak, A.; Cabaj, J. Graphene Quantum Dots-Based Electrochemical Biosensor for Catecholamine Neurotransmitters Detection. *Electroanalysis* **2018**, *30*, 1781–1790. [CrossRef]

33. Faridnouri, H.; Ghourchian, H.; Hashemnia, S. Direct electron transfer enhancement of covalently bound tyrosinase to glassy carbon via Woodward's reagent K. *Bioelectrochemistry* **2011**, *82*, 1–9. [CrossRef] [PubMed]

34. Wang, J.; Martinez, T.; Yaniv, D.R.; McCormick, L.D. Scanning tunneling microscopic investigation of surface fouling of glassy carbon surfaces due to phenol oxidation. *J. Electroanal. Chem. Interfacial Electrochem.* **1991**, *313*, 129–140. [CrossRef]

35. Desimoni, E.; Brunetti, B. Presenting Analytical Performances of Electrochemical Sensors. Some Suggestions. *Electroanalalysis* **2013**, *25*, 1645–1651. [CrossRef]

36. Yao, H.; Li, S.; Tang, Y.; Chen, Y.; Chen, Y.; Lin, X. Selective oxidation of serotonin and norepinephrine over eriochrome cyanine R film modified glassy carbon electrode. *Electrochim. Acta* **2009**, *54*, 4607–4612. [CrossRef]

37. Mazloum-Ardakani, M.; Beitollahi, H.; Sheikh-Mohseni, M.A.; Naeimi, H.; Taghavinia, N. Novel nanostructure electrochemical sensor for electrocatalytic determination of norepinephrine in the presence of high concentrations of acetaminophene and folic acid. *Appl. Catal. A* **2010**, *378*, 195–201. [CrossRef]

38. Lu, L.P.; Wang, S.Q.; Lin, X.Q. Fabrication of layer-by-layer deposited multilayer films containing DNA and gold nanoparticle for norepinephrine biosensor. *Anal. Chim. Acta* **2004**, *519*, 161–166. [CrossRef]

39. Mazloum-Ardakani, M.; Sheikh-Mohseni, M.A.; Abdollahi-Alibeik, M.; Benvidi, A. Electrochemical sensor for simultaneous determination of norepinephrine, paracetamol and folic acid by a nanostructured mesoporous material. *Sens. Actuators B* **2012**, *171*, 380–386. [CrossRef]

sensors

Article

A Crosstalk- and Interferent-Free Dual Electrode Amperometric Biosensor for the Simultaneous Determination of Choline and Phosphocholine

Rosanna Ciriello and Antonio Guerrieri *

Dipartimento di Scienze, Università degli Studi della Basilicata, Viale dell'Ateneo Lucano 10, 85100 Potenza, Italy; rosanna.ciriello@unibas.it
* Correspondence: antonio.guerrieri@unibas.it; Tel.: +39-0971-205460

Abstract: Choline (Ch) and phosphocholine (PCh) levels in tissues are associated to tissue growth and so to carcinogenesis. Till now, only highly sophisticated and expensive techniques like those based on NMR spectroscopy or GC/LC- high resolution mass spectrometry permitted Ch and PCh analysis but very few of them were capable of a simultaneous determination of these analytes. Thus, a never reported before amperometric biosensor for PCh analysis based on choline oxidase and alkaline phosphatase co-immobilized onto a Pt electrode by co-crosslinking has been developed. Coupling the developed biosensor with a parallel sensor but specific to Ch, a crosstalk-free dual electrode biosensor was also developed, permitting the simultaneous determination of Ch and PCh in flow injection analysis. This novel sensing device performed remarkably in terms of sensitivity, linear range, and limit of detection so to exceed in most cases the more complex analytical instrumentations. Further, electrode modification by overoxidized polypyrrole permitted the development of a fouling- and interferent-free dual electrode biosensor which appeared promising for the simultaneous determination of Ch and PCh in a real sample.

Keywords: choline analysis; phosphocholine analysis; choline oxidase; alkaline phosphatase; enzyme immobilization; overoxidized polypyrrole; electropolymerized non-conducting polymer; dual electrode biosensor; simultaneous determination; flow injection analysis

check for **updates**

Citation: Ciriello, R.; Guerrieri, A. A Crosstalk- and Interferent-Free Dual Electrode Amperometric Biosensor for the Simultaneous Determination of Choline and Phosphocholine. *Sensors* **2021**, *21*, 3545. https://doi.org/10.3390/s21103545

Academic Editor: Ajeet Kaushik

Received: 5 May 2021
Accepted: 14 May 2021
Published: 19 May 2021

Publisher's Note: MDPI stays neutral with regard to jurisdictional claims in published maps and institutional affiliations.

1. Introduction

Choline (Ch), a vitamin-like nutrient found in many common foods, is essential for several biological functions in human body [1]. Three different pathways are involved in its In Vivo metabolism [1]. Its phosphorylation by choline kinase, an enzyme widely distributed in mammalian tissues, gives phosphocholine (PCh), the precursor involved in the Kennedy pathway for the biosynthesis of phosphatidylcholine. On the other hand, mainly in liver and kidney, Ch is irreversibly oxidized to betaine, a methyl donor involved in the biosynthesis of methionine. Last, but not less important, Ch is acetylated, essentially in cholinergic neurons, by acetylcholine transferase to acetylcholine (ACh), an important and well-known neurotransmitter.

The Ch uptake and its metabolism are tissue-dependent so the levels of Ch and its metabolites are related to tissue growth and plausibly to carcinogenesis [2,3]. In fact, Ch levels change significantly in malignant tissue transformations [3] while abnormal PCh levels were found in many human cancers and transformed cell lines [2]; even several chemicals, well-known to be carcinogenic, appear to stimulate the formation of PCh in several cell types [2]. As a proof of these findings, abnormal levels of Ch and PCh were found in some human tumors [4], mainly in breast cancers [3,5–7]; in addition, some studies pointed out similar features also for non-Hodgkin's lymphoma [8] and endometrial cancers [9]. Too, abnormal levels were also observed for other diseases like chronic pancreatitis [10], polycystic ovary syndrome [11], and some serious neurological diseases [12,13], thus pointing out the relevance of Ch and PCh levels in clinical diagnostics. Remarkably, Ch and PCh levels are related due to the Kennedy pathway, so often the ratio of Ch/PCh levels, rather than their absolute levels, is considered as a diagnostic and prognostic marker.

While many analytical methods have been developed for Ch analysis, few approaches were dedicated to PCh and even fewer for the simultaneous determination of both analytes. Pomfret et al. [14] described a procedure able to isolate and detect Ch and PCh in biological tissues and fluids. Unfortunately, the approach appears quite complex, laborious, and time consuming since requiring a preliminary HPLC separation of analytes, further isolation by TLC, chemical conversion of PCh in Ch for at least 24 h, and final quantitation by GC-MS after Ch derivatization. Koc et al. [15] circumvented all these separation procedures by using a LC separation coupled to electrospray ionization—isotope dilution mass spectrometry (LC/ESI-IDMS). This method permitted the analysis of Ch, PCh, and other related compounds in e.g., mouse liver and rat fetal brain, but required the use of deuterium-labeled internal standards of all the analytes, lengthy time analysis, and much more expensive equipment. Bioimaging of PCh, cholesterol, and galactosylceramide on rat cerebellar cortex was possible by time-of-flight secondary-ion-mass-spectrometry (TOF-SIMS) [16], but unfortunately, simultaneous detection of Ch was not reported. A hydrophilic interaction liquid chromatography combined with high resolution ESI mass spectrometry (HILIC LC-MS/MS) method has been proposed for the analysis of Ch, PCh, and other related Ch compounds in egg yolks [17] and rat livers [18]. Nevertheless, this approach requires the use of deuterated standards and a complex as expensive equipment like QTRAP™ MS which heavily limit its use in clinical laboratories; fortunately, Mimmi et al. [19,20] showed that the same approach could be pursued with less MS demand, but 1D NMR was required as a complementary analytical technique. Finally, an ultrahigh performance liquid chromatography (UPLC)—tandem mass spectroscopy was developed for the simultaneous analysis of Ch, PCh, and other involved neurotransmitters and metabolites [21] but this approach was validated only for zebrafish embryos and larvae.

Certainly, NMR spectroscopy represents today one of the most used approach for human cancer characterization, diagnostic and prognostic analysis in clinical environment [4] and hence for the simultaneous analysis of Ch and PCh in those tissues. Further, 1D and 2D proton NMR has been used for the biochemical characterization of perchloric acid extracts of metastatic lymph nodes of breast cancer patients [5]. ^{31}P edited ^{1}H NMR spectroscopy permitted quantification of Ch, PCh, and other analytes in human brain tumor extracts [22]. ^{1}H high resolution magic angle spinning NMR spectroscopy allowed metabolic profiling of Ch and PCh in human lung cancer tissues [23]. High field (7T) ^{31}P NMR imaging coupled to ^{1}H NMR spectroscopy demonstrated useful for the non-invasive determination of important biomarkers in human breast [7] and prostate [24] cancers but determination of Ch and PCh required two techniques and cannot be considered a simultaneous determination. ^{1}H high resolution NMR spectroscopy was recently described as able to analyze Ch and PCh in fine needle aspirate biopsies of breast cancer lesions [25]. In Vivo ^{31}P NMR imaging allowed metabolic profiling of human breast cancer xenografts [26] but Ch analysis was not possible. NMR approaches were also described for the analysis of Ch and PCh in solutions [27], in liver tissues [28], as well as in ovarian cancer diagnostic [29] but this latter required peculiar signal data processing [30,31] instead of the conventional Fourier transform usually found in commercial NMR data stations.

Even if representing the best actual approach in clinical studies and analysis, NMR spectroscopy is not without drawbacks when applied for Ch and PCh simultaneous analysis. As an example, water-suppressed ^{1}H NMR spectroscopy is required for total choline containing compounds analysis [4]. In addition, the water signal is used as an internal reference for the absolute quantification of metabolites so additional measurements are required for the acquisition of this internal reference [32]. Further, ^{1}H NMR spectroscopy is not able to resolve Ch and PCh signals as well as those between PCh and related phospholipid compounds [4,7] even in simple biological fluids like urine [33] so that the use of ^{13}C labeling is required [3]; a similar failure was observed also for ^{31}P NMR when decoupling of PCh and phosphoethanolamine signals is required [4]. Even in the most capable approaches (as above described), finding or using a "blank" tissue (i.e., tissues with cells which are the normal counterparts) for comparing cancer spectra is not straightforward [4]: this is possible for breast, primary liver, prostate, and glioma cancers but not for sarcomas, lymphomas, melanomas, and head or neck carcinomas where the background contains skeletal muscle, which of course cannot be considered the normal counterpart; moreover, for primary brain tumors, where the single voxel localization technique is used to circumvent interferent signals from normal brain parts, the tumor is usually surrounded by edema, so in NMR scans, it is difficult to establish where the cancer ends and the edema begins [4].

As surely it is evident, the above approaches, whatever well-established in some cases, all show a restricted or at least complicated practicability since requiring complex and expensive instruments and thus skilled operators or/and time-consuming and complicate sample pre-treatments, e.g., do not satisfy the requirements for a simple, fast, specific, and accurate analysis, usually necessary in clinical and medical fields. In this respect, immobilized enzyme sensors, i.e., biosensors, can play undoubtedly an important role since compacts, easy in their use, almost cheap and producing analytical responses quickly, with high sensitivity and specificity, thus not requiring any sample pre-treatment. Unfortunately, enzymatic approaches for Ch and PCh simultaneous analysis are still not described till now. Masoom et al. [34] described a method based on immobilized enzyme reactors (IMERs) and amperometric detection of the released hydrogen peroxide according to the following enzymatic reactions with both analytes:

$$phosphocholine + H_2O \xrightarrow{ALP} > choline + phosphate$$

$$choline + 2O_2 + H_2O \xrightarrow{ChO} > betaine + 2H_2O_2$$

Accordingly, using a choline oxidase (ChO)-based IMER or an alkaline phosphatase (ALP)-based IMER coupled in series with a ChO-IMER, they were able to analyze Ch and PCh, respectively, but a simultaneous determination of both analytes was not reported. An HPLC procedure coupled to an acetylcholine esterase/ChO-IMER followed by amperometric detection permitted the analysis of Ch and its metabolites in rat brain and body fluids [35] as well as in mouse tissue [36]. Nevertheless, PCh analysis necessitated sample pre-treatment and incubation with ALP before analysis, whereas its quantification came from the increase of free Ch peak after ALP hydrolysis so that the simultaneous determination of both analytes was unachievable.

In the field of Ch biosensing, we have already described a fast response and sensitive amperometric biosensor for Ch and ACh analysis [37]. Based on ChO and acetylcholine esterase (AChE) immobilized onto a Pt electrode through co-crosslinking with bovine serum albumin by glutaraldehyde, the immobilization procedure there developed proved rapid, easy but robust and simply adaptable for any working electrode configuration, even for conventional HPLC flow cells, thus permitting, e.g., the development of a novel HPLC detector for Ch and ACh determination in brain tissue homogenates [38]. Further, this latter can easily be modifiable in dual electrode configuration, thus permitting, after proper anti-fouling and anti-interference electrode modification (see below), the simultaneous detection of both analytes without the need of any chromatographic separation [39]. The developed biosensor proved even useful for the realization of a disposable device aimed at a rapid screening of AChE activity in soil extract [40] or to the development of novel clinical assays for serum cholinesterase activity [41] or Ch detection in biological fluids from patients on hemodialysis [42].

Therefore, in the present paper, it will be described for the first time the development of a novel amperometric biosensor for PCh analysis based on ALP and ChO co-immobilized onto a Pt electrode by co-crosslinking; more importantly, it will be demonstrated that a simultaneous determination of Ch and PCh can be achieved by a crosstalk-free dual electrode amperometric biosensor based on ChO and ChO-ALP enzyme immobilized onto Pt electrodes by co-crosslinking. Electropolymerized non-conducting films with built-in permselectivity [39,41–54] have widely demonstrated to be able, in real sample analysis (e.g., blood, serum), to protect the sensor surface from electrode fouling effects and to remove, with unmatched efficiency, common interfering species in glucose, Ch, and Ach, as well as in lysine determination. Accordingly, electrode modification by overoxidized polypyrrole permitted us to develop a fouling- and interferent-free dual electrode biosensor which appeared promising for the simultaneous determination of Ch and PCh in the clinical sample.

2. Materials and Methods

2.1. Materials

Choline chloride, phosphocholine chloride calcium salt tetrahydrate, choline oxidase (EC 1.1.3.17 from *Alcaligenes species*, 14 U/mg of solid), alkaline phosphatase (EC 3.1.3.1 from bovine intestinal mucosa, 13 DEA units/mg of solid, product code P7640), alkaline phosphatase (EC 3.1.3.1 from bovine calf intestinal, 2,000 DEA units, product code P7923), alkaline phosphatase (EC 3.1.3.1 from shrimp, 1,300 DEA units, product code P9088), ascorbic acid, *L*-cysteine, glutaraldehyde (grade II, 25% aqueous solution), and bovine serum albumin (fraction V) were bought from Sigma Chemical Co. (St. Louis, MO, USA). Alkaline phosphatase (EC 3.1.3.1 from bovine intestinal mucosa, 22,990 U/mL, product code 79835), uric acid, and pyrrole came from Aldrich (Steinheim, Germany). Analytical reagent grade chemicals were used as other materials. Double distilled-deionized water was used to prepare all solutions. Choline chloride was dried under vacuum over P_2O_5 for at least 3 days and stored in a vacuum desiccator at 4 °C. When required, phosphocholine was purified by using a Strata-X-C 33 μm polymeric strong cation exchange tube (Phenomenex, Castel Maggiore (BO), Italy). Pyrrole was purified by vacuum distillation (62 °C) and stored at 4 °C. Choline and phosphocholine stock solutions were stored at 4 °C: their dilute solutions, pyrrole and interferant solutions were prepared just before their use.

2.2. Apparatus

Batch electrochemical, rotating disk electrode (RDE) experiments were performed using an AMEL (Milan, Italy) Model 466 polarographic analyzer coupled to a JJ instruments mod. CR65OS Yt chart recorder. The electrochemical cell was a conventional three-electrode system consisting of an Ag/AgCl, KCl satd. reference electrode, a Pt counter electrode, and a Pt RDE. The RDE was a CTV101 Speed Control Unit, EDI 101 Rotating Disc Electrode (Radiometer, Copenhagen, Denmark).

Flow injection experiments were performed using a Gilson Minipuls 3 peristaltic pump (Gilson Medical Electronics, Villiers-Le-Bel, France) and a seven-port injection valve (Rheodyne mod. 7725, Cotati, CA, USA) with a 20 μL sample loop. The electrochemical detector was an EG&G Model 400 (Princeton Applied Research, Princeton, NJ, USA) including a thin-layer electrochemical cell with a single Pt disk (3 mm diameter) working electrode and an Ag/AgCl, 3 M NaCl reference electrode; for dual electrodes experiments, a dual Pt disk (both 3 mm diameter) working electrode (parallel configuration) was used; in both cases, two thin layer flow cell dual gaskets (Bioanalytical Systems, West Lafayett, IN, USA) of 0.004-inch thickness were used. A PEEK tubing (0.25 mm ID, 150 cm length) was used to connect the sample injection valve to the electrochemical detector. A Kipp & Zonen (Delft, The Netherlands) mod. BD 12 Flatbed recorder was used for flow injection signal chart recording. The flow injection setup is illustrated in Scheme S1 in Supplementary Materials.

Controlled electrochemical deposition of polypyrrole film onto the single or dual Pt electrodes was carried on a conventional three electrode cell equipped with an Ag/AgCl, KCl satd. reference electrode and a Pt counter electrode by using an EG&G model 263A potentiostat/galvanostat equipped with a M270 electrochemical research software (EG&G

Princeton Applied Research, Princeton, NJ, USA) version 4.23 for data control and acquisition.

2.3. Methods

2.3.1. Electrode Modification

Before each electrode modification, the Pt disk working electrode of the thin-layer electrochemical cell was preliminary cleaned by few drops of hot nitric acid, washed with bidistilled water, then polished to a mirror finish by alumina (0.05 μm particles), and finally extensive washed and sonicated in bidistilled water. To check the presence of electrode surface impurities as well as for electrode conditioning, the electrodes were cycled in 0.5 M sulfuric acid solution between -0.210 and $+ 1.190$ V (vs. Ag/AgCl, KCl satd.) at 200 mV/s until a steady state cyclic voltammogram was obtained, followed by copious rinsing with double distilled-deionized water.

Platinum disk working electrodes were modified by electrosynthesized polypyrrole (PPy) films by potentiostatically growth at $+ 0.7$ V (vs. Ag/AgCl, KCl satd.) from a 0.1 M KCl supporting electrolyte solution containing 0.4 M pyrrole; a deposition charge of 300 mC/cm^2 was typically used permitting the growth of an approximately 0.7 μm thick film [55]. The so-obtained Pt/PPy modified electrodes were successively overoxidized at $+ 0.7$ V (vs. Ag/AgCl, KCl satd.) in phosphate buffer (pH 7, I 0.1 M) for about 6 h until a steady-state background current was observed; the resultant overoxidized Pt/PPy electrodes (Pt/oPPy) were then washed with double distilled-deionized water and air-dried at room temperature.

2.3.2. Biosensor Preparation

Choline oxidase (ChO) was immobilized onto bare Pt or Pt/oPPy electrodes as follow. Then, 1.00 mg of ChO (corresponding to 14 units) and 16 mg of bovine serum albumin (BSA) were carefully dissolved avoiding air bubble formation into 200 μL of phosphate buffer (pH 6.5, I 0.1 M). The resulting solution was divided in two identical aliquots of 100 μL. An aliquot was mixed with 50 μL of phosphate buffer (pH 6.5, I 0.1 M) and 15 μL of 2.5% glutaraldehyde (GLU) solution (25% GLU solution diluted 1:10 with phosphate buffer pH 7.4, I 0.1 M) and used for ChO biosensor preparation. Alkaline phosphatase (ALP) and ChO were co-immobilized onto bare Pt or Pt/oPPy electrodes as follow. The remaining aliquot of the previous prepared ChO-BSA solution was mixed with 50 μL of ALP (22,990 U/mL) and 15 μL of 2.5% GLU solution and used for ChO-ALP biosensor preparation.

ChO and ChO-ALP biosensors were prepared as follow. Three μL of the enzyme solution from the appropriate aliquots described above were carefully pipetted onto the bare Pt or Pt/oPPy disk working electrode surface and meticulously spread out to cover completely the electrode surface avoiding air bubble formation. After that, the so modified electrode was left to crosslinking and air-drying at room temperature for few minutes. To avoid any cross-contamination from weakly bounded ALP onto ChO biosensor in dual electrodes experiments, both electrodes were successively cured with 10 μL of BSA solution (16 mg of BSA into 300 μL of phosphate buffer pH 6.5, I 0.1 M) to saturate any free glutaraldehyde residues left onto the enzyme membrane. Before their first use, the biosensors were preliminarily soaked in the background electrolyte for few minutes to allow removing of weakly bound or adsorbed enzyme and swelling of the enzyme layer. When not in use, the biosensor was stored in phosphate buffer, pH 6.5, I 0.1 M, at 4 °C in the dark.

2.3.3. Electrochemical Measurements

The detection potential in all the electrochemical measurements was chosen as the lowest potential value, showing maximum sensitivity and pH independence towards hydrogen peroxide oxidation and, hence, towards analyte detection.

The detection potential in batch electrochemical, RDE experiments was + 0.7 V (vs. Ag/AgCl, KCl satd.) and the rotation rate was 1000 rpm. A borate buffer (pH 9.0, *I* 0.1 M) was used as supporting electrolyte and for sample analysis. Solutions were air saturated and the temperature was ambient.

In flow injection electrochemical measurements, the detection potential was +0.7 V (vs. Ag/AgCl, 3 M NaCl). Unless otherwise specified, in flow injection experiments, the carrier stream was a borate buffer (pH 9.0, *I* 0.1 M), the sample injection volume was 20 μL, and flow rate was 1 mL/min; solutions and carrier stream were air saturated and the temperature was ambient.

3. Results and Discussion

3.1. Enzyme Immobilization and Choice of ALP Source

Choline oxidase (ChO) was immobilized and ChO and alkaline phosphatase (ALP) were co-immobilized onto bare Pt or modified Pt electrodes by co-crosslinking [56], i.e., by crosslinking with glutaraldehyde (GLU) through the use of an inert protein, bovine serum albumin (BSA), sweetening the crosslinker reactivity. This valuable immobilization approach allowed us to obtain an immobilized enzyme layer with high biocomponent activity and good mechanical properties by simply casting a small amount of the proper co-crosslinking enzyme solution onto the electrode surface [37]. Obviously, the pH of enzyme immobilization solution and the inert protein and crosslinker concentrations have strong impact on the efficiency of enzyme immobilization and their catalytic properties so that their influence was studied and optimized. In agreement with a previous finding [37], the use of a phosphate buffer (pH 6.5, *I* 0.1 M) assured a good catalytic and mechanical stability of the immobilized enzyme layer while reducing the co-crosslinking reaction time; similarly, a BSA concentration of about 50 mg/mL and a GLU concentration of nearly 0.2% permitted high enzyme loadings, reasonable gelation times while reducing undesired denaturation effects. Finally, a volume V_c of 3 μL of the enzyme solution cast onto the electrode surface assured the formation of a strongly adherent, thin enzymatic membrane on the top of the working electrode with high permeation and low response time towards both analytes and the produced hydrogen peroxide (for a deepening about the influence of pH, BSA, GLU concentrations, and V_c, as well as for a morphological study see references [37,42,53]).

In agreement with previous findings [37–42,54], ChO from *Alcaligenes species* (the almost unique available commercial source) proved successful for the realization of Ch biosensors. ALP, vice versa, comes from different sources and distinctive commercial preparations (see Section 2) so the choice of a satisfactory ALP source for successful PCh biosensing required the preparation of different Pt/ChO-ALP biosensors at different ratios of their enzyme activities and their preliminary analytical characterization in batch electrochemical, rotating disk electrode (RDE) experiments. In this context, absolute and relative analyte sensitivities, response time, operational and long-term stability of the relevant biosensors were the analytical performances here studied to select the optimal enzyme source (see Table S1 in Supplementary Materials for a relevant survey about). Almost all the enzyme sources gave biosensors with low absolute and relative PCh sensitivity respect to Ch one. Interesting, increasing the ALP to ChO activity ratio to increase the relative PCh sensitivity towards Ch did not give the expected result since the PCh absolute sensitivity remained almost the same. This suggested that the PCh response poorly depends on ChO loading (i.e., it is saturating) while Ch response is ChO loading limiting. Accordingly, tweaking the ALP to ChO activity ratio was unsuccessful for optimizing the performances of most biosensors. Anyway, among the different enzyme sources, ALP from bovine intestinal mucosa (22,990 U/mL, product code 79835, see Section 2) permitted the fabrication of Pt/ChO-ALP biosensors with the highest Ch and PCh absolute sensitivities, suitable sensitivity ratio, and high operational and long-term stability; notably, Pt/ChO-ALP biosensors as here developed showed enzyme layers so stable under stirred or flowing solutions and response times (as evaluated in batch RDE experiments) so low for both analytes (about

1 s) to allow their use in flow injection analysis without any distortion of the relevant flow injection sample peaks. Accordingly, this enzyme source was selected for further experiments where ChO and ALP were co-immobilized onto the Pt working electrode of a typical thin-layer electrochemical cell (see Scheme S1 in Supplementary Materials for further details).

3.2. Influence of Buffer Composition and Its pH

Preliminary, the effect of buffer composition was studied by using different buffers. In agreement with previous findings [54], phosphate and borate buffers showed the highest responses towards Ch at Pt/ChO-ALP electrodes; anyway, the PCh used in our experiments comes from its calcium salt (see Section 2), so phosphate buffers were excluded to avoid any calcium phosphate precipitation in solution.

The influence of pH on the Pt/Cho-ALP biosensor responses towards Ch and PCh was investigated in the pH range 6–10 using a universal buffer (acetate/borate/tris) at fixed ionic strength (I 0.1 M) as supporting electrolyte (note that the influence of pH on Pt/ChO biosensor was studied and reported elsewhere [37]). As Figure S1 in Supplementary Materials shows, a characteristic bell-shaped curve was observed for both analytes with a maximum located at about pH 9; notably, the behavior observed at the Pt/Cho-ALP biosensor towards Ch agrees with that observed at a Pt/ChO biosensor [37], suggesting that the immobilized ALP does not induce any significant change on ChO kinetics. Further, the observed maximum for immobilized ALP agrees with the optimal pH reported for its soluble form [57,58], indicating that the enzyme membrane does not influence significantly the ionization process of the immobilized enzyme.

Finally, the effect of magnesium ions on PCh response at the Pt/Cho-ALP biosensor was also investigated since an enhancement of ALP activity by Mg^{++} was reported elsewhere [59]. On contrast, in the present case, no PCh response increase was observed using magnesium chloride up to 0.5 mM so this salt was not used in the supporting electrolyte.

Accordingly, a borate buffer at pH 9 was used in all the experiments.

3.3. Analytical Performances of the Dual Electrode Biosensor

Due to the enzymatic detection approach here used, where PCh needs firstly to be converted to Ch for its detection (see the enzyme reactions in the Introduction section), a single Pt/ChO-ALP biosensor can just perform an integrated determination of both analytes. The feasibility of a simultaneous determination of Ch and PCh by flow injection analysis has been here studied developing a dual electrode amperometric biosensor (see Scheme 1 and Scheme S1 in Supplementary Materials for further details), provided that the Pt/ChO-ALP biosensor gives additivity of Ch and PCh responses and crosstalk side effects due to e.g., hydrogen peroxide detection by both the electrodes are absent (vide infra).

In this respect, Figure 1 shows typical flow injection responses observed at the dual electrode, amperometric biosensor due to replicate injections of standard solutions of both analytes. Flow injection peaks showed peak shapes and widths quite similar to those observed for common electroactive substances (e.g., hydrogen peroxide) at bare Pt electrode, thus suggesting that the enzymatic conversion is much faster than flowing sample and the presence of the enzymatic membranes onto the electrode surface does not hinder the permeation of both substrates and the product of the enzymatic reaction.

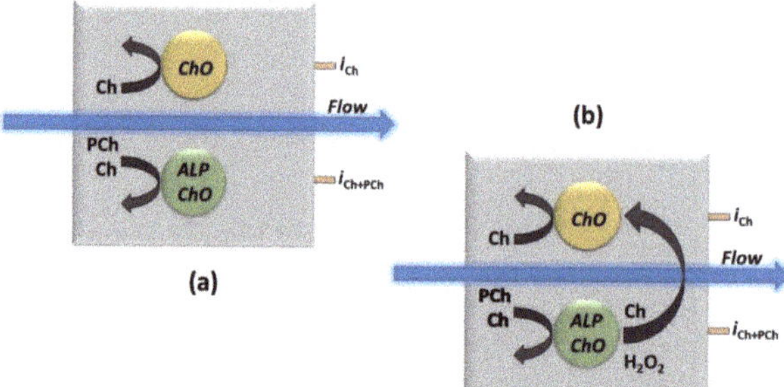

Scheme 1. The approach for the simultaneous determination by the dual electrode (parallel configuration) amperometric biosensor thin-layer cell (as shown in Scheme S1 in Supplementary Materials) used in the flow injection experiments for Ch and PCh analysis (**a**) and the example of a crosstalk effect in measurements (**b**). ChO and ALP/ChO refer to ChO and ALP/ChO biosensors, respectively, the black curved arrows (left side in both sketches) represent the enzymatic detection, the blue ones, the carrier flow direction while i_{Ch} and i_{PCh} the currents for Ch and PCh measurements, respectively, as measured at their respective electrical connections. The larger black curved arrow (right side in right sketch) represents the potential crosstalk effect due to hydrogen peroxide detection from the side electrode.

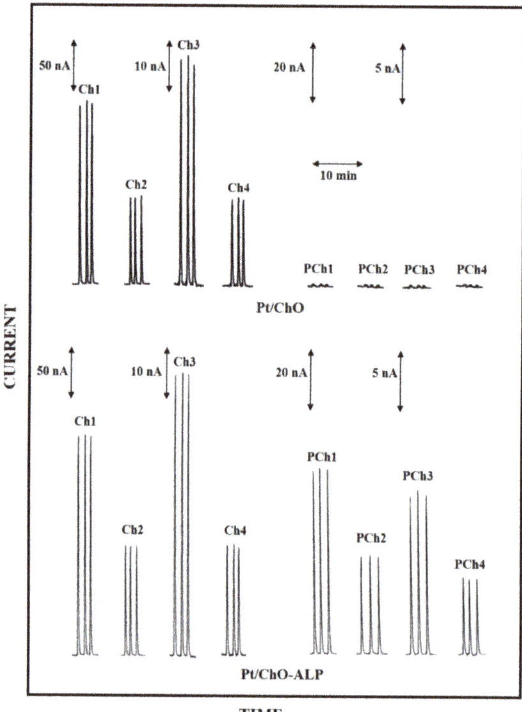

Figure 1. Typical flow injection responses at the dual electrode biosensor for replicate injections of Ch 0.1 mM (Ch1), 0.05 mM (Ch2), 0.025 mM (Ch3), 0.01 mM (Ch4) and PCh 0.5 mM (PCh1), 0.25 mM (PCh2), 0.1 mM (PCh3), 0.05 mM (PCh4). Upper and lower traces refer to responses at Pt/ChO and Pt/Cho-ALP biosensors, respectively. Carrier solution: borate buffer (pH 9, *I* 0.1 M); flow rate 1 mL/min; injection volume 20 µL. Other experimental conditions as described in Section 2.

Due to the detection scheme, Ch responses were observed at both the enzyme modified electrodes (i.e., Pt/ChO and Pt/ChO-ALP, left side upper and lower traces in Figure 1, respectively) while PCh responses were observed only at the Pt/ChO-ALP biosensor (right side lower traces in Figure 1), whereas they are practically absent at the Pt/ChO electrode (right side upper traces in Figure 1) if not due to injection artefacts (vide infra). As Figure 1 demonstrates, the dual electrode amperometric biosensors are practically crosstalk-free (i.e., no PCh responses were observed at the parallel Pt/ChO electrode due to e.g., hydrogen peroxide lateral diffusion from the side Pt/ChO-ALP biosensor). Further, the occurrence of somewhat cross-contamination from weakly bounded ALP onto ChO biosensor during their use (and thus generating crosstalk) was avoided by saturating any free GLU residues left onto the enzyme membrane by BSA (a provision used for biosensor preparation, see Section 2). Indeed, in some cases, a crosstalk effect was observed for some PCh batches, with spurious responses at the Pt/ChO biosensor, which increased with the PCh concentration increase: fortunately, this was only due to traces of Ch in some commercial preparations, as verified by purifying PCh standard solution by a strong cation exchange column, so care should be used in selecting and using PCh standards.

Figure 2 shows typical calibration curves at the dual electrode biosensor. As can be seen, all plots displayed linear and saturated responses at low and high concentrations, respectively, for both analytes and for both biosensors, confirming the kinetic behavior expected for an enzyme catalysis. Michaelis–Menten data fitting (correlation coefficients better than 0.997) gave apparent kinetic constants reported in Table 1. As Figure 2 and Table 1 show, ALP co-immobilization had a detrimental effect on ChO maximal response (i.e., the maximal current I'_{max} for Ch at Pt/ChO-ALP was about half with respect to Pt/ChO) but the apparent Michaelis–Menten constants K'_m were almost the same suggesting that the ChO loading in ChO-ALP membrane is lower, e.g., the co-immobilization of ALP lowers the efficiency in ChO immobilization. Finally, I'_{max} for PCh was even lower while K'_m was higher. Fortunately, all these kinetic features will not preclude the analytical performances of both biosensors, as it will show below.

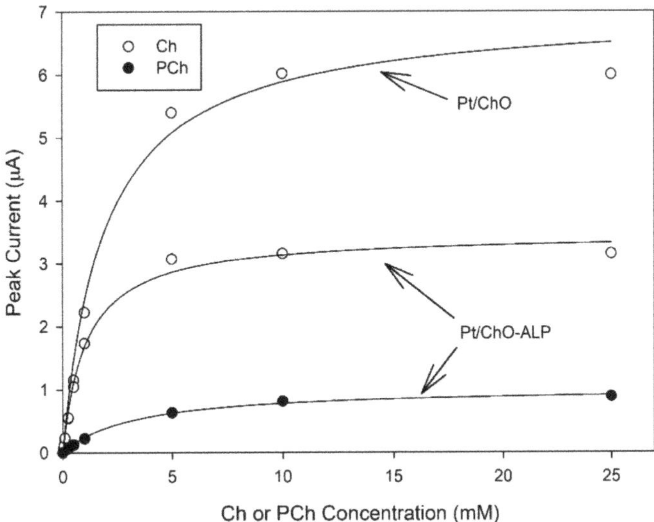

Figure 2. Calibration curves at the dual electrode biosensor for replicate injections of Ch and PCh. Continuous lines refer to Michaelis-Menten fitting of data. Carrier solution: borate buffer (pH 9, I 0.1 M); flow rate 1 mL/min; injection volume 20 µL. Other experimental conditions as described in the Section 2.

Table 1. Analytical performances at the (unmodified) dual electrode biosensor [1].

Analyte	Biosensor	K'_m [2] (mM)	I'_{max} [3] (µA)	Sensitivity (µA/mM)	Linear Range (µM)	LOD [4] (µM/pmol)
PCh	Pt/ChO-ALP	3.6 ± 0.2	1.03 ± 0.02	0.216	7–1000	6.9/139
Ch	Pt/ChO-ALP	1.19 ± 0.07	3.6 ± 0.1	2.03	0.7–1000	0.68/13.6
Ch	Pt/ChO	2.1 ± 0.3	7.3 ± 0.3	2.13	0.7–1000	0.65/13.0

[1] Carrier solution: borate buffer (pH 9, I 0.1 M); flow rate 1 mL/min; injection volume 20 µL. Other experimental conditions as described in Section 2; [2] K'_m: apparent Michaelis-Menten constant; [3] I'_{max}: apparent maximal current; [4] LOD: relative (µM) and absolute (pmol injected) limit of detection calculated at a signal-to-noise ratio of 3.

Linear parts of calibration curves from Figure 1 are shown in Figure S2: as can be seen, the dual electrode biosensor gave excellent linear responses for both analytes and at both the electrodes. Linear fitting of their data (correlation coefficients better than 0.999) gave the sensitivities reported in Table 1: sensitivities towards Ch were practically identical for both the sensors and the responses were linear for more than 3 decades of Ch concentrations (see Table 1). Sensitivity towards PCh detection (see Table 1) was almost a tenth than Ch but sufficiently higher to permit PCh detection as expected in real sample (see below); of course, the lower sensitivity reduced the PCh linear range (see Table 1) due to a higher detection limit (see below). Lastly, the within-a-day coefficients of variations (CV) for replicate (n = 5) PCh injections were 4.3%, 2.1%, and 1.9% at 250, 25, and 5 µM PCh levels, respectively. At the same, the Pt/ChO-ALP electrode, CVs for replicate (n = 5) Ch injections were 2.9%, 2.3%, and 2.5% at 100, 10, and 1 µM PCh levels, respectively, whereas at the Pt/ChO electrode, were 1.3%, 3.0%, and 3.4% at the same Ch levels.

Figure 3 shows the dual electrode biosensor responses for replicate injection of Ch and PCh at concentration levels close to their detection limits. As can be seen, the responses are mainly limited slightly by a signal drift (almost due to the peristaltic pump used in the flow injection setup) but significantly by the presence of injection artefacts on analyte determination (see responses in figure where B refers to replicate injections of the carrier solution) due to the change of pressure occurring at the injection step: both signal limitations were included in the noise floor analysis. In this respect, Table 1 reports the observed limit of detection (LOD) for Ch and PCh at a signal-to-noise ratio of 3. Please note that in Table 1 and whenever it applies, the absolute limit of detection expresses the minimum amount of analyte detectable, while the relative limit of detection refers to the minimum concentration of analyte detectable. Of course, these values can be significantly lowered by using a pulseless pumping system and an improved sample injector as proved elsewhere [38]. Nevertheless, the present LOD values are much better than those reported using the complex and expensive NMR spectroscopy techniques. In fact, for Ch, they are typically in the millimolar range [22,32], sometimes requiring scan times up to 10 h; only the high sophisticated 1H–14N HSQC detection technique [27] gives (estimated) LOD of about 2 µM using a 16 min scan time (reported LOD for Ch is 1 mM at a S/N = 1700), which is still higher than the present LODs. Analytical approaches using the enzimatic conversion, as in the present case, but using enzyme reactors and electrohemical detection, described higher LODs [34]. Further, neither LODs for PCh, neither a simultaneous determination with Ch were reported [34–36]. LC/MS standard approaches [14,15,17,18] stated in the best cases both absolute and relative LODs, comparable with those reported in Table 1; anyway, more complex and expensive LC/MS techniques [21] reported LODs down to nanomolar range. Even if not so sensitive as the last technique, the present approach shows linear ranges and LODs more than adequate for real sample analysis (see references [14,15,17,18,21,22,27,32,34–36] for sample and analyte ranges).

In flow injection approaches using an amperometric biosensor as in the present case, flow rate is well-known to rule the apparent enzyme kinetics and the measured currents [60]. In this respect, Figure S3 reports the flow rate dependences for Ch and PCh at the dual amperometric biosensor. For both analytes, the biosensor responses decreased, increasing the flow rate, leveling off starting from about 1 mL/min. This behavior, already reported and rationalized elsewhere (see for example references [37–39,42]), is anticipated since increasing the flow rate reduces the enzymatic conversion efficiency and/or diffuses away from the enzyme membrane (and hence from the Pt electrode) that produced hydrogen peroxide. Figure S3 apparently shows better analytical performances at low flow rate since of the higher responses: unfortunately, injection artefacts increased as well (see Figure S4), thus nullifying the increased response of analyte. Indeed, while at 0.2 mL/min, LOD for PCh was 6.5 μM (corresponding to 130 pmol injected), LODs for Ch at the same flow rate were 0.69 μM (13.7 pmol injected) and 0.78 μM (15.6 pmol injected) at Pt/ChO and Pt/ChO-ALP electrodes, respectively, i.e., comparable with those reported in Table 1. Accordingly, a flow rate of 1 mL/min was selected since the onset of the flow rate independence shown in Figure S3.

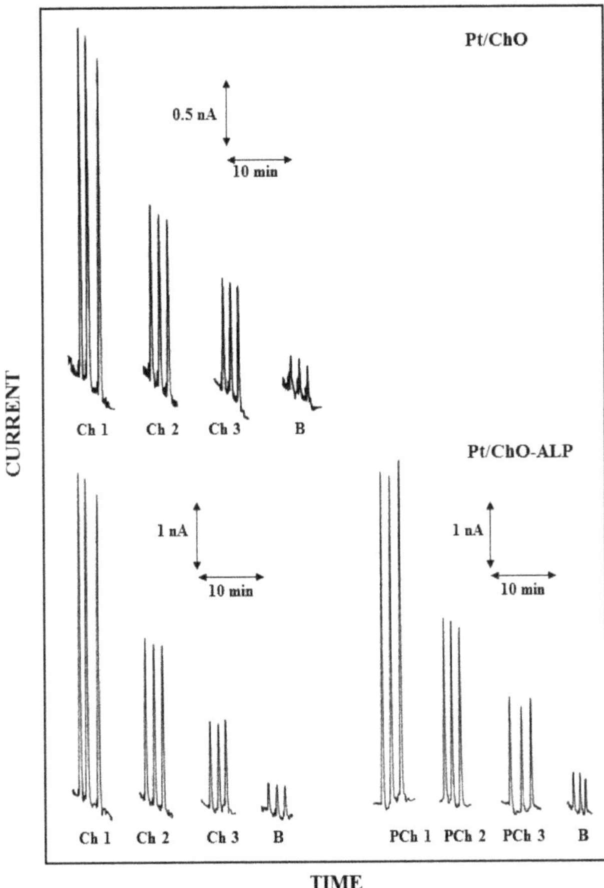

Figure 3. Typical flow injection responses at the dual electrode biosensor for replicate injections of Ch 2.5 μM (Ch1), 1.25 μM (Ch2), 0.625 μM (Ch3) and PCh 25 μM (PCh1), 12.5 μM (PCh2), 6.25 μM (PCh3); B refers to responses due to replicate injections of the carrier solution, i.e., the buffer used for preparing sample solutions. Upper and lower traces refer to responses at Pt/ChO and Pt/ChO-ALP biosensors, respectively. Carrier solution: borate buffer (pH 9, *I* 0.1 M); flow rate 1 mL/min; injection volume 20 μL. Other experimental conditions as described in Section 2.

3.4. Simultaneous Determination of Ch and PCh at the Dual Electrode Biosensor

Even if the present dual electrode biosensor demonstrated excellent analytical performances for single analyte detection, a simultaneous determination of both analytes requires that all the molecular events (i.e., enzyme catalysis/electrochemical detection) occurring for Ch detection does not interfere with those similarly involved for PCh and vice versa. In the best case, the amperometric responses for both analytes should be linearly related to their concentration levels, in their respective linear ranges, as described by the following relations:

$$I_{BI} = K_1 \, [Ch] + K_2 \, [PCh] \tag{1}$$

$$I_M = K_3 \, [Ch] \tag{2}$$

where I_{BI} and I_M are the amperometric responses at Pt/ChO-ALP and Pt/ChO electrodes, respectively, [Ch] and [PCh] are the concentration levels in the sample mixture of both analytes, while K_1, K_2, and K_3 are the respective analyte sensitivities. To validate the additivity of analyte responses, several Ch and PCh mixtures were analyzed using the present dual electrode biosensor for a concentration range as listed in Table 1 in the linear range column. In this respect, Figure 4 shows the 3d plot of the observed responses at the Pt/ChO electrode, while Figure 5 shows some relevant slices of the 3d plot at some constant analyte levels.

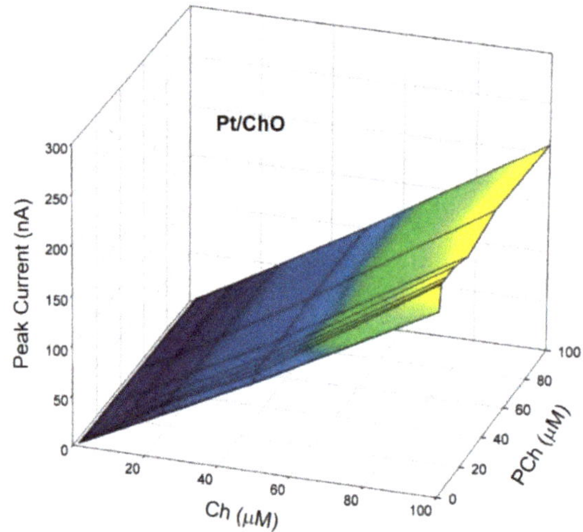

Figure 4. Calibration 3d plot at the Pt/ChO electrode of the dual electrode biosensor for replicate injections of mixtures of Ch and PCh. Carrier solution: borate buffer (pH 9, *I* 0.1 M); flow rate 1 mL/min; injection volume 20 μL. Other experimental conditions as described in Section 2.

Due to the specificity towards Ch and the lack of any crosstalk effects, the amperometric responses shown in Figure 4 for Pt/ChO of the present dual electrode biosensor are linearly related to Ch but independent of PCh. Accordingly, the detector response is represented by a plane with a slope equal to Ch sensitivity at the Pt/ChO electrode, parallel to the PCh concentration axis and with zero intercept. Particularly, Figure 5 (upper plot) shows the linear dependence of responses towards Ch whatever the PCh level in the sample mixture while the constant behaviors observed in the lower plots in Figure 5 certified the independence of Ch responses from PCh whatever the Ch level. Accordingly, the responses of the dual electrode biosensor at the Pt/ChO detector can be described by Equation (2).

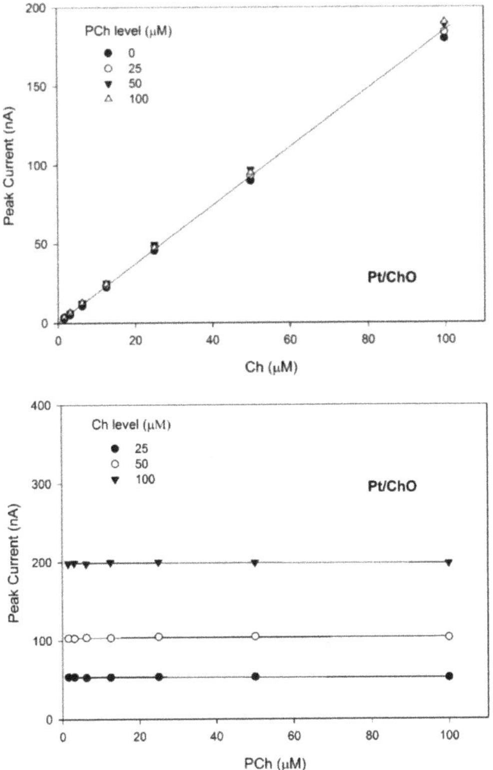

Figure 5. Calibration lines at the Pt/ChO electrode of the dual electrode biosensor for replicate injections of mixtures of Ch and PCh at several concentration levels of PCh (upper plot) and Ch (lower plot). Continuous lines refer to linear fitting of data. Carrier solution: borate buffer (pH 9, I 0.1 M); flow rate 1 mL/min; injection volume 20 µL. Other experimental conditions as described Section 2.

On the other hand, Figure 6 shows the 3d plot of the observed responses at the Pt/ChO-ALP electrode, while Figure 7 shows some relevant slices of the 3d plot at some constant analyte levels. As Figure 6 shows, the amperometric responses at the Pt/ChO-ALP electrode are linearly and additively dependent from both analytes: now the plane representing the detector response has its lower vertex at zero and tilted as the sensitivities toward Ch and PCh. Particularly, both plots in Figure 7 confirmed the additivity of responses for both analytes in the relevant linear ranges reported in Table 1. According to these experimental findings, Equation (1) can be used to describe the responses of Ch and PCh at the Pt/ChO-ALP electrode of the present biosensor; in particular, this relation can be simplified to:

$$I_{BI} = K_1 [Ch] + A_2 \tag{3}$$

for Ch determination (Figure 7, upper plot), and to:

$$I_{BI} = A_1 + K_2 [PCh] \tag{4}$$

for PCh determination (Figure 7, lower plot), A_1 and A_2 representing the constant values of Ch and PCh, respectively.

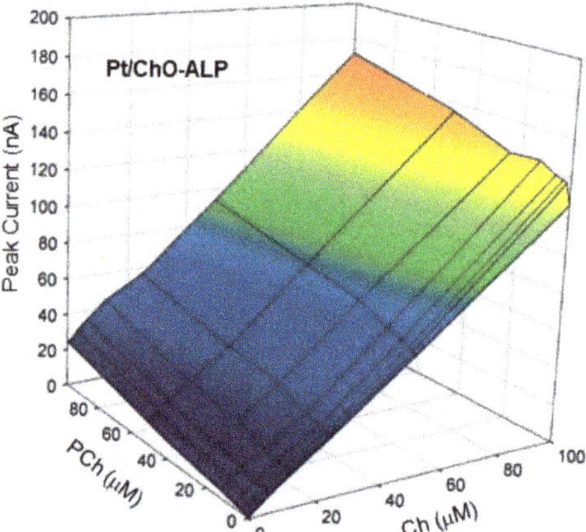

Figure 6. Calibration 3d plot at the Pt/ChO-ALP electrode of the dual electrode biosensor for replicate injections of mixtures of Ch and PCh. Carrier solution: borate buffer (pH 9, *I* 0.1 M); flow rate 1 mL/min; injection volume 20 μL. Other experimental conditions as described in Section 2.

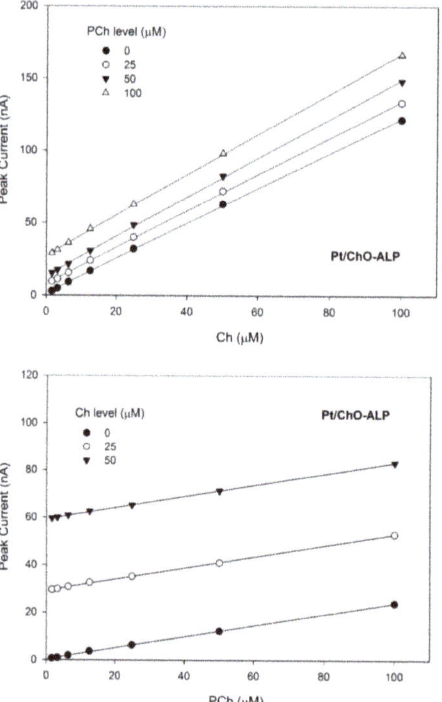

Figure 7. Calibration lines at the Pt/ChO-ALP electrode of the dual electrode biosensor for replicate injections of mixtures of Ch and PCh at several concentration levels of PCh (upper plot) and Ch (lower plot). Continuous lines refer to linear fitting of data. Carrier solution: borate buffer (pH 9, *I* 0.1 M); flow rate 1 mL/min; injection volume 20 μL. Other experimental conditions as described in Section 2.

Due to these experimental results, proper calibration of the Pt/ChO electrode (i.e., the determination of K_3 in Equation (2)) can permit to analyze Ch levels in the sample whatever the PCh concentration. Similarly, calibration of the Pt/ChO-ALP electrode toward Ch and PCh (i.e., the determination of K_1 and K_2 in Equation (1)) can allow PCh determination in Ch/PCh mixtures since the Ch contribution to the overall response can be calculated by the simultaneous response at the Pt/ChO electrode.

Finally, it should be noted that the behavior of the dual electrode biosensors has been studied even for Ch and PCh concentrations greater than their respective linear range upper limits, i.e., when the enzyme kinetics of both enzymes are not linear and saturating (see Figure 2 and Table 1 for the relevant analyte levels). This characterization, of course, had no relevance for the analytical studies of the proposed biosensor due to the non-linear responses so that was not here described. Nevertheless, a significant depression of Ch responses was observed for high PCh levels, thus suggesting ChO inhibition by PCh, an inhibition phenomenon never reported for this enzyme. A detailed study about this behavior is reported elsewhere.

3.5. Interference-Free Performances of the Dual Electrode Biosensor

Even if the enzymatic transduction assures specific biorecognition of Ch and PCh, unfortunately the succeeding electrochemical detection is poorly selective since any endogenous electroactive compounds in the sample could give an amperometric response if oxidable at the potential of hydrogen peroxide detection. Of course, this side effect at least biases the analyte responses. As a proof of this deleterious effect, Figure S5 compares typical flow injection responses of Ch and PCh at the dual electrode biosensor with those due to representative endogenous electroactive compounds, namely ascorbate (AA), urate (AU), and cysteine (CYS) at their respective physiological levels in biological fluids. As can be seen, the co-occurrence of these compounds generates significant bias on analyte response, even many times higher than e.g., the Ch response (compare with AU responses). Even if lowering the flow rate is expected to increase the biosensor response (see above) while decreasing the faradic currents due to interferent oxidation, the bias introduced in the analyte measurement is still dramatic even at low flow rate (as Figure S6 shows), so an application to real sample analysis appears impracticable using this biosensor configuration.

The usual approach in amperometric biosensors to overcomes this adverse effect is the use of discrete membranes (like polycarbonate, cellulose, and so) onto the electrode surface, selectively limiting the diffusion of endogenous electroactive compounds present in the sample. In addition to be scarcely effective, the use of these relatively macroscopic and thick membranes limits also, usually by distortion, the analyte response and precludes any miniaturization of the sensing device [48–52]. Electropolymerized non-conducting films with built-in permselectivity [39,41–54,61] have extensively demonstrated to protect the sensor surface from electrode fouling effects (e.g., due to high molecular weight proteins in sample) and to remove, with unmatched efficiency, common electroactive interfering species in glucose, Ch, and Ach, as well as in lysine determination in real sample analysis (e.g., blood, serum, dialysates, tissue homogenate). In this approach, a suitable monomer (e.g., o-aminophenol, pyrrole, 2-naphthol) is electrooxidized at the Pt electrode surface so forming in situ a polymeric, non-conducting, strong adherent film with adverse permselectivity towards many endogenous electroactive compounds; apart from minimizing these undesired, interferent effects, this approach easily permits the deposition of very thin-film (so not distorting the analyte response), even with multilayer facilities and, due to the electrochemical deposition technique there used, permits the miniaturization of any sensing device.

Accordingly, in the present case, both the electrodes of the dual electrode biosensor were modified, before enzyme immobilization, by filming the bare Pt electrodes with overoxidized polypyrrole (oPPy): the resulting electrodes, here described as Pt/oPPy/ChO and Pt/oPPy/ChO-ALP electrodes, proved highly stable in stirred or flowing solution since modified with a thin film quite adherent to the enzyme membrane and to the underneath

Pt surface. The high selectivity of the electrochemical detection obtained here by this approach is shown by Figure 8, where low Ch and PCh levels are compared to responses due to some endogenous electroactive compounds at their physiological concentrations; as the figure demonstrates, the modification with oPPy dramatically reduces the interferent effect due to these substances (compare Figure 8 with Figure S5 where this electrode modification was not there used): in particular, the interference effect due to AA and CYS is practically suppressed whereas the interference signal due to AU is even much more suppressed. Similar interference-free performances were also observed at low flow rate (e.g., 0.2 mL/min, see Figure S7). Notably, also the undesired artefacts due to sample injection were significantly reduced (compare responses B in Figure 8 with those already shown in Figure 3): undoubtedly, the oPPy modification makes somewhat the electrode immunes from the dielectric constant changes due to the injected buffer and/or pressure changes during sample injection. At the same time, the presence of oPPy film onto the Pt electrode does not limit or reduce the analytical signal, i.e., the oxidation of the enzymatic produced hydrogen peroxide, as a comparison of Figure 8 with Figure 3 (where oPPy modification is absent) shows. In this respect, Table 2 (and Table S2 for low flow rates) summarizes the analytical performances of the dual electrode biosensor modified by oPPy: with respect to data already reported in Table 1, sensitivities were just a bit lower that the unmodified case while linear ranges were practically identical; on the contrary, absolute and relative LODs were significantly lower as a consequence of the reduced noise (e.g., injection artefacts) in measurements, confirming that the performances of the biosensor were quite adequate for real sample analysis even at low analyte levels.

Table 2. Analytical performances at the oPPy modified dual electrode biosensor [1].

Analyte	Biosensor	Sensitivity (μA/mM)	Linear Range (μM)	LOD [2] (μM/pmol)
PCh	Pt/oPPy/ChO-ALP	0.197	3–1000	2.7/53.6
Ch	Pt/oPPy/ChO-ALP	1.30	0.3–1000	0.19/3.9
Ch	Pt/oPPy/ChO	1.39	0.3–1000	0.28/5.6

[1] Carrier solution: borate buffer (pH 9, I 0.1 M); flow rate 1 mL/min; injection volume 20 μL. Other experimental conditions as described in Section 2. [2] LOD: relative (μM) and absolute (pmol injected) limit of detection calculated at a signal-to-noise ratio of 3.

A quantitative mode to describe the interference effect due to endogenous electroactive compounds relies on the use of the analyte bias factor, i.e., the concentration (or the amount) of the analyte (Ch or PCh in the present case) which generates a response equal to those of interferent species at their maximal physiological levels. Accordingly, Table 3 reports the tested bias values on Ch and PCh measurements as due to AA, AU, and CYS interferences. As can be seen, in the worst cases, the interferent effects due to the investigated interferents mimic an analyte response (bias) corresponding to 0.25 and 1.6 μM for Ch and PCh, respectively. All the observed biases were comparable if not lower than LODs performed for Ch and PCh: in particular, the biases introduced in analyte measurements were comparable to the observed artefacts due to injection so noise due sample analysis cannot be distinguished from interference response. Of course, this assured that the present biosensor, as modified by oPPy, is certainly specific towards Ch and PCh at least for the endogenous interference compounds here tested.

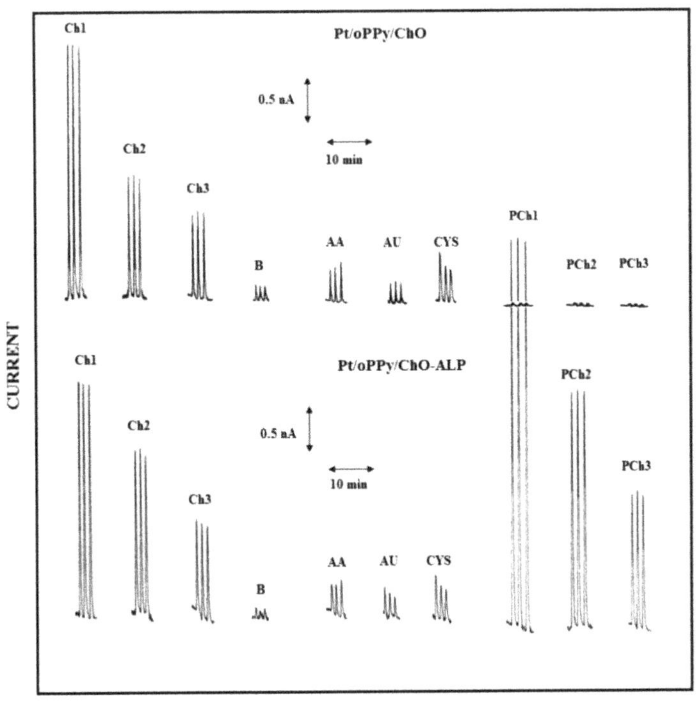

Figure 8. Typical flow injection responses at the oPPy modified dual electrode biosensor for replicate injections of Ch 2.5 µM (Ch1), 1.25 µM (Ch2), 0.625 µM (Ch3), PCh 25 µM (PCh1), 12.5 µM (PCh2), 6.25 µM (PCh3) compared to responses due to AA 0.1 mM, AU 0.5 mM, and CYS 0.2 mM; B refers to responses due to replicate injections of the carrier solution, i.e., the buffer used for preparing sample solutions. Upper and lower traces refer to responses at Pt/oPPy/ChO and Pt/oPPy/ChO-ALP biosensors, respectively. Carrier solution: borate buffer (pH 9, *I* 0.1 M); flow rate 1 mL/min; injection volume 20 µL. Other experimental conditions as described in Section 2.

Table 3. Biases on Ch and PCh measurements by interferents at the oPPy modified dual electrode biosensor [1].

Analyte	Biosensor	Bias (µM) For:		
		AA (0.1 mM)	AU (0.5 mM)	CYS (0.2 mM)
PCh	Pt/oPPy/ChO-ALP	1.5	1.2	1.6
Ch	Pt/oPPy/ChO-ALP	0.19	0.15	0.22
Ch	Pt/oPPy/ChO	0.2	0.14	0.25

[1] Carrier solution: borate buffer (pH 9, *I* 0.1 M); flow rate 1 mL/min; injection volume 20 µL. Other experimental conditions as described in Section 2.

The analytical performances of the here described dual electrode biosensor demonstrated quite appealing in terms of sensitivities, linear ranges, LODs, and high specificity so application to real sample analysis appears promising: unfortunately, biological sample like biological fluids and tissues, where significant Ch and PCh levels are expected (see the Introduction section), were not available, particularly during these last times. Anyway, the proposed dual electrode biosensor was tested for the simultaneous determination of Ch and PCh in biologically-like synthetic sample (i.e., bovine serum albumin solutions enriched with endogenous electroactive compounds) spiked with known amount of several Ch/PCh mixtures: according to a paired Student's *t*-test (1—α = 0.95 level), the found analyte levels were not significantly different from those observed for Ch/PCh standards so excluding any matrix effects in analyte quantitation due to high molecular proteins in the sample. Further, the successfully application of alike biosensors in high complex sample like foods and biological fluids [38,40–45,48,51–54] could comfort about a successfully application even for the more complex sample.

3.6. Stability of the Dual Electrode Biosensor

The stability of the biosensor was studied either at short or long time. The operational stability was tested by repetitive injection of Ch/PCh mixtures under the continuous run of the flow injection system: during all the working day (8—12 h), no significant difference in responses was observed for both the analyte and the within-a-day CVs were those already here reported (see above).

Storage and long-term stability of the dual electrode biosensor was studied by discontinuously monitoring the Ch and PCh responses of a biosensor continuously used over about one month and stored in buffer at 4 °C in the dark when not in use: no significant loss in sensitivity was observed during more than 30 days, after an initial sensitivity drop of about 30% observed gradually during the first 5 days.

4. Conclusions

Ch and PCh levels in tissues are associated with tissue growth and so to carcinogenesis. Very few methods have been reported till now demonstrating capable of Ch or PCh analysis and even fewer of a simultaneous determination of both analytes, while the ratio of Ch/PCh levels, rather than their absolute levels, is useful as a diagnostic and prognostic marker in these pathologies. Since all these methods rely on highly sophisticated and expensive analytical techniques (NMR spectroscopy or GC/LC-high resolution mass spectrometry), there is the necessity to design and develop more simple, fast, specific, and accurate analytical approaches, as usually necessary in clinical and medical fields.

This paper demonstrates that biosensing could permit the development of simple, easy to fabricate, and easy to use analytical devices, satisfying the above goals. By co-crosslinking of ChO or ChO-ALP enzymes onto the Pt surface of a simple electrochemical thin-layer cell, a novel PCh biosensor has been developed, a biosensor never reported before. Further, coupling the developed biosensor with a parallel ChO sensor in a dual electrode electrochemical cell, a crosstalk-free dual electrode biosensor has been also developed, permitting the simultaneous determination of Ch and PCh in flow injection analysis. This novel sensing device performed remarkably in terms of sensitivity, linear range, and limit of detection, thus exceeding in most cases the more complex analytical instrumentations.

To circumvent matrix effects in real sample analysis due to high molecular weight proteins and endogenous electroactive compounds generating deleterious fouling effects onto the electrodes and unpredictable bias on analyte responses, respectively, electropolymerized non-conducting films with built-in permselectivity have been used to increase the selectivity of the electrochemical detection. Thus, electrode modification by oPPy permitted the development a fouling- and interferent-free dual electrode biosensor which appeared surely promising for the simultaneous determination of Ch and PCh in a real sample like biological fluids and tissues. Work in this direction is in progress in the present laboratory.

Supplementary Materials: The following are available online at https://www.mdpi.com/article/10.3390/s21103545/s1, Scheme S1. Schematic diagram of the flow injection setup used in the experiments showing in particular the electrochemical cell and the relevant dual electrode biosensor used in the amperometric measurements (the reference electrode and the flow cell gaskets were omitted for sake of figure clarity). Table S1. Comparison of main analytical performances of choline oxidase (ChO)/alkaline phosphatase (ALP) biosensors produced with different ALP sources. Figure S1. Normalized sensitivities towards Ch and PCh *vs* pH at a typical rotating disk Pt/ChO-ALP electrode. Supporting electrolyte: acetate/borate/tris buffer (*I* 0.1 M). Rotation rate: 1000 rpm. Experimental condition as described in the Materials and Methods section of the paper. Figure S2. Linear parts of the calibration curves at the dual electrode biosensor for replicate injections of Ch and PCh as shown in Figure 1 (see the relevant section of the paper). Continuous lines refer to linear fitting of data (correlation coefficients better than 0.999). Carrier solution: borate buffer (pH 9, *I* 0.1 M); flow rate 1 mL/min; injection volume 20 µL. Other experimental conditions as described in the Materials and Methods section of the paper. Figure S3. Normalized peak currents vs flow rate plots for replicate injections of Ch (2.5 µM) and PCh (25 µM) at Pt/ChO (left) and Pt/ChO-ALP (right) dual electrode biosensor. Carrier solution: borate buffer (pH 9, I 0.1 M); injection volume 20 µL. Other experimental conditions as described in the Materials and Methods section of the paper. Figure S4. Typical flow injection responses at the dual electrode biosensor for replicate injections of Ch 2.5 µM (Ch1), 1.25 µM (Ch2), 0.625 µM (Ch3) and PCh 25 µM (PCh1), 12.5 µM (PCh2), 6.25 µM (PCh3); B refers to responses due to replicate injections of the carrier solution, i.e., the buffer used for preparing sample solutions. Upper and lower traces refer to responses at Pt/ChO and Pt/Cho-ALP biosensors, respectively. Carrier solution: borate buffer (pH 9, I 0.1 M); flow rate 0.2 mL/min; injection volume 20 µL. Other experimental conditions as described in the Materials and Methods section of the paper. Figure S5. Comparison of flow injection responses at the dual electrode biosensor for replicate injections of Ascorbate (AA, 0.1 mM), Urate (AU, 0.5 mM), Cysteine (CYS, 0.2 mM), Ch (0.1 mM), and PCh (1 mM) at a flow rate of 1 mL/min. Upper and lower traces refer to responses at Pt/ChO and Pt/Cho-ALP biosensors, respectively. Carrier solution: borate buffer (pH 9, I 0.1 M); injection volume 20 µL. Other experimental conditions as described in the Materials and Methods section of the paper. Figure S6. Comparison of flow injection responses at the dual electrode biosensor for replicate injections of Ascorbate (AA, 0.1 mM), Urate (AU, 0.5 mM), Cysteine (CYS, 0.2 mM), Ch (0.1 mM), and PCh (1 mM) at a flow rate of 0.2 mL/min. Upper and lower traces refer to responses at Pt/ChO and Pt/Cho-ALP biosensors, respectively. Carrier solution: borate buffer (pH 9, I 0.1 M); injection volume 20 µL. Other experimental conditions as described in the Materials and Methods section of the paper. Figure S7. Typical flow injection responses at the oPPy modified dual electrode biosensor for replicate injections of Ch 2.5 µM (Ch1), 1.25 µM (Ch2), 0.625 µM (Ch3), PCh 25 µM (PCh1), 12.5 µM (PCh2), 6.25 µM (PCh3) compared to responses due to AA 0.1 mM, AU 0.5 mM, and CYS 0.2 mM; B refers to responses due to replicate injections of the carrier solution, i.e., the buffer used for preparing sample solutions. Upper and lower traces refer to responses at Pt/oPPy/ChO and Pt/oPPy/ChO-ALP biosensors, respectively. Carrier solution: borate buffer (pH 9, I 0.1 M); flow rate 0.2 mL/min; injection volume 20 µL. Other experimental conditions as described in the Materials and Methods section of the paper. Table S2. Analytical performances at the oPPy modified dual electrode biosensor[1] at low flow rate.

Author Contributions: Conceptualization, A.G.; methodology, R.C. and A.G; investigation, R.C., and A.G. writing—original draft preparation, A.G.; writing—review and editing, A.G.; visualization, A.G.; supervision, A.G.; project administration, A.G.; funding acquisition, A.G. All authors have read and agreed to the published version of the manuscript.

Funding: The Authors gratefully acknowledge financial support from Ministero dell'Istruzione, dell'Università e della Ricerca Italiana (MIUR).

Institutional Review Board Statement: Not applicable.

Informed Consent Statement: Not applicable.

Data Availability Statement: Not applicable.

Acknowledgments: This work comes from the experimental contributions by Vitina Gruosso, Tiziana Notarangelo and Rossella Fucci theses (2006–2007 academic year) which are fully acknowledge for their experimental skills.

Conflicts of Interest: The authors declare no conflict of interest.

References

1. Zeisel, S.H. Choline deficiency. *J. Nutr. Biochem.* **1990**, *1*, 332–349. [CrossRef]
2. Kiss, Z. Regulation of Mitogenesis by Water-Soluble Phospholipid Intermediates1Abbreviations: PCho—phosphocholine; PEtn—phosphoethanolamine; PtdCholphosphatidylcholine; PtdEtn—phosphatidylethanolamine; CK—choline kinase; EK— ethanolamine kinase; PLD—phospholipase D. *Cell. Signal.* **1999**, *11*, 149–157. [CrossRef]
3. Katz-Brull, R.; Margalit, R.; Bendel, P.; Degani, H. Choline metabolism in breast cancer; H-, C- and P-NMR studies of cells and tumors. *Magn. Reson. Mater. Biol. Phys. Med.* **1998**, *6*, 44–52. [CrossRef] [PubMed]
4. Negendank, W. Studies of human tumors by MRS: A review. *NMR Biomed.* **1992**, *5*, 303–324. [CrossRef]

5. Sharma, U.; Mehta, A.; Seenu, V.; Jagannathan, N.R. Biochemical characterization of metastatic lymph nodes of breast cancer patients by in vitro 1H magnetic resonance spectroscopy: A pilot study. *Magn. Reson. Imaging* **2004**, *22*, 697–706. [CrossRef]

6. Sitter, B.; Lundgren, S.; Bathen, T.F.; Halgunset, J.; Fjosne, H.E.; Gribbestad, I.S. Comparison of HR MAS MR spectroscopic profiles of breast cancer tissue with clinical parameters. *NMR Biomed.* **2006**, *19*, 30–40. [CrossRef]

7. Klomp, D.W.J.; van de Bank, B.L.; Raaijmakers, A.; Korteweg, M.A.; Possanzini, C.; Boer, V.O.; van de Berg, C.A.T.; van de Bosch, M.A.A.J.; Luijten, P.R. 31 P MRSI and 1 H MRS at 7 T: Initial results in human breast cancer. *NMR Biomed.* **2011**, *24*, 1337–1342. [CrossRef]

8. Arias-Mendoza, F.; Smith, M.R.; Brown, T.R. Predicting treatment response in non-Hodgkin's lymphoma from the pretreatment tumor content of phosphoethanolamine plus phosphocholine1. *Acad. Radiol.* **2004**, *11*, 368–376. [CrossRef]

9. Cheng, S.-C.; Chen, K.; Chiu, C.-Y.; Lu, K.-Y.; Lu, H.-Y.; Chiang, M.-H.; Tsai, C.-K.; Lo, C.-J.; Cheng, M.-L.; Chang, T.-C.; et al. Metabolomic biomarkers in cervicovaginal fluid for detecting endometrial cancer through nuclear magnetic resonance spectroscopy. *Metabolomics* **2019**, *15*, 146. [CrossRef]

10. Sun, G.; Wang, J.; Zhang, J.; Ma, C.; Shao, C.; Hao, J.; Zheng, J.; Feng, X.; Zuo, C. High-resolution magic angle spinning 1H magnetic resonance spectroscopy detects choline as a biomarker in a swine obstructive chronic pancreatitis model at an early stage. *Mol. Biosyst.* **2014**, *10*, 467–474. [CrossRef]

11. Vonica, C.L.; Ilie, I.R.; Socaciu, C.; Moraru, C.; Georgescu, B.; Farcaş, A.; Roman, G.; Mureşan, A.A.; Georgescu, C.E. Lipidomics biomarkers in women with polycystic ovary syndrome (PCOS) using ultra-high performance liquid chromatography-quadrupole time of flight electrospray in a positive ionization mode mass spectrometry. *Scand. J. Clin. Lab. Investig.* **2019**, *79*, 437–442. [CrossRef]

12. Walter, A.; Korth, U.; Hilgert, M.; Hartmann, J.; Weichel, O.; Hilgert, M.; Fassbender, K.; Schmitt, A.; Klein, J. Glycerophosphocholine is elevated in cerebrospinal fluid of Alzheimer patients. *Neurobiol. Aging* **2004**, *25*, 1299–1303. [CrossRef]

13. Senaratne, R.; Milne, A.M.; MacQueen, G.M.; Hall, G.B.C. Increased choline-containing compounds in the orbitofrontal cortex and hippocampus in euthymic patients with bipolar disorder: A proton magnetic resonance spectroscopy study. *Psychiatry Res. Neuroimaging* **2009**, *172*, 205–209. [CrossRef]

14. Pomfret, E.A.; DaCosta, K.-A.; Schurman, L.L.; Zeisel, S.H. Measurement of choline and choline metabolite concentrations using high-pressure liquid chromatography and gas chromatography-mass spectrometry. *Anal. Biochem.* **1989**, *180*, 85–90. [CrossRef]

15. Koc, H.; Mar, M.-H.; Ranasinghe, A.; Swenberg, J.A.; Zeisel, S.H. Quantitation of Choline and Its Metabolites in Tissues and Foods by Liquid Chromatography/Electrospray Ionization-Isotope Dilution Mass Spectrometry. *Anal. Chem.* **2002**, *74*, 4734–4740. [CrossRef]

16. Nygren, H.; Börner, K.; Hagenhoff, B.; Malmberg, P.; Månsson, J.-E. Localization of cholesterol, phosphocholine and galactosylceramide in rat cerebellar cortex with imaging TOF-SIMS equipped with a bismuth cluster ion source. *Biochim. Biophys. Acta Mol. Cell Biol. Lipids* **2005**, *1737*, 102–110. [CrossRef]

17. Zhao, Y.-Y.; Xiong, Y.; Curtis, J.M. Measurement of phospholipids by hydrophilic interaction liquid chromatography coupled to tandem mass spectrometry: The determination of choline containing compounds in foods. *J. Chromatogr. A* **2011**, *1218*, 5470–5479. [CrossRef]

18. Xiong, Y.; Zhao, Y.-Y.; Goruk, S.; Oilund, K.; Field, C.J.; Jacobs, R.L.; Curtis, J.M. Validation of an LC-MS/MS method for the quantification of choline-related compounds and phospholipids in foods and tissues. *J. Chromatogr. B* **2012**, *911*, 170–179. [CrossRef]

19. Mimmi, M.C.; Picotti, P.; Corazza, A.; Betto, E.; Pucillo, C.E.; Cesaratto, L.; Cedolini, C.; Londero, V.; Zuiani, C.; Bazzocchi, M.; et al. High-performance metabolic marker assessment in breast cancer tissue by mass spectrometry. *Clin. Chem. Lab. Med.* **2011**, *49*, 317–324. [CrossRef]

20. Mimmi, M.C.; Finato, N.; Pizzolato, G.; Beltrami, C.A.; Fogolari, F.; Corazza, A.; Esposito, G. Absolute quantification of choline-related biomarkers in breast cancer biopsies by liquid chromatography electrospray ionization mass spectrometry. *Anal. Cell. Pathol.* **2013**, *36*, 71–83. [CrossRef]

21. Santos-Fandila, A.; Vázquez, E.; Barranco, A.; Zafra-Gómez, A.; Navalón, A.; Rueda, R.; Ramírez, M. Analysis of 17 neurotransmitters, metabolites and precursors in zebrafish through the life cycle using ultrahigh performance liquid chromatography-tandem mass spectrometry. *J. Chromatogr. B Anal. Technol. Biomed. Life Sci.* **2015**, *1001*, 191–201. [CrossRef]

22. Loening, N.M.; Chamberlin, A.M.; Zepeda, A.G.; Gonzalez, R.G.; Cheng, L.L. Quantification of phosphocholine and glycerophosphocholine with31P edited1H NMR spectroscopy. *NMR Biomed.* **2005**, *18*, 413–420. [CrossRef]

23. Rocha, C.M.; Barros, A.S.; Gil, A.M.; Goodfellow, B.J.; Humpfer, E.; Spraul, M.; Carreira, I.M.; Melo, J.B.; Bernardo, J.; Gomes, A.; et al. Metabolic profiling of human lung cancer tissue by1H high resolution magic angle spinning (HRMAS) NMR spectroscopy. *J. Proteome Res.* **2010**, *9*, 319–332. [CrossRef]

24. Luttje, M.P.; Italiaander, M.G.M.; Arteaga de Castro, C.S.; van der Kemp, W.J.M.; Luijten, P.R.; van Vulpen, M.; van der Heide, U.A.; Klomp, D.W.J. 31 P MR spectroscopic imaging combined with 1 H MR spectroscopic imaging in the human prostate using a double tuned endorectal coil at 7T. *Magn. Reson. Med.* **2014**, *72*, 1516–1521. [CrossRef]

25. Pearce, J.M.; Mahoney, M.C.; Lee, J.-H.; Chu, W.-J.; Cecil, K.M.; Strakowski, S.M.; Komoroski, R.A. 1H NMR analysis of choline metabolites in fine-needle-aspirate biopsies of breast cancer. *Magn. Reson. Mater. Phys. Biol. Med.* **2013**, *26*, 337–343. [CrossRef]

26. Esmaeili, M.; Moestue, S.A.; Hamans, B.C.; Veltien, A.; Kristian, A.; Engebråten, O.; Maelandsmo, G.M.; Gribbestad, I.S.; Bathen, T.F.; Heerschap, A. In Vivo 31 P magnetic resonance spectroscopic imaging (MRSI) for metabolic profiling of human breast cancer xenografts. *J. Magn. Reson. Imaging* **2015**, *41*, 601–609. [CrossRef]

27. Mao, J.; Jiang, L.; Jiang, B.; Liu, M.; Mao, X. 1H–14N HSQC detection of choline-containing compounds in solutions. *J. Magn. Reson.* **2010**, *206*, 157–160. [CrossRef]

28. Mao, J.; Jiang, L.; Jiang, B.; Liu, M.; Mao, X. A Selective NMR Method for Detecting Choline Containing Compounds in Liver Tissue: The 1 H–14 N HSQC Experiment. *J. Am. Chem. Soc.* **2010**, *132*, 17349–17351. [CrossRef] [PubMed]

29. Belkić, D.; Belkić, K. In Vivo magnetic resonance spectroscopy for ovarian cancer diagnostics: Quantification by the fast Padé transform. *J. Math. Chem.* **2017**, *55*, 349–405. [CrossRef]

30. Belkić, D.; Belkić, K. Mathematical optimization of In Vivo NMR chemistry through the fast Padé transform: Potential relevance for early breast cancer detection by magnetic resonance spectroscopy. *J. Math. Chem.* **2006**, *40*, 85–103. [CrossRef]

31. Belkić, D.; Belkić, K. The potential for practical improvements in cancer diagnostics by mathematically-optimized magnetic resonance spectroscopy. *J. Math. Chem.* **2011**, *49*, 2408–2440. [CrossRef]

32. Özdemir, M.S.; De Deene, Y.; Fieremans, E.; Lemahieu, I. Quantitative proton magnetic resonance spectroscopy without water suppression. *J. Instrum.* **2009**, *4*, P06014. [CrossRef]

33. Morniroli, D.; Dessì, A.; Giannì, M.L.; Roggero, P.; Noto, A.; Atzori, L.; Lussu, M.; Fanos, V.; Mosca, F. Is the body composition development in premature infants associated with a distinctive nuclear magnetic resonance metabolomic profiling of urine? *J. Matern. Neonatal Med.* **2019**, *32*, 2310–2318. [CrossRef]

34. Masoom, M.; Roberti, R.; Binaglia, L. Determination of phosphatidylcholine in a flow injection system using immobilized enzyme reactors. *Anal. Biochem.* **1990**, *187*, 240–245. [CrossRef]

35. Klein, J.; Gonzalez, R.; Köppen, A.; Löffelholz, K. Free choline and choline metabolites in rat brain and body fluids: Sensitive determination and implications for choline supply to the brain. *Neurochem. Int.* **1993**, *22*, 293–300. [CrossRef]

36. Murai, S.; Saito, H.; Shirato, R.; Kawaguchi, T. An improved method for assaying phosphocholine and glycerophosphocholine in mouse tissue. *J. Pharmacol. Toxicol. Methods* **2001**, *46*, 103–109. [CrossRef]

37. Guerrieri, A.; De Benedetto, G.E.; Palmisano, F.; Zambonin, P.G. Amperometric sensor for choline and acetylcholine based on a platinum electrode modified by a co-crosslinked bienzymic system. *Analyst* **1995**, *120*, 2731–2736. [CrossRef]

38. Guerrieri, A.; Palmisano, F. An Acetylcholinesterase/Choline Oxidase-Based Amperometric Biosensor as a Liquid Chromatography Detector for Acetylcholine and Choline Determination in Brain Tissue Homogenates. *Anal. Chem.* **2001**, *73*, 2875–2882. [CrossRef]

39. Guerrieri, A.; Lattanzio, V.; Palmisano, F.; Zambonin, P.G. Electrosynthesized poly(pyrrole)/poly(2-naphthol) bilayer membrane as an effective anti-interference layer for simultaneous determination of acethylcholine and choline by a dual electrode amperometric biosensor. *Biosens. Bioelectron.* **2006**, *21*, 1710–1718. [CrossRef]

40. Guerrieri, A.; Monaci, L.; Quinto, M.; Palmisano, F.; Aldridge, W.N.; van der Hoff, G.R.; Zoonen, P.; van Tran-Minh, C.; Dumschat, C.; Muller, H.; et al. A disposable amperometric biosensor for rapid screening of anticholinesterase activity in soil extracts. *Analyst* **2002**, *127*, 5–7. [CrossRef]

41. Ciriello, R.; Lo Magro, S.; Guerrieri, A. Assay of serum cholinesterase activity by an amperometric biosensor based on a co-crosslinked choline oxidase/overoxidized polypyrrole bilayer. *Analyst* **2018**, *143*, 920–929. [CrossRef] [PubMed]

42. Guerrieri, A.; Ciriello, R.; Crispo, F.; Bianco, G. Detection of choline in biological fluids from patients on haemodialysis by an amperometric biosensor based on a novel anti-interference bilayer. *Bioelectrochemistry* **2019**, *129*, 135–143. [CrossRef] [PubMed]

43. Centonze, D.; Guerrieri, A.; Malitesta, C.; Palmisano, F.; Zambonin, P.G. Interference-free glucose sensor based on glucose-oxidase immobilized in an overoxidized non-conducting polypyrrole film. *Fresenius. J. Anal. Chem.* **1992**, *342*, 729–733. [CrossRef]

44. Centonze, D.; Guerrieri, A.; Malitesta, C.; Palmisano, F.; Zambonin, P.G. An in-situ electro-synthesized poly-o-phenylenediamine/ glucose oxidase amperometric biosensor for flow injection determination of glucose in serum. *Ann. Chim.* **1992**, *82*, 219–234.

45. Palmisano, F.; Centonze, D.; Guerrieri, A.; Zambonin, P.G. An interference-free biosensor based on glucose oxidase electrochemically immobilized in a non-conducting poly(pyrrole) film for continuous subcutaneous monitoring of glucose through microdialysis sampling. *Biosens. Bioelectron.* **1993**, *8*, 393–399. [CrossRef]

46. Centonze, D.; Guerrieri, A.; Palmisano, F.; Torsi, L.; Zambonin, P.G. Electrochemically prepared glucose biosensors: Kinetic and faradaic processes involving ascorbic acid and role of the electropolymerized film in preventing electrode-fouling. *Fresenius. J. Anal. Chem.* **1994**, *349*, 497–501. [CrossRef]

47. Palmisano, F.; Guerrieri, A.; Quinto, M.; Zambonin, P.G. Electrosynthesized Bilayer Polymeric Membrane for Effective Elimination of Electroactive Interferents in Amperometric Biosensors. *Anal. Chem.* **1995**, *67*, 1005–1009. [CrossRef]

48. Guerrieri, A.; De Benedetto, G.E.; Palmisano, F.; Zambonin, P.G. Electrosynthesized non-conducting polymers as permselective membranes in amperometric enzyme electrodes: A glucose biosensor based on a co-crosslinked glucose oxidase/overoxidized polypyrrole bilayer. *Biosens. Bioelectron.* **1998**, *13*, 103–112. [CrossRef]

49. Ciriello, R.; Cataldi, T.R.I.; Centonze, D.; Guerrieri, A. Permselective behavior of an electrosynthesized, nonconducting thin film of poly(2-naphthol) and its application to enzyme immobilization. *Electroanalysis* **2000**, *12*, 825–830. [CrossRef]

50. Guerrieri, A.; Ciriello, R.; Centonze, D. Permselective and enzyme-entrapping behaviours of an electropolymerized, non-conducting, poly(o-aminophenol) thin film-modified electrode: A critical study. *Biosens. Bioelectron.* **2009**, *24*, 1550–1556. [CrossRef]

51. Guerrieri, A.; Ciriello, R.; Cataldi, T.R.I. A novel amperometric biosensor based on a co-crosslinked l-lysine-α-oxidase/overoxidized polypyrrole bilayer for the highly selective determination of l-lysine. *Anal. Chim. Acta* **2013**, *795*, 52–59. [CrossRef] [PubMed]
52. Ciriello, R.; Cataldi, T.R.I.; Crispo, F.; Guerrieri, A. Quantification of l-lysine in cheese by a novel amperometric biosensor. *Food Chem.* **2015**, *169*, 13–19. [CrossRef] [PubMed]
53. Ciriello, R.; De Gennaro, F.; Frascaro, S.; Guerrieri, A. A novel approach for the selective analysis of L-lysine in untreated human serum by a co-crosslinked L-lysine-α-oxidase/overoxidized polypyrrole bilayer based amperometric biosensor. *Bioelectrochemistry* **2018**, *124*, 47–56. [CrossRef] [PubMed]
54. Ciriello, R.; Guerrieri, A. Assay of Phospholipase D Activity by an Amperometric Choline Oxidase Biosensor. *Sensors* **2020**, *20*, 1304. [CrossRef]
55. Fortier, G.; Brassard, E.; Bélanger, D. Optimization of a polypyrrole glucose oxidase biosensor. *Biosens. Bioelectron.* **1990**, *5*, 473–490. [CrossRef]
56. Kennedy, J.F.; White, C.A. Principles of immobilization of enzymes. In *Handbook of Enzyme Biotechnology*; Wiseman, A., Ed.; Ellis Horwood Lim.: Binghamton, NY, USA; John Wiley & Sons: Hoboken, NJ, USA, 1985; pp. 147–207.
57. Fernley, H.N. Mammalian alkaline phosphatase. In *The Enzymes*; Boyer, P.D., Ed.; Academic Press: New York, NY, USA; London, UK, 1971; pp. 417–477.
58. Latner, A.L.; Parsons, M.; Skillen, A.W. Isoelectric focusing of alkaline phosphatases from human kidney and calf intestine. *Enzymologia* **1971**, *40*, 1–7.
59. Armstrong, A.R. Purification of the active phosphatase found in dog faeces. *Biochem. J.* **1935**, *29*, 2020–2022. [CrossRef]
60. Guerrieri, A.; Ciriello, R.; Bianco, G.; De Gennaro, F.; Frascaro, S. Allosteric Enzyme-Based Biosensors-Kinetic Behaviours of Immobilised L-Lysine-α-Oxidase from Trichoderma viride: pH Influence and Allosteric Properties. *Biosensors* **2020**, *10*, 145. [CrossRef]
61. Carbone, M.E.; Ciriello, R.; Guerrieri, A.; Salvi, A.M. XPS investigation on the chemical structure of a very thin, insulating, film synthesized on platinum by electropolymerization of o-aminophenol (oAP) in aqueous solution at neutral pH. *Surf. Interface Anal.* **2014**, *46*, 1081–1085. [CrossRef]

Article

Employment of 1-Methoxy-5-Ethyl Phenazinium Ethyl Sulfate as a Stable Electron Mediator in Flavin Oxidoreductases-Based Sensors

Maya Fitriana [1,2], Noya Loew [3,4], Arief Budi Witarto [2], Kazunori Ikebukuro [1], Koji Sode [3,*] and Wakako Tsugawa [1,*]

[1] Department of Biotechnology and Life Science, Graduate School of Engineering, Tokyo University of Agriculture and Technology, 2-24-16 Naka-cho, Koganei, Tokyo 184-8588, Japan; s206376w@st.go.tuat.ac.jp (M.F.); ikebu@cc.tuat.ac.jp (K.I.)

[2] Faculty of Biotechnology, Sumbawa University of Technology, Jl. Raya Olat Maras, Batu Alang, Moyo Hulu, Sumbawa Besar 84371, Indonesia; witarto@gmail.com

[3] Joint Department of Biomedical Engineering, The University of North Carolina at Chapel Hill and North Carolina State University, Chapel Hill, NC 27599, USA; noya-loew@rs.tus.ac.jp

[4] Department of Pure and Applied Chemistry, Faculty of Science and Technology, Tokyo University of Science, 2641 Yamazaki, Noda, Chiba 278-8510, Japan

* Correspondence: ksode@email.unc.edu (K.S.); tsugawa@cc.tuat.ac.jp (W.T.); Tel.: +1-919-966-3550 (K.S.); +81-42-388-7027 (W.T.)

Received: 21 April 2020; Accepted: 13 May 2020; Published: 15 May 2020

Abstract: In this paper, a novel electron mediator, 1-methoxy-5-ethyl phenazinium ethyl sulfate (mPES), was introduced as a versatile mediator for disposable enzyme sensor strips, employing representative flavin oxidoreductases, lactate oxidase (LOx), glucose dehydrogenase (GDH), and fructosyl peptide oxidase (FPOx). A disposable lactate enzyme sensor with oxygen insensitive *Aerococcus viridans*-derived engineered LOx (*Av*LOx), with A96L mutant as the enzyme, was constructed. The constructed lactate sensor exhibited a high sensitivity (0.73 ± 0.12 µA/mM) and wide linear range (0–50 mM lactate), showings that mPES functions as an effective mediator for *Av*LOx. Employing mPES as mediator allowed this amperometric lactate sensor to be operated at a relatively low potential of +0.2 V to 0 V vs. Ag/AgCl, thus avoiding interference from uric acid and acetaminophen. The lactate sensors were adequately stable for at least 48 days of storage at 25 °C. These results indicated that mPES can be replaced with 1-methoxy-5-methyl phenazinium methyl sulfate (mPMS), which we previously reported as the best mediator for *Av*LOx-based lactate sensors. Furthermore, this study revealed that mPES can be used as an effective electron mediator for the enzyme sensors employing representative flavin oxidoreductases, GDH-based glucose sensors, and FPOx-based hemoglobin A1c (HbA1c) sensors.

Keywords: 1-methoxy-5-ethyl phenazinium ethyl sulfate; disposable enzyme sensor; lactate oxidase; glucose dehydrogenase; fructosyl peptide oxidase; electrochemical enzyme sensor; biomedical engineering

---

1. Introduction

Electrochemical enzyme sensors are the most widely studied form of biosensors due to their simple construction and achievable adequate performance. As is commonly known, in electrochemical enzyme sensors, the substrate (or analyte) is oxidized by a redox enzyme, which results in a product and a reduced cofactor of the enzyme in the reductive half reaction; in other words, electrons are transferred from the substrate to the cofactor of the enzyme [1]. For example, lactate is oxidized by the flavoenzyme lactate oxidase (LOx) harboring flavin mononucleotide (FMN) as a cofactor [2–6]. This results in pyruvate and reduced flavin. In first-generation electrochemical enzyme sensors, the reduced cofactor is reoxidized, and molecular oxygen (O_2) is reduced to hydrogen peroxide (H_2O_2) in the subsequent oxidative half reaction [7–9]. The resulting H_2O_2 is then oxidized at the surface of the electrode, which is held at an appropriate potential; thus, the electrons are transferred to the electrode and can be measured as an electrochemical current signal. However, the oxidation of H_2O_2 requires a high potential ($\geq$+0.6 V vs. Ag/AgCl [1,9]). At this high potential, oxidation of other redox substances in blood can occur, which leads to an erroneous increase in the response current and can cause serious problems [1]. In second-generation electrochemical biosensors, this problem is solved by utilizing an artificial electron mediator to transfer electrons from the reduced cofactor of the enzyme to the electrode. Mediators are preferentially used, especially in disposable strip-type enzyme sensors, because they allow a lower application potential than what is necessary to oxidize H_2O_2, which leads to fewer errors due to redox interference.

Second-generation electrochemical biosensors for lactate are currently commercially available for several applications, including self-monitoring [10] and point-of-care testing [11] of blood lactate. Blood lactate levels are a relevant parameter in clinical diagnosis and sports medicine, and the availability of disposable enzyme sensors for the determination of blood lactate is increasing.

LOx derived from *Aerococcus viridans* (*Av*LOx) is widely used in lactate enzyme sensors [12–19], including in commercially available lactate sensor strips [11]. However, LOx utilizes O_2 as an electron acceptor, which causes inherent problems in second-generation sensors and results in a lower sensor signal. This problem is common to second-generation biosensors utilizing oxidases and can be solved by utilizing dehydrogenases, which do not utilize O_2 as an electron acceptor. Our group recently engineered *Av*LOx to have a very low reactivity to O_2, thus converting the enzyme into a dehydrogenase [18]. By utilizing this engineered *Av*LOx, an A96L mutant, the influence of O_2 is minimized. Thus, utilizing this mutant enzyme can increase the accuracy of lactate sensors.

Attempts are continuing to improve the performance of second-generation lactate sensors, including improving their accuracy and shelf life. In addition to the enzyme, the mediator is a key element in second-generation electrochemical enzyme sensors, so the choice of mediator to be employed in enzyme sensors is one way to improve their performance. When choosing the mediator, the preference of the enzyme for the mediator as its electron acceptor must to be considered, as it affects the efficiency of the electron transfer from the cofactor of the enzyme to the electrode surface and thus determines the suitability of the mediator for sensor construction. Studies have shown that the availability of a mediator may be influenced by electrostatic interactions between the mediator and the enzyme surface or by steric hindrance between the mediator and the enzyme [19–23]. To avoid interference from the oxidation of redox substances present in the sample and thus minimize errors in the response of the biosensor, the biosensor should be operated at a low potential, which can be achieved with a mediator with a low redox potential [24]. Another parameter to be considered in the search for a good mediator is its stability under the sensor fabrication and storage of sensors, where the mediators will be exposed under several conditions, i.e., at room temperature, high humidity, and under the light illumination.

In a recent study, we reported on the mediator preference of *Av*LOx [19]. The light stable form of phenazinium methyl sulfate (PMS), a popular redox dye in spectrometric assays, 1-methoxy-5-methyl phenazinium methyl sulfate (mPMS), was shown to be a more effective mediator for *Av*LOx than potassium ferricyanide [19], a popular mediator commonly used in commercial enzyme sensors. Hexaammine ruthenium(III) chloride, a mediator that is currently gaining attention in the development of enzyme sensors due to its stability, i.e., to light irradiation, and low redox potential (-0.11 V vs. Ag/AgCl [25]) compared to potassium ferricyanide (redox potential +0.23 V vs. Ag/AgCl [26]), was not utilized as an electron acceptor by *Av*LOx [19]. The lactate sensors employing mPMS as the mediator showed a higher sensitivity and a wider linear range than those employing potassium ferricyanide as the mediator. Due to the low redox potential of mPMS (-0.11 V vs. Ag/AgCl [27]), the lactate sensors can be operated at a low potential. Therefore, employing mPMS as the mediator for lactate sensors based on *Av*LOx improves the accuracy of the lactate sensors by avoiding interference from redox substances in the blood [19]. A commercially available blood lactate sensor based on LOx that employs mPMS in combination with hexaammine ruthenium(III) chloride supports these findings [10]. In this commercial lactate sensor, both the effectivity of mPMS as the primary electron acceptor for LOx and the low potential of hexaammine ruthenium(III) chloride are exploited by fabricating lactate sensor strips with a low amount of mPMS to act as the primary electron acceptor and mediator between the enzyme and secondary electron acceptor and a high amount of hexaammine ruthenium(III) chloride to act as the secondary electron acceptor and mediator between the primary electron acceptor and the electrode [10].

Despite its many advantages, however, hexaammine ruthenium(III) chloride contains the rare metal ruthenium, the use of which, considering preservation and sustainability, should be avoided in the future. The organic mediator mPMS is easily synthesized and does not contain any rare elements. Unfortunately, mPMS is not stable in acidic to neutral solutions [27–29]. The low stability of mPMS can shorten the shelf life of enzyme sensors employing this mediator.

In 2018, 1-methoxy-5-ethyl phenazinium ethyl sulfate (mPES) became commercially available. This new mediator is stable in solution over a wide pH range [29,30]. Therefore, *Av*LOx-based lactate sensors containing mPES are expected to have a longer shelf life than corresponding sensors containing mPMS. The similar structure and similar electrochemical characteristics of mPES and mPMS suggest that sensors with mPES as the mediator can be developed with negligible interference from redox substances and thus similarly high accuracy. However, the utilization of mPES as a mediator in enzyme sensors has yet to be reported.

In this study, disposable, strip-type enzyme sensors based on representative flavin oxidoreductases, *Av*LOx, *Aspergillus flavus*-glucose dehydrogenase (*Af*GDH), and *Phaeosphaeria nodorum*-fructosyl peptide oxidase (*Pn*FPOx), employing mPES as an electron mediator, were constructed to assess the universal applicability of mPES as a mediator in enzyme sensor strips.

2. Materials and Methods

2.1. Materials and Apparatus

Sucrose, Tween 20, and silica bead desiccant were purchased from Kanto Chemical Co., Inc. (Tokyo, Japan). Sodium L-lactate was purchased from Sigma-Aldrich (St. Louis, MO, USA). D(+)-Glucose was purchased from Wako Pure Chemical Industries Ltd. (Osaka, Japan). Fructosyl valine was purchased from Santa Cruz Biotech, Inc. (Dallas, TX, USA). L(+)-Ascorbic acid was purchased from Nacalai Tesque, Inc. (Kyoto, Japan). Uric acid and 4'-hydroxyacetanilide were purchased from Tokyo Chemical Industry Co., Ltd. (Tokyo, Japan), while mPES was kindly provided by Dojindo Laboratories (Kumamoto, Japan). All other chemicals were of reagent grade.

Screen-printed carbon electrodes (SPCEs, working electrode [WE]: carbon 2.4 mm^2; reference electrode [RE]: silver/silver chloride [Ag/AgCl]; counter electrode [CE]: carbon) were kindly supplied by i-SENS (Seoul, Republic of Korea). All electrochemical measurements were carried out using a VersaSTAT4 potentiostat from Princeton Applied Research (Princeton, NJ, USA).

2.2. Enzyme Preparation and Activity Evaluation

The *Av*LOx A96L mutant was produced by following the methods of Hiraka et al. [18] with the following minor modifications. The cells of *E. coli* harboring the *Av*LOx A96L mutant were grown in 100 mL medium using 500 mL baffle flasks and incubated at 30 °C and 120 rpm for 36 h. The dye-mediated lactate dehydrogenase activities were measured in 20 mM PPB (pH 7.0) containing 0.06 mM 2,6-dichloroindophenol (DCIP), 4 mM phenazine methosulfate (PMS), and various concentrations of sodium L-lactate. The dehydrogenase activities were determined by monitoring the reduced DCIP at 600 nm based on the molar absorption coefficient of DCIP at pH 7.0 (16.3 mM^{-1} cm^{-1}). One unit of dehydrogenase activity was defined as the amount of enzyme necessary to catalyze the reduction of 1 μmol DCIP per minute at 25 °C. These assays were performed in triplicate. Production and activity evaluation of the engineered glucose dehydrogenase V149C/G190C mutant derived from *Aspergillus flavus* (*Af*GDH CC mutant) and fructosyl peptide oxidase N56A mutant derived from *Phaeosphaeria nodorum* (*Pn*FPOx N56A mutant) were carried out according to the methods by Sakai et al. [31] and Kim et al. [32], respectively.

2.3. Fabrication of Lactate Sensors and Evaluation of the Sensor Response

A mixture of *Av*LOx, mPES, sucrose, and Tween 20 in 100 mM PPB (pH 7.0) was prepared. One microliter of the mixture, containing 1 U enzyme, 100 mM mediator, 4% sucrose, and 2% Tween 20, was deposited onto the WE of an SPCE. The mixture was then dried onto the SPCE at room temperature (RT) at less than 1% relative humidity (RH) for 3 h. This resulted in an amount of 1 U enzyme, 100 nmol mPES, 40 μg sucrose, and 20 μg Tween 20 per sensor strip. A spacer (75 μm thickness) and a cover were attached to the dry modified SPCEs to form a μL-volume capillary space above the electrodes, resulting in disposable lactate sensors. Figure 1 shows the configuration of SPCE used in this study and the procedure to prepare lactate enzyme sensor. Chronoamperometry (CA) measurements were carried out to evaluate the response currents toward lactate by loading 1.0 μL lactate (0–50 mM) onto the sensor strip, which was connected to the potentiostat. A potential of +0.2 V vs. Ag/AgCl was applied 60 s after the addition of lactate, and the response currents were monitored for another 60 s. The sensitivity, the linear range, the limit of detection (LOD), the limit of quantitation (LOQ), and reproducibility based on the relative standard deviation (RSD) value were calculated as follows.

$$\text{Sensitivity } (\mu A/mM) = \text{the slope of the calibration curves obtained from plotting lactate concentrations versus response currents.}$$

$$\text{Linear range } (mM) = \text{the range of lactate concentration obtained from the linear line of the calibration curves of lactate concentrations versus response currents.}$$

$$\text{LOD } (mM) = \frac{3.3 \times \text{standard deviation of background current (0 mM lactate)}}{\text{slope (sensitivity of lactate sensor)}}$$

$$\text{LOQ } (mM) = \frac{10 \times \text{standard deviation of background current (0 mM lactate)}}{\text{slope (sensitivity of lactate sensor)}}$$

$$\text{RSD } (\%) = \frac{100 \times \text{standard deviation of particular point (i.e., for currents obtained with 2 mM lactate)}}{\text{mean of response current with 2 mM lactate (N = 3)}} 100\%$$

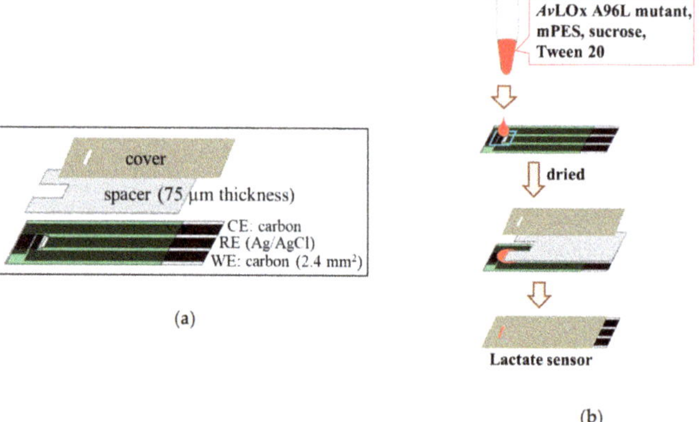

Figure 1. Illustration of screen-printed carbon electrode (SPCE) used in this study; (**a**) components of the SPCE and (**b**) construction of lactate sensor.

2.4. Evaluation of Response Currents of Lactate Sensors with Different Application Potentials

Lactate sensors were fabricated according to the method described in Section 2.3. The response currents were evaluated by applying a potential of 0 V, +0.05 V, +0.1 V, or +0.2 V vs. Ag/AgCl.

2.5. Evaluation of Storage Stability of Lactate Sensors

Lactate sensors were fabricated by following the method described in Section 2.3 but with 5 U enzyme per strip. The lactate sensors were divided into several sets, containing a minimum of 18 sensors each. Each set was wrapped in aluminum foil. All sets were put in a dark plastic box containing silica bead desiccant and stored at 25 °C for up to 50 days. After 2 days (prestorage), the response currents of the first set of lactate sensors were evaluated with 0-50 mM lactate, and the obtained data set was assigned as a control (Day 0). Subsequent sets of lactate sensors were evaluated after 12, 28, and 48 more days of storage (Day 12, Day 28, and Day 48).

2.6. Evaluation of Lactate Sensors toward Interferences

Lactate sensors were fabricated according to the method described in Section 2.3. The sensors were evaluated using mixtures of 2 mM or 10 mM lactate containing 0.17 mM ascorbic acid, 0.1 mM uric acid, or 0.3 mM acetaminophen. As with pure lactate solutions, 1.0 µL of the mixture was loaded onto the lactate sensor, and the response current was monitored.

2.7. Employing mPES as an Electron Mediator for Other Enzyme Sensors

The corresponding enzyme sensors were fabricated utilizing the enzymes *Af*GDH or *Pn*FPOx (0.1 U per sensor strip) according to the same method described in Section 2.3. The response currents were evaluated by loading solutions of 0-50 mM glucose or fructosyl valine onto the respective sensor strip based on *Af*GDH or *Pn*FPOx, respectively, and potential was applied 90 s after loading the solution.

3. Results

3.1. Construction of a Disposable Lactate Sensor

In this study, the versatility of mPES was demonstrated by constructing a disposable type lactate enzyme sensor strip, employing SCPE as the representative disposable electrode, widely utilized in the commercially available glucose enzyme sensor strips. A solution containing mediator and enzyme was dried onto a film electrode, and a spacer and cover were attached to complete the disposable lactate sensor. The novel, stable mPES [29] was employed as the mediator, and O_2 insensitive *Av*LOx A96L mutant was employed as the enzyme [18]. When the sample lactate solution was loaded into the sensor strip, the LOx and the mediator dissolved, and the enzyme reaction began. The time between the start of the enzyme reaction and application of the potential (start of the electrochemical reaction) is known as the waiting time. The waiting time was optimized to 60 s to achieve complete oxidation of lactate by the enzyme and thus full turnover. The response currents observed after application of the potential were generated by oxidation of the reduced mediator at the electrode; the initial amount of the reduced mediator was proportional to the initial amount of lactate in the sample.

The response currents toward 0–50 mM lactate were monitored amperometrically at an operation potential of +0.2 V vs. Ag/AgCl (operation potential was determined from cyclic voltammograms of the mediator, Supplementary Material Figure S1). An initial high response current was observed, which gradually decreased with time until it reached a steady state. In Figure 2a, the time courses of the response currents, or amperograms, are shown. The current at 10 s after potential application was plotted against the lactate concentrations to obtain the calibration curve (Figure 2b). The observed response currents increased linearly depending on the lactate concentration. A wide linear range of up to 50 mM lactate was obtained. The sensitivity of the lactate sensor, defined as the slope of the linear range, was 0.73 ± 0.12 µA/mM. The lactate sensor also showed good reproducibility at each tested lactate concentration (RSD < 7%) (Table 1). The LOD and LOQ was 0.5 mM and 1.8 mM, respectively. These results show that the lactate sensor was constructed successfully and that lactate can be measured with this lactate sensor.

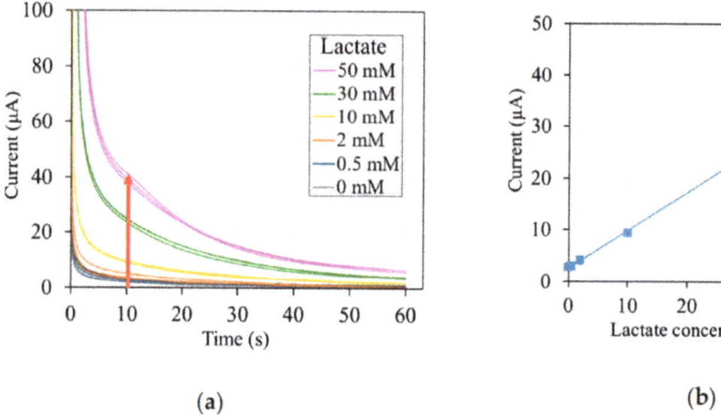

Figure 2. Response currents of lactate sensors; (**a**) time courses and (**b**) calibration curve (linear regression: y = 0.73x + 2.5; R^2 > 0.99). The arrow in (**a**) indicates increasing lactate concentration. Sensor composition: 1 U *Av*LOx A96L, 100 nmol mPES, 40 µg sucrose, and 20 µg Tween 20. Waiting time: 60 s. Applied potential: +0.2 V vs. Ag/AgCl. N = 3. Sampling point for the calibration curve: at 10 s.

Table 1. Properties of biosensors employing 1-methoxy-5-ethyl phenazinium ethyl sulfate (mPES) in this study.

Enzyme	Amount of Enzyme (U/strip)	Linear Range (0 mM to)	Sensitivity (µA/mM)	R^2	LOD (mM)	RSD (%)
*Av*LOx A96L mutant	1	50 mM [a]	0.73 ± 0.12	>0.99	0.5	<7
*Af*GDH CC mutant	0.1	50 mM [b]	0.25	>0.99	2.6	<3 [d]
*Pn*FPOx N56A mutant	0.1	10 mM [c]	0.55	>0.99	1.1	<8 [e]

Analyte: [a] lactate, [b] glucose, and [c] fructosyl valine. LOD: limit of detection; RSD: relative standard deviation. [d] for 10 mM, [e] for 0.5 and 10 mM.

3.2. Optimization of Operation Potentials

To optimize the operation potential, response currents were evaluated with different applied potentials, from 0 to +0.2 V (vs. Ag/AgCl) (Figure 3). A clear dependency of the response current on the lactate concentration was observed at all potentials. However, at 0 V applied, the linear range extended only up to 30 mM (Figure 3a). At higher operating potentials, the linear range extended up to 50 mM (Figure 3b–d). This is evidenced in the fact that the linear regression of the response currents obtained has the same slope and a high R^2 value with and without the response current value obtained with 50 mM lactate included when a potential higher than 0 V was applied (continuous vs. dashed lines in Figure 3b–d). When 0 V was applied, the response current value obtained with 50 mM lactate significantly deviated from the linear regression line obtained with response current values up to 30 mM lactate (Figure 3a).

A linear range of up to 30 mM lactate, which was obtained at all tested operation potentials, is more than enough to cover the standard range of blood lactate levels in humans, which can increase to a maximum of 25 mM [22]. The lactate sensor showed a good sensitivity of approximately 0.7 µA/mM, defined as the slope of the linear calibration, for the range of up to 30 mM lactate for all evaluated potentials, indicating that the sensor can be used at any operating potential between 0 V and +0.2 V vs. Ag/AgCl for measurements of up to 30 mM lactate. If higher lactate concentrations are expected, an operation potential of +0.05 V or higher should be used.

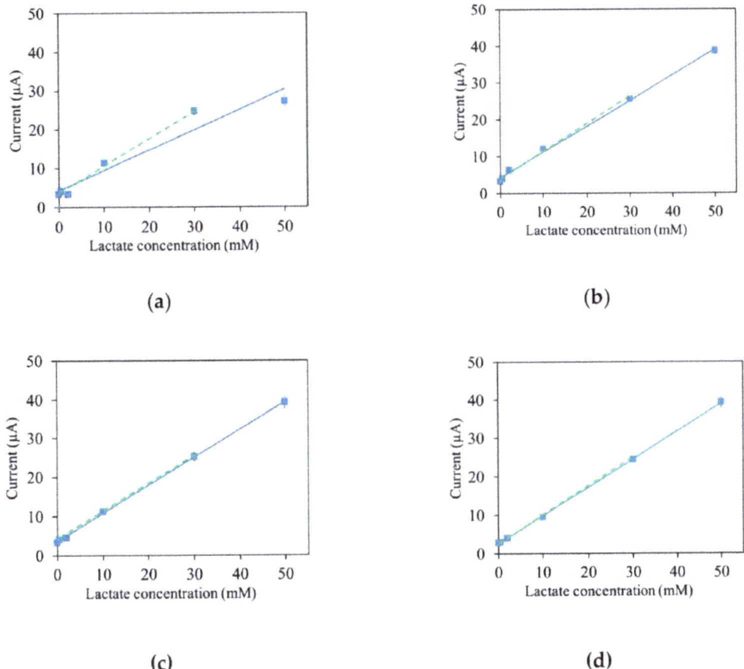

Figure 3. Response currents at different application potentials: (**a**) 0 V, (**b**) +0.05 V, (**c**) +0.1 V, and (**d**) +0.2 V vs. Ag/AgCl. Blue continuous line: linear regression for the range of up to 50 mM; green dashed line: linear regression for the range of up to 30 mM. Sensitivity for the range up to 50 mM: 0.52 µA/mM (R^2 = 0.93) in (**a**) and 0.69–0.73 µA/mM (R^2 > 0.99) in (**b–d**). Sensitivity for the range up to 30 mM: 0.69–0.73 µA/mM (R^2 > 0.99) in (**a–d**). Sensor composition: 1 U *AvLOx* A96L, 100 nmol mPES, 40 µg sucrose, and 20 µg Tween 20. Waiting time: 60 s. N = 3. Calibration curves at sampling point: 10 s.

3.3. Interference of Common Redox Substances

The response of the lactate sensors in the presence of possibly interfering redox substances was evaluated. For this, mixtures of 2 mM and 10 mM lactate with either ascorbic acid, uric acid, or acetaminophen were loaded into the sensor strips. In the presence of ascorbic acid, the observed response currents were higher than the corresponding signals of samples without ascorbic acid (Figure 4). Student's unpaired *t*-test showed a *p* value of < 0.05, indicating a significant difference. In the presence of uric acid or acetaminophen, there was no significant difference in the response currents compared to corresponding currents in the absence of redox substances (p > 0.05). Measurements of samples with 0 mM lactate and 10 mM interferent were also carried out, with corresponding results (Supplementary Material Figure S2). These results suggest that the lactate sensors are affected by ascorbic acid but not by uric acid or acetaminophen.

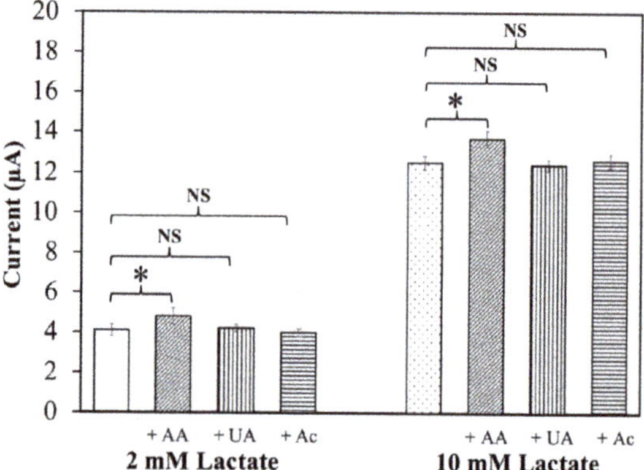

Figure 4. Response currents of lactate sensors to 2 mM or 10 mM lactate in the absence and presence of redox substances: 0.17 mM ascorbic acid (AA), 0.1 mM uric acid (UA) or 0.3 mM acetaminophen (Ac). * Student's unpaired *t*-test value $p < 0.05$ (suggesting a significant difference between data); NS: $p > 0.05$ (suggesting no significant differences between data). Sensor composition: 1 U *AvLOx* A96L, 100 nmol mPES, 40 µg sucrose, and 20 µg Tween 20. Waiting time: 60 s. Applied potential: +0.2 V vs. Ag/AgCl. N = 3. Sampling point: 10 s.

3.4. Storage Stability of the Lactate Sensors

The storage stability of the lactate sensors was evaluated with sensor strips with 5 U of *AvLOx* A96L per sensor, to avoid the case that enzyme inactivation would be the major factor to show a decrease in the sensor performance. The stability of the sensor with 1 U of *AvLOx* A96L per sensor is shown in Supplementary Material Figure S3, where no significant difference was observed from those of the sensor with 5 U of enzyme. The lactate sensors were stored for up to 2 + 48 days at 25 °C in the dark, with the first 2 days designated as prestorage. Sets of sensors were evaluated at 0, 12, 28, and 48 days of main storage (Figure 5). The sensitivity for calibration of up to 30 mM lactate was approximately 0.4 µA/mM at 0 days of main storage. The lower sensitivity compared to the sensors in Sections 3.1 and 3.2 can be explained by the increased amount of protein. However, the sensitivity is still acceptably high. The detection limit of the sensor also did not change during the storage for 48 days at 25 °C (Supplementary Material Table S1).

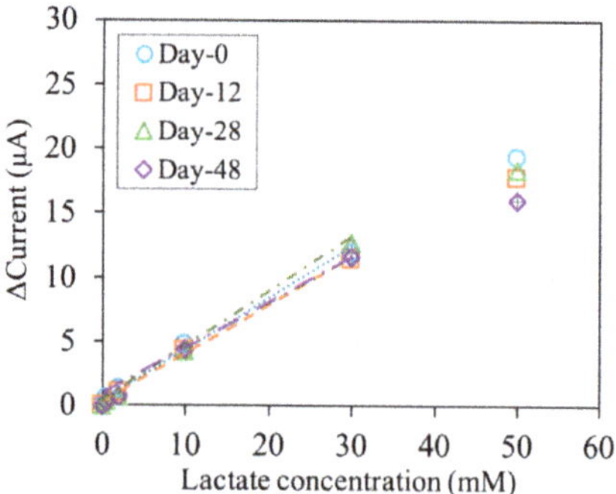

Figure 5. Calibration curves of lactate sensors stored at 25 °C in the dark, evaluated after storage for 2 days (prestorage) + 0, 12, 28, or 48 days. Linear regression was determined for up to 30 mM lactate; sensitivities were 0.38–0.42 µA/mM ($R^2 > 0.99$) for all calibration curves. Sensor composition: 5 U *AvLOx* A96L, 100 nmol mPES, 40 µg sucrose, and 20 µg Tween 20. Waiting time: 60 s. Applied potential: +0.2 V vs. Ag/AgCl. N = 3. Sampling point: 10 s.

The sensitivity for calibration of up to 30 mM lactate did not change significantly for 48 days of storage (Figure 5). The response current obtained with 50 mM lactate, however, decreased slightly with storage time. To support these findings, electrode strips containing only mediator were prepared and evaluated using a mixture of *Av*LOx A96L mutant and lactate (Supplementary Material Figure S4). The excellent storage stability of the mediator electrode strips at 25 °C and good storage stability at 45 °C suggests that the slight decrease in sensitivity to 50 mM lactate might be due to the enzyme. Therefore, this sensor can be used for measurements of up to 30 mM lactate after up to 48 days of storage at 25 °C in the dark without loss of sensitivity. For measurements of higher lactate concentrations, new lactate sensors should be used.

3.5. mPES as an Electron Mediator for Other Enzyme Sensors

Finally, to investigate the usability of mPES as a mediator for other enzymes, enzyme sensor strips were constructed utilizing mPES and the diagnostic enzymes *Af*GDH and *Pn*FPOx. The responses of the sensors based on *Af*GDH and *Pn*FPOx toward glucose and fructosyl valine, respectively, were evaluated. The response currents at 10 s were plotted against the analyte concentration, and calibration curves were obtained (Figure 6).

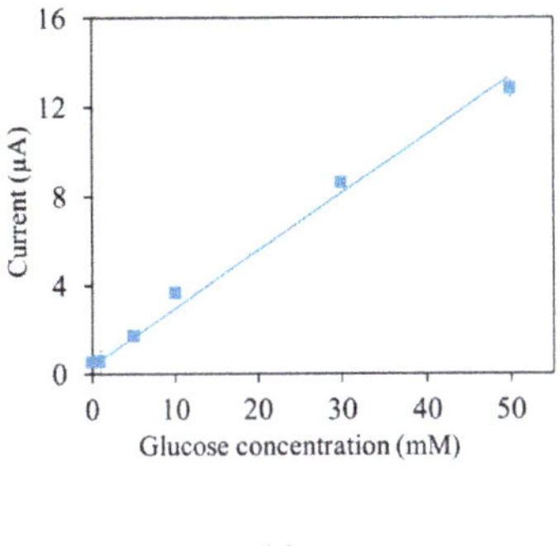
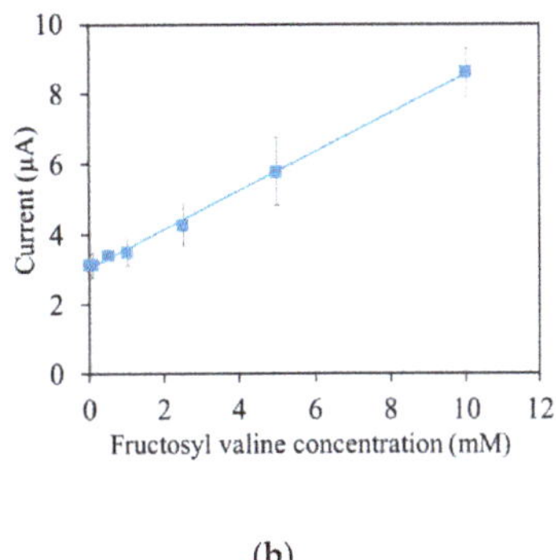

(a)   (b)

Figure 6. Calibration curves of biosensors utilizing mPES and (a) *Af*GDH CC mutant (linear regression: y = 0.25x + 0.65; R^2 = 0.99) or (b) *Pn*FPOx N56A mutant (linear regression: y = 0.55x + 3; R^2 > 0.99). Sensor composition: 0.1 U enzyme, 100 nmol mPES, 40 µg sucrose, and 20 µg Tween 20. Waiting time: 90 s. Applied potential: +0.2 V vs. Ag/AgCl. N = 3. Sampling point: 10 s.

A clear linear dependency of the response currents on the analyte concentration was observed, with a linear range of up to 50 mM glucose for the sensor strips containing *Af*GDH and up to 10 mM fructosyl valine for the sensor strips containing *Pn*FPOx. These results show that both *Af*GDH- and *Pn*FPOx-based enzyme sensor strips were constructed successfully utilizing mPES as a mediator.

4. Discussion

In this study, a novel electron mediator, mPES, was introduced as a versatile mediator for disposable enzyme sensor strips, employing representative flavin oxidoreductases, *Av*LOx, *Af*GDH, and *Pn*FPOx.

In our previous work, an oxygen insensitive mutant, *Av*LOx A96L, was constructed, and it was demonstrated the employment of this mutant was the solution to overcome the inherent problem of oxidase that the sensor signal is affected by oxygen [18]. However, we did not conclude the appropriate mediator for the lactate sensor employing *Av*LOx A96L. In the other achievement, we reported the investigation of electron mediators for enzyme sensors employing *Av*LOx, where we concluded that mPMS was the best [19]. However, mPMS is recognized as an unstable mediator, especially in neutral condition [27–29]. Therefore, we have been investigating an alternate mediator for LOx-based enzyme sensors, and we came across the use of mPES, which is the derivative of mPMS, but is more chemically stable than mPMS.

The properties of the sensor strips investigated in this study are summarized in Table 1. All three sensors have a linear range that sufficiently covers the range needed for the measurement of blood lactate levels, blood glucose levels, and HbA1c values. All three sensors also have good sensitivity, a low LOD value, a high R^2 value for the linear regression, and a low RSD of the measurements. These parameters indicate the high accuracy of the sensors. In this study, to demonstrate the versatility of mPES, a SCPE was employed as the representative disposable electrode, which is widely utilized in the commercially available glucose enzyme sensor strips. Current results did not indicate any obstacle issues of mPES to be used for sensors with another electrode material, such as gold, platinum, and palladium, which are conventionally utilized for disposable sensor strips.

The method used in these enzyme sensors based on monitoring of lactate (*Av*LOx A96L), glucose (*Af*GDH), and fructosyl amino acid (*Pn*FPOx) acid is an end-point assay, as has been utilized for commercially available glucose sensors for self-monitoring of blood glucose. Considering the amount of target molecule in 1 μl sample, 0.1 U of enzyme will be theoretically enough to oxidize an entire sample within the setting time for incubation (60 s for lactate and 90 s for glucose and fructosyl amino acid). In these types of enzyme sensors, the enzyme reaction reaches its equilibrium relatively fast, usually before the potential is applied. The response current is an indicator for the concentration of the reduced mediator at equilibrium. For example, in the current experiment with the sample volume of 1 μL with 50 mM of lactate sample, there are 50 nmol of lactate and 1 U of *Av*LOx A96L. Considering the definition of 1 U unit of dehydrogenase activity was defined as the amount of enzyme necessary to catalyze the reduction of 1 μmol (1000 nmol) DCIP in 60 s at 25 °C, which corresponds to the oxidation of 500 nmol of lactate in 60 s, excess amount of *Av*LOx A96L existed on the electrode in this experiment. Usually, enzyme sensor strips contain an excess amount of enzyme, considering the shelf life of a sensor, as the enzyme gradually inactivates during the preservation. We chose 1 U or 5 U of *Av*LOx A96L considering further investigation of mediator stability during the preservation, where enzyme stability would not be the limiting steps. For the investigation of the availability of mPES for *Af*GDH and *Pn*FPOx, where we did not plan a stability investigation, 0.1 U of enzymes were employed with 90 s for enzyme reaction, which provided enough time to arrive at end points for each experiment.

In addition to the characteristics of the enzyme, the characteristics of the mediator are a key factor in determining the properties of second-generation electrochemical biosensors. A commonly considered characteristic of the mediator is its redox potential, which determines the possible operating potentials of the sensor and thus influences the accuracy in terms of sensitivity to redox substance-type interference. Here, mPES has a low redox potential, which results in a low operating potential of +0.2 V. An even lower operating potential can be used with mPES as a mediator, although this might come at the cost of a narrower linear range (Figure 3).

A low operating potential generally means few interferences by redox substances that can be directly oxidized at the electrode. Three common redox substances in blood are ascorbic acid, uric acid, and acetaminophen, with the redox potential approximately +0.16 V, +0.45 V [33], and +0.2 V [34] (vs Ag/AgCl), respectively. In interference tests using the *Av*LOx-based sensors with 2 mM and 10 mM lactate, which are within and above the physiological range of lactate concentration in blood, respectively, we could show that the operating potential of +0.2 V is sufficiently low to eliminate interference by uric acid and acetaminophen (Figure 4). This was further supported by the fact that the sensors showed a response similar to the background (0 mM lactate) to 10 mM uric acid or acetaminophen (Supplementary Material Figure S2a).

However, a small, yet significant, bias in the response current is observed in the presence of ascorbic acid at an operating potential of +0.2 V (Figure 4), despite the potential being lower than the redox potential of ascorbic acid. Furthermore, the response of the sensor strips to 10 mM ascorbic acid was almost as high as that to 10 mM lactate (Figure S2a). Moreover, the response current did not decrease much at lower operating potentials, although a large decrease in the direct oxidation of a substance with the redox potential of ascorbic acid would be expected (Figure S2b). Reports have shown that ascorbic acid interferes by reducing the mediator rather than by being oxidized at the electrode [35–39], as exploited by a commercially available portable analyzer measuring reducing compounds, mainly ascorbic acid, by their ability to reduce the mediator, which is then quantified by oxidizing the mediator at the electrode [40]. Therefore, it is feasible that ascorbic acid can reduce mPES, which is then oxidized at the electrode and thus leads to a false signal.

Another characteristic of the mediator to be considered is its stability. This is especially the case in disposable sensor strips that operate with the end-point principle, meaning that the enzyme reaction is completed before the electrode reaction begins. In contrast to the kinetic principle, in which the mediator is recycled between the enzyme and electrode, for the end-point principle, a large amount of mediator is needed, as no recycling occurs. Thus, the storage stability of the sensor strips relies significantly on the storage stability of the mediator in the strips. Reports show that mPES is more stable than similar mediators, such as mPMS, over a wide range of pH values [29,30]. At highly alkaline conditions, decomposition of mPES due to ring-opening may occur [30]. Generally, neutral pH values are maintained during the fabrication and use of enzyme sensor strips, so that instability due to extreme pH conditions is not of concern. In this study, we investigated the storage stability of *Av*LOx-based sensor strips containing mPES. For this investigation, sensor strips containing more enzyme were used to avoid a decrease in the response current due to enzyme degradation, which led to a lower initial sensitivity due to the higher amount of protein. No significant decrease in the sensitivity in the range of 0–30 mM lactate was observed after 48 days of storage at 25 °C (Figure 5). The mediator itself proved to be stable at temperatures of at least 45 °C (Figure S4). Therefore, employing mPES as a mediator should be a successful step toward the development of sensor strips with an extended shelf life.

These results indicated that mPES is a promising organic mediator for strip-type disposable second-generation electrochemical biosensors, thanks to the superior characteristics of mPES, that is, the low redox potential, compatibility with various diagnostically relevant enzymes, and especially the high stability. The most common mediator, potassium ferricyanide, has a high redox potential (+0.23 V vs. Ag/AgCl [26]). Therefore, the amperometric enzyme sensor with potassium ferricyanide should be operated at high potential, otherwise consequently the signal will suffer from the oxidation current of electrochemically active ingredient potentially existing in samples. Potassium ferricyanide is also a labile molecule, which is easily inactivated during the preservation, especially under light exposure. Hexaammine ruthenium(III) has similar redox potential (−0.11 V vs. Ag/AgCl [25]) to mPES, however, hexaammine ruthenium(III) is not available as the mediator for native *Av*LOx and *Af*GDH as reported [19,20]. In addition, the use of the rare metal element, ruthenium, which has limited existence in the earth, is not recommended as the mediator for a disposable sensor. These inherent issues of using potassium ferricyanide or hexaammine ruthenium(III) as mediators will be overcome by substituting with mPES. PMS and mPMS have been widely utilized as the mediators of a variety of enzyme electrochemical investigations, thanks to their low redox potential (−0.11 V vs. Ag/AgCl [27]). Considering that PMS is a labile molecule toward light exposure, mPMS was developed, however mPMS is not stable at neutral to alkaline condition. mPES keeps low redox potential (−0.14 V vs. Ag/AgCl, Supplementary Material Figure S1) stability under light exposure, but with improved stability over a wide pH range. This study demonstrated that mPES can be used as the electron mediator of a variety of oxidoreductases, and thus, it is the most promising mediator for further studies of electrochemical sensors.

5. Conclusions

In this study, a novel electron mediator, mPES, was introduced for disposable enzyme sensor strips, employing representative flavin oxidoreductases, *Av*LOx, *Af*GDH, and *Pn*FPOx, by demonstrating versatility of this mediator including applying potential for the amperometric enzyme sensors, impact of potential redox active ingredients, storage stability, and availability as the electron acceptor for various enzymes. The successfully constructed lactate sensor had a high sensitivity (0.73 ± 0.12 μA/mM) and a wide linear range (0–50 mM lactate). Thanks to the redox potential of mPES, the sensor was operated by applying a low operation potential of +0.2 V vs. Ag/AgCl, or even at low as low as 0 V, without significant loss in sensitivity, and consequently, no interference was observed in the presence of uric acid and acetaminophen. Ascorbic acid reacts with mPES, and thus, precautions need to be implemented to account for ascorbic acid in the sample. Furthermore, the lactate sensors were stable for at least 48 days of storage at 25 °C in the dark, which is unusually long for an organic mediator without the addition of stabilizing agents. The compatibility of mPES with other representative diagnostic enzymes, such as GDH and FPOx, was also approved. Thus, compared to potassium ferricyanide, or hexaammine ruthenium(III) chloride, mPES might be a good alternative mediator to consider for future developments of second-generation electrochemical enzyme sensor strips.

Supplementary Materials: The following are available online at http://www.mdpi.com/1424-8220/20/10/2825/s1.

Author Contributions: Conceptualization, M.F., N.L., K.S. and W.T.; methodology and validation, M.F., N.L., K.S. and W.T.; formal analysis, M.F., N.L. and W.T.; investigation, M.F.; resources, N.L., K.S. and W.T.; data curation, M.F., N.L., K.S. and W.T.; writing—original draft preparation, M.F.; writing—review and editing, N.L., A.B.W., K.S. and W.T.; visualization, M.F.; supervision, N.L., A.B.W., K.I., W.T. and K.S.; project administration, M.F., K.I., K.S. and W.T. All authors have read and agreed to the published version of the manuscript.

Funding: This research received no external funding.

Acknowledgments: The authors thank Dojindo Laboratories for their kindness to provide mPES and i-SENS Inc. (Seoul, Republic of Korea) for their kindness to provide the SPCE strips. The authors also thank Kentaro Hiraka, Department of Biotechnology, Graduate School of Engineering, Tokyo University of Agriculture and Technology, for helpful discussions. The authors thank Junko-Okuda Shimazaki, Post-Doctoral Research Associate at Joint Department of Biomedical Engineering, The University of North Carolina at Chapel Hill and North Carolina State University, Chapel Hill, NC, USA. The first author (M.F.) acknowledges the Indonesia Endowment Fund for Education, Ministry of Finance of Indonesia (LPDP RI) for providing a scholarship.

Conflicts of Interest: The authors declare no conflict of interest.

References

1. Monteiro, T.; Almeida, G. Electrochemical Enzyme Biosensors Revisited: Old Solutions for New Problems. *Crit. Rev. Anal. Chem.* **2018**, *49*, 44–66. [CrossRef] [PubMed]

2. Duncan, J.D.; Wallis, J.O.; Azari, M.R. Purification and properties of *Aerococcus viridans* lactate oxidase. *Biochem. Biophys. Res. Commun.* **1989**, *164*, 919–926. [CrossRef]

3. Maeda-Yorita, K.; Aki, K.; Sagai, H.; Misaki, H.; Massey, V. L-lactate oxidase and L-lactate monooxygenase: Mechanistic variations on a common structural theme. *Biochim.* **1995**, *77*, 631–642. [CrossRef]

4. Leiros, I.; Wang, E.; Rasmussen, T.; Oksanen, E.; Repo, H.; Petersen, S.B.; Heikinheimo, P.; Hough, E. The 2.1 Å structure of *Aerococcus viridans* L-lactate oxidase (LOX). *Acta Crystallogr. Sect. F Struct. Boil. Cryst. Commun.* **2006**, *62*, 1185–1190. [CrossRef]

5. Umena, Y.; Yorita, K.; Matsuoka, T.; Kita, A.; Fukui, K.; Morimoto, Y. The crystal structure of L-lactate oxidase from *Aerococcus viridans* at 2.1 Å resolution reveals the mechanism of strict substrate recognition. *Biochem. Biophys. Res. Commun.* **2006**, *350*, 249–256. [CrossRef]

6. Li, S.J.; Umena, Y.; Yorita, K.; Matsuoka, T.; Kita, A.; Fukui, K.; Morimoto, Y. Crystallographic study on the interaction of L-lactate oxidase with pyruvate at 1.9 Å resolution. *Biochem. Biophys. Res. Commun.* **2007**, *358*, 1002–1007. [CrossRef]

7. Karube, I.; Matsunaga, T.; Teraoka, N.; Suzuki, S. Microbioassay of phenylalanine in blood sera with a lactate electrode. *Anal. Chim. Acta* **1980**, *119*, 271–276. [CrossRef]

8. Matsunaga, T.; Karube, I.; Teraoka, N.; Suzuki, S. Determination of cell numbers of lactic acid producing bacteria by lactate sensor. *Appl. Microbiol. Biotechnol.* **1982**, *16*, 157–160. [CrossRef]

9. Clark, L.C., Jr.; Noyes, L.K.; Grooms, T.A.; Moore, M.S. Rapid micromeasurement of lactate on whole blood. *Crit. Care. Med.* **1984**, *12*, 461–464. [CrossRef]

10. Lactate Pro 2. Available online: https://www.pmda.go.jp/PmdaSearch/ivdDetail/ResultDataSetPDF/100639_25A2X00001000013_A_01_03 (accessed on 9 May 2020).

11. Nova StatStrip®Lactate Test Strips-For Use Only with the Nove StatStrip Lactate Hospital Meter. Available online: https://www.woodleyequipment.com/docs/product_insert_lactate_meter_test_strip.pdf (accessed on 9 May 2020).

12. Spohn, U.; Narasaiah, D.; Gorton, L. Reagentless Hydrogen Peroxide and L-Lactate Sensors Based on Carbon Paste Electrodes modified with different peroxidases and lactate oxidases. *J. Prakt. Chem.* **1997**, *339*, 607–614. [CrossRef]

13. Moser, I. Biosensor arrays for simultaneous measurement of glucose, lactate, glutamate, and glutamine. *Biosens. Bioelectron.* **2002**, *17*, 297–302. [CrossRef]

14. Thomas, N.; Lähdesmäki, I.; Parviz, B. A contact lens with an integrated lactate sensor. *Sens. Actuators B Chem.* **2012**, *162*, 128–134. [CrossRef]

15. Taurino, I.; Reiss, R.; Richter, M.; Fairhead, M.; Thöny-Meyer, L.; De Micheli, G.; Carrara, S. Comparative study of three lactate oxidases from *Aerococcus viridans* for biosensing applications. *Electrochim. Acta* **2013**, *93*, 72–79. [CrossRef]

16. Andrus, L.P.; Unruh, R.; Wisniewski, N.A.; McShane, M.J. Characterization of Lactate Sensors Based on Lactate Oxidase and Palladium Benzoporphyrin Immobilized in Hydrogels. *Biosensors* **2015**, *5*, 398–416. [CrossRef] [PubMed]

17. Bollella, P.; Sharma, S.; Cass, A.E.G.; Antiochia, R.; Sharma, S. Microneedle-based biosensor for minimally-invasive lactate detection. *Biosens. Bioelectron.* **2019**, *123*, 152–159. [CrossRef] [PubMed]

18. Hiraka, K.; Kojima, K.; Lin, C.-E.; Tsugawa, W.; Asano, R.; La Belle, J.; Sode, K. Minimizing the effects of oxygen interference on L-lactate sensors by a single amino acid mutation in *Aerococcus viridans* L-lactate oxidase. *Biosens. Bioelectron.* **2018**, *103*, 163–170. [CrossRef]

19. Loew, N.; Fitriana, M.; Hiraka, K.; Sode, K.; Tsugawa, W. Characterization of electron mediator preference of *Aerococcus viridans*-derived lactate oxidase for use in disposable enzyme sensor strips. *Sens. Mater.* **2017**, *29*, 1703–1711. [CrossRef]

20. Okurita, M.; Suzuki, N.; Loew, N.; Yoshida, H.; Tsugawa, W.; Mori, K.; Kojima, K.; Klonoff, D.C.; Sode, K. Engineered fungus derived FAD-dependent glucose dehydrogenase with acquired ability to utilize hexaammineruthenium(III) as an electron acceptor. *Bioelectrochemistry* **2018**, *123*, 62–69. [CrossRef]

21. Nieh, C.-H.; Tsujimura, S.; Shirai, O.; Kano, K. Electrostatic and steric interaction between redox polymers and some flavoenzymes in mediated bioelectrocatalysis. *J. Electroanal. Chem.* **2013**, *689*, 26–30. [CrossRef]

22. Loew, N.; Tsugawa, W.; Nagae, D.; Kojima, K.; Sode, K. Mediator Preference of Two Different FAD-Dependent Glucose Dehydrogenases Employed in Disposable Enzyme Glucose Sensors. *Sensors* **2017**, *17*, 2636. [CrossRef]

23. Tsuruoka, N.; Sadakane, T.; Hayashi, R.; Tsujimura, S. Bimolecular Rate Constants for FAD-Dependent Glucose Dehydrogenase from *Aspergillus terreus* and Organic Electron Acceptors. *Int. J. Mol. Sci.* **2017**, *18*, 604. [CrossRef] [PubMed]

24. Chaubey, A.; Malhotra, B. Mediated biosensors. *Biosens. Bioelectron.* **2002**, *17*, 441–456. [CrossRef]

25. Metzker, G.; De Aguiar, I.; Martins, S.C.; Schultz, M.S.; Vasconcellos, L.C.; Franco, U.W. Electrochemical and chemical aspects of ruthenium(II) and (III) ammines in basic solution: The role of the ruthenium(IV) species. *Inorg. Chim. Acta* **2014**, *416*, 142–146. [CrossRef]

26. O'Reilly, J.E. Oxidation-reduction potential of the ferro-ferricyanide system in buffer solutions. *Biochim. Biophys. Acta (BBA) Bioenerg.* **1973**, *292*, 509–515. [CrossRef]

27. Kimura, Y.; Niki, K. Electrochemical oxidation of nicotinamide-adenine dinucleotide (NADH) by modified pyrolytic graphite electrode. *Anal. Sci.* **1985**, *1*, 271–274. [CrossRef]

28. Jahn, B.; Jonasson, N.S.W.; Hu, H.; Singer, H.; Pol, A.; Good, N.M.; Camp, H.J.M.O.D.; Martinez-Gomez, N.C.; Daumann, L.J. Understanding the chemistry of the artificial electron acceptors PES, PMS, DCPIP and Wurster's Blue in methanol dehydrogenase assays. *J. Boil. Inorg. Chem.* **2020**, *25*, 199–212. [CrossRef]

29. Methoxy PES. Available online: https://www.dojindo.co.jp/products/M470/ (accessed on 9 May 2020).

30. Yomo, T.; Sawai, H.; Urabe, I.; Yamada, Y.; Okada, H. Synthesis and characterization of 1-substituted 5-alkylphenazine derivatives carrying functional groups. *JBIC J. Boil. Inorg. Chem.* **1989**, *179*, 293–298. [CrossRef]

31. Sakai, G.; Kojima, K.; Mori, K.; Oonishi, Y.; Sode, K. Stabilization of fungi-derived recombinant FAD-dependent glucose dehydrogenase by introducing a disulfide bond. *Biotechnol. Lett.* **2015**, *37*, 1091–1099. [CrossRef]

32. Kim, S.; Nibe, E.; Tsugawa, W.; Kojima, K.; Ferri, S.; Sode, K. Construction of engineered fructosyl peptidyl oxidase for enzyme sensor applications under normal atmospheric conditions. *Biotechnol. Lett.* **2011**, *34*, 491–497. [CrossRef]
33. Kamel, M.M.; Abdalla, E.M.; Ibrahim, M.; Temerk, Y. Electrochemical studies of ascorbic acid, dopamine, and uric acid at a dl-norvaline-deposited glassy carbon electrode. *Can. J. Chem.* **2014**, *92*, 329–336. [CrossRef]
34. Wangfuengkanagul, N.; Chailapakul, O. Electrochemical analysis of acetaminophen using a boron-doped diamond thin film electrode applied to flow injection system. *J. Pharm. Biomed. Anal.* **2002**, *28*, 841–847. [CrossRef]
35. Pournaghi-Azar, M.; Ojani, R. Catalytic oxidation of ascorbic acid by some ferrocene derivative mediators at the glassy carbon electrode. Application to the voltammetric resolution of ascorbic acid and dopamine in the same sample. *Talanta* **1995**, *42*, 1839–1848. [CrossRef]
36. Murthy, A.S.; Sharma, J. Benzoquinone modified electrode for sensing NADH and ascorbic acid. *Talanta* **1998**, *45*, 951–956. [CrossRef]
37. Pournaghi-Azar, M.H.; Ojani, R. Attempt to incorporate ferrocenecarboxylic acid into polypyrrole during the electropolymerization of pyrrole in chloroform: Its application to the electrocatalytic oxidation of ascorbic acid. *J. Solid State Electrochem.* **1999**, *3*, 392–396. [CrossRef]
38. Pournaghi-Azar, M.H.; Ojani, R. Electrochemistry and electrocatalytic activity of polypyrrole/ferrocyanide films on a glassy carbon electrode. *J. Solid State Electrochem.* **2000**, *4*, 75–79. [CrossRef]
39. Raoof, J.B.; Ojani, R.; Beitollahi, H.; Hossienzadeh, R. Electrocatalytic Determination of Ascorbic Acid at the Surface of 2,7-Bis(ferrocenyl ethyl)fluoren-9-one Modified Carbon Paste Electrode. *Electroanalysis* **2006**, *18*, 1193–1201. [CrossRef]
40. VC Sensor for Pocket Chem VC. Available online: http://www.arkray-vc.com/spec/# (accessed on 9 May 2020).

 *sensors*

Article

Assay of Phospholipase D Activity by an Amperometric Choline Oxidase Biosensor

Rosanna Ciriello * and Antonio Guerrieri 

Dipartimento di Scienze, Università degli Studi della Basilicata, Viale dell'Ateneo Lucano 10, 85100 Potenza, Italy; antonio.guerrieri@unibas.it
* Correspondence: rosanna.ciriello@unibas.it; Tel.: +39-0971-205944

Received: 16 January 2020; Accepted: 25 February 2020; Published: 27 February 2020

Abstract: A novel electrochemical method to assay phospholipase D (PLD) activity is proposed based on the employment of a choline biosensor realized by immobilizing choline oxidase through co-crosslinking on an overoxidized polypyrrole film previously deposited on a platinum electrode. To perform the assay, an aliquot of a PLD standard solution is typically added to borate buffer containing phosphatidylcholine at a certain concentration and the oxidation current of hydrogen peroxide is then measured at the rotating modified electrode by applying a detection potential of +0.7 V vs. SCE. Various experimental parameters influencing the assay were studied and optimized. The employment of 0.75% (v/v) Triton X-100, 0.2 mM calcium chloride, 5 mM phosphatidylcholine, and borate buffer at pH 8.0, ionic strength (I) 0.05 M allowed to achieve considerable current responses. In order to assure a controlled mass transport and, at the same time, high sensitivity, an electrode rotation rate of 200 rpm was selected. The proposed method showed a sensitivity of 24 $(nA/s) \cdot (IU/mL)^{-1}$, a wide linear range up to 0.33 IU/mL, fast response time and appreciable long-term stability. The limit of detection, evaluated from the linear calibration curve, was 0.005 IU/mL (S/N = 3). Finally, due to the presence of overoxidized polypyrrole film characterized by notable rejection properties towards electroactive compounds, a practical application to real sample analysis can be envisaged.

Keywords: choline biosensor; amperometric detection; overoxidized polypyrrole film; phospholipase D assay; phosphatidylcholine

1. Introduction

Phospholipase D (PLD) is a lipolytic enzyme which in vivo hydrolyses the terminal phosphodiester bond of phospholipids, generating phosphatidic acid (PA) and a free polar head group [1]. In addition, in the presence of a primary alcohol, it is able to catalyze the in vitro transphosphatidylation of the naturally abundant phosphatidylcholine [2]. PLD has been isolated and sequenced from a variety of sources and, as recently reviewed [3], can be regarded as a key component of many cellular and physiological processes.

In mammalian systems PLD is involved in cellular processes including membrane trafficking, cell proliferation, protein secretion, and metabolic regulation [1]. It has been associated with important pathological disorders such as thrombotic and neurodegenerative diseases, and several cancers [3]. Abnormalities in PLD activity in many human cancers have been noticed, highlighting a possible PLD role as a diagnostic biomarker and even as a target for drug discovery [4].

In plants, functional studies on PLD have shown its possible effects on membrane deterioration in response to stress injuries, senescence, aging, and pathogenesis [5].

In yeast, PLD activity has been shown to be involved mainly in sporulation and adaptation of nutrient utilization [6]. Bacterial PLD may act as virulence determinant and can mimic some reactions in cell [7]. Among hydrolytic enzymes from bacterial source, PLD from *Streptomyces Chromofuscus* is characterized by a notable transesterification capacity. This property can be exploited for the preparation of less abundant natural phospholipids, such as phosphatidylserine, phosphatidylglycerol, and novel artificial phospholipids of interest in the field of pharmaceutical and food industries, starting from crude or purified phosphatidylcholine [8,9]. Moreover, this bacterial PLD has been used as a model system for the mammalian enzyme since it mimics some reactions in cells [10]. A full understanding on the activation and inhibition mechanisms of this enzyme may be useful for in vivo studies on mammalian systems.

A wide variety of assay procedures have been realized for the determination of the activity of PLD enzymes. They can be substantially divided into two classes depending on the employment of exogenously-provided substrates, for in vitro measurements, or endogenous substrates for in vivo measurements [11]. In both cases the most used substrates are certainly radiolabeled phospholipids (radiolabeled acyl chains, phosphorus, or head group of the phospholipid). The radio-assay procedure has generally been reported to be fast and highly sensitive allowing to detect PLD in samples where its activity is very low, such as brain areas and peripheral tissues of mouse [12]. Despite these advantages, radio-assay is quite laborious and requires the employment of expensive and potentially health hazardous compounds.

Other methods have been proposed based on titrimetric or pH-stat techniques [13], chromatographic technique [14], conductimetry [15], infrared spectroscopy [16] and nuclear magnetic resonance [17]. These methods, while being viable alternatives to radio-assay, are not suited for routine analysis. In this context, enzyme coupled assays have gain increasing importance. They are based on the quantification of choline released during the hydrolysis of phosphatidylcholine by PLD. The released choline is measured by the formation of a colored or fluorescent complex, upon the sequential reactions of choline oxidase and peroxidase enzymes [18].

In order to bypass the requirement of coupled enzymes, synthetic substrates of PLD have been used such as phosphatidyl p-nitrophenol [19]. A graphene-based nano-assembly has been proposed as a fluorescence biosensor for sensitive detection of PLD activity by employing a fluorescein-labeled phospholipid [20]: even if quite interesting, since miming a cellular membrane, this approach requires labeling, a fluorescence assay and was not applied to any real sample where endogenous substances could generate fluorescence signals interfering with the PLD assay. Nanoprobes for fluorescence detection and imaging of PLD in cell lysate were also realized based as well on rhodamine B-labeled phospholipids [21,22].

Alternative spectrophotometric methods were based on the determination of phosphatidic acid through iron (III) complexation [23] or through Ca^{2+} complexation with 8-hydroxyquinoline [24]. In this last case, a direct and continuous PLD assay was achieved exploiting the chelation-enhanced fluorescence property of 8-hydroxyquinoline.

The aim of the present work was to realize an assay method sensitive, although nonradioactive, and easy to realize, without requiring the use of synthetic phospholipids, derivatization reaction, and instrumentations particularly expensive and difficult to handle. We developed an enzyme coupled procedure, based on a choline amperometric biosensor, able to provide real time measurements and suitable for high throughput assays. Indeed, electrochemical biosensors are clearly advantageous since they can overcome some limitations of the spectroscopic methods: they require a minimum sample pretreatment and can be used also for the analysis of turbid or colored samples.

The employment of choline amperometric biosensors for PLD assay is almost unexplored. Vrbova et al. [25] reported a PLD assay method based on a Clark-type electrode modified by a nylon net co-immobilizing choline oxidase and catalase. Besides the limitations typical of oxygen sensors, another drawback was surely the immobilization technique adopted which was quite complex and time consuming, requiring an incubation period of the enzymatic solution of 4 days.

The device we propose is surely easier to realize while assuring high enzyme stability. The approach we adopted is based on the detection of hydrogen peroxide which requires high applied potential thus causing severe interference problems from electroactive endogenous compounds presents in real samples. The modification of the Pt electrode by overoxidized polypyrrole, before choline oxidase (ChO) immobilization by co-crosslinking, could successfully solve this problem. This film is characterized by notable rejection properties towards interferents, as already demonstrated in the case of other enzymes and analytes [26–28].

The resulting device could be used in general to assay other enzymes than PLD which release choline as well. Indeed, we have already employed this biosensor to assay cholinesterase activity in human serum samples [29]. Having demonstrated the validity of such a sensor, in this work we wanted to extent its employment to the analysis of PLD by substantially modifying the reaction environment. Particularly, being PLD activity dependent on whether the substrate is readily accessible and on the presence of divalent cations at a certain concentration [3], a high sensor response was assured only upon having opportunely optimized the experimental conditions. As it will be shown, notable performances were achieved.

2. Materials and Methods

2.1. Materials

Choline oxidase (EC 1.1.3.17 from Alcaligenes species, 12 U/mg of solid), serum bovine albumin (fraction V), choline chloride, glutaraldehyde (grade II, 25% aqueous solution), L-α-phosphatidylcholine (Type XI-E: from fresh egg yolk, approx. 99%, 100 mg/mL solution in chloroform) and Triton X-100 were obtained from Sigma (Sigma Chemical Co.; St. Louis, MO, USA). Phospholipase D (EC 3.1.4.4., from *Streptomyces Chromofuscus*, 50,000 U/mL) was purchased from Biomol Research Laboratories Inc. (Butler Pike, Plymouth Meeting, PA, USA). Pig lung comes from a local butcher shop.

The above chemicals were of analytical reagent grade and were used without further purification with the exception of pyrrole (Aldrich, Steinheim, Germany) which was purified by distillation under vacuum by fixing the temperature at 62 °C. Choline chloride before its use was left to dry in a vacuum desiccator for 3 days in the presence of P2O5.

Choline stock solutions were prepared in double distilled-deionized water and stored in the dark at 4 °C. Fresh L-α-phosphatidylcholine standard solutions were prepared daily by evaporating chloroform and then dissolving the required amounts in 50 mM borate buffer (pH 8.0) containing 0.75% (v/v) Triton X-100 and 0.2 mM calcium chloride. Phospholipase D solutions were prepared in borate buffer pH 8.0 unless otherwise stated. The PLD activity is expressed in International Units (IU) (1 IU releases 1 µmol of choline per minute) throughout the work.

2.2. Spectrophotometric Assay

Before use, PLD activity was assayed by an established colorimetric method using a reagent kit available from Biomol (Phospholipase D Assay Catalog SE-301). The assay comprises three reaction steps. In the first step, PLD catalyzes the hydrolysis of phosphatidylcholine to choline and phosphatidic acid. In the second step, the oxidation of choline catalyzed by choline oxidase produces two hydrogen peroxides. Peroxidase catalyzes the third step, in which the two hydrogen peroxides, phenol and 4-aminoantipyrine react to produce a quinonimine dye, with a millimolar extinction coefficient of 12.2 mM^{-1} cm^{-1} at 500 nm. Measurement of the absorbance at 500 nm can then be used to calculate the amount of choline produced by the phospholipase reaction.

2.3. Apparatus

To perform electrochemical experiments, an EG&G (Princeton Applied Research, Princeton, NJ, USA) model 263 A potentiostat/galvanostat was used. Data acquisition was accomplished through a M270 electrochemical research software (EG&G) version 4.23.

A conventional three electrode cell was used consisting of a platinum counter electrode and a saturated calomel reference electrode (SCE). The working electrode was a platinum disk (2 mm diameter) embedded in a poly(tetrafluoroethylene) (PTFE) body. Experiments under controlled mass transport were performed by a CTV101 Speed Control Unit, EDI 101 Rotating Disc Electrode (RDE) (Radiometer Copenhagen). A rotation rate of 200 rpm was typically adopted unless otherwise stated.

Spectrophotometric measurements were performed using a single beam LKB Biochrom Ultrospec II 4050 UV/Visible spectrophotometer (Biochrom Ltd., Cambridge, UK).

2.4. Biosensor Preparation

In order to assure a reproducible electrode surface, before each modification Pt substrates were subjected to a cleaning procedure comprising a first step of dipping in hot nitric acid for few minutes. Then the electrode was polished by mechanical abrasion employing an alumina slurry with particles of 0.05 μm. Finally, to remove any adsorbed alumina, the electrode was copiously rinsed with bi-distilled water and sonicated for several minutes.

Polypyrrole films (PPy) were electrosynthesized potentiostatically using a 0.4 M pyrrole solution in 0.1 M KCl electrolyte. A selected potential of +0.7 V versus SCE was applied and polymer growth was stopped when a deposition charge of 300 mC/cm^2 was passed. The formed film was then immersed in a phosphate buffer (pH = 7.0, I = 0.1 M) and overoxidized by applying a fixed potential of +0.7 V versus SCE. After a time of at least 7 h, a steady-state background current was reached and the overoxidation process was stopped. The resulting modified electrode (Pt/PPyox) was rinsed with bi-distilled water and air-dried at room temperature before enzyme immobilization.

Choline oxidase immobilization was carried out by following a procedure already reported [30]: 16 mg of bovine serum albumin (BSA) and 1 mg of ChO were dissolved in 300 μL of a phosphate buffer solution (I = 0.1 M, pH = 6.5); 30 μL of 2.5% glutaraldehyde solution (25% glutaraldehyde solution diluted 1:10 with phosphate buffer) were then added; 3 μl of the resulting solution were pipetted onto the Pt/PPyox working electrode surface and spread out to completely cover the electrode surface avoiding air bubble formation. Finally, the modified electrode was air-dried at room temperature and soaked in the background electrolyte to remove weakly bound or adsorbed enzyme. The enzyme electrode was stored in phosphate buffer, pH 6.5, I 0.1 M, at 4 °C in the dark.

2.5. Electrochemical Measurement

A detection potential of +0.7 V versus SCE was adopted to perform all the electrochemical measurements. PLD assay was performed through batch addition experiments: the modified electrode was immersed in a borate buffer (I = 0.05 M, pH = 8.0) solution stirred at 200 rpm and containing calcium chloride 0.2 mM and Triton X-100 0.75% (v/v); an aliquot of a phosphatidylcholine solution, and then an aliquot of a standard PLD solution were added into the cell and the current variation with time was recorded. A temperature of 37 °C was maintained for the duration of the assay by a thermostatic bath.

2.6. Real Sample Analysis

Extraction of PLD from pig lung was performed accordingly to a procedure described elsewhere [31]. Briefly, about 500 g of pig lung was minced and homogenized in sucrose buffer at 4 °C, centrifuged and stored at −80 °C until its use. Just before analysis, the thawed homogenized was diluted in buffer to reach the desired PLD activity, stirred, further centrifuged, and the desired supernatant aliquot was added to the electrochemical cell for PLD assay.

3. Results and Discussion

In the present work, we have proposed an assay method based on the employment of a modified Pt/PPyox/ChO electrode, realized following the fabrication steps shown in Scheme 1, which is immersed in a thermostated solution of phosphatidylcholine and kept under a fixed rotation rate. Then, to perform the assay, an aliquot of a standard PLD solution is injected in the cell which causes the release of free choline (Ch). The immobilized ChO catalyzes the oxidation of Ch to betaine with the subsequent production of hydrogen peroxide. At the electrode surface, opportunely polarized, hydrogen peroxide undergoes progressive oxidation producing a current signal which increases linearly with time. PLD activity in the cell solution can be evaluated from the slope of the current signal increase, as shown in Scheme 1, after having opportunely calibrated the sensor.

Following are the reactions involved to produce the sensor response

$$\text{phosphatidylcholine} + 2H_2O \xrightarrow{PLD} \text{phosphatidic acid} + \text{choline} \tag{1}$$

$$\text{choline} + 2O_2 + H_2O \xrightarrow{ChO} \text{betaine} + 2H_2O_2 \tag{2}$$

$$H_2O_2 \rightarrow O_2 + 2H^+ + 2e^- \tag{3}$$

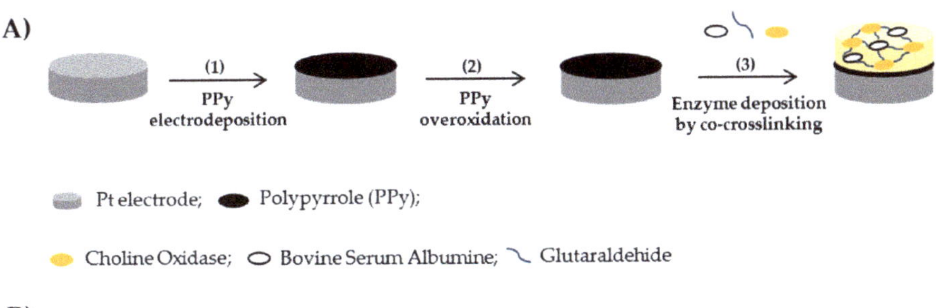

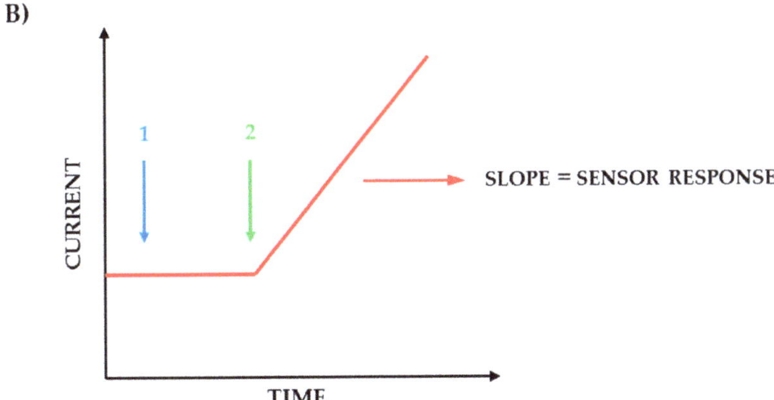

Scheme 1. Fabrication steps of the biosensor (**A**) and current signal acquisition (**B**).

PLDs from the culture broth of *Streptomyces* are commercially available or can be prepared by fermentation from some *Actinomycetes* strains [36,37]. Imamura and Horiuti provided information on the purification of PLD from Streptomyces Chromofuscus secreted into the growth medium [38]. They partially characterized this bacterial PLD furnishing a molecular weight of 57 kDa and an isoelectric point of pH 5.1. To optimize the enzymatic reaction, alkaline pH values are required in the range 8–9 and the presence of detergent (Triton X-100, deoxycholate), Ca^{2+} or both of them revealed able to significantly increase the velocity [38]. Indeed, like the eukaryotic enzymes PLD from Streptomyces Chromofuscus requires Ca^{2+} ion for activity whereas, unlike other phospholipases, does not exhibit "interfacial activation" or "surface dilution" [39,40]. This means that this enzyme does not show preference for substrate in the form of micellar short chain phospholipids rather than monomeric ones [39] and its specific activity does not depend on the mole fraction of substrate in a micelle or bilayer surface [40]. Nevertheless, considering that the catalytic reaction involves vesicle substrates, the transfer of PLD from aqueous solution to the lipid-water interface is a required step for the reaction to take place.

The notable performances of the choline oxidase biosensor towards choline detection have already been summarized in our previous work [29]. Fast response time, high sensitivity, wide linear range, and significant long-term stability were achieved. Furthermore, the presence of the underlying overoxidized polypyrrole film guaranteed a notable rejection of interferent compounds, electroactive at the high overpotential required to detect H_2O_2, thus allowing the application of the sensor to assay Cholinesterase in real samples as complex as human serum [29]. Electrochemical, XPS, and ESEM characterizations of these biosensors were already described elsewhere [28–30,32–35]. Based on these important achievements, in the present work we wanted to employ such a biosensor to develop a novel PLD assay method.

The binding mechanism that has been proposed for the enzyme is a two-step ping-pong like ordered one: the substrate combines with the enzyme giving a covalent phosphatidyl-enzyme moiety with release of choline; then, water attacking to the covalent adduct induces the release of phosphatidic acid [41]. The cleavage of the P-O bond is caused by the nucleophilic attack of water on the distal phosphate ester bond. It is worth noting that, in the presence of a primary alcohol at a high concentration, the phosphatidyl-enzyme intermediate can be decomposed to a different phospholipid.

It is evident that PLD activity is affected by various experimental variables, such as substrate accessibility, presence of a bication at a proper concentration, and a surfactant able to solubilize at the right extent the substrate, which need a careful investigation in addition to the more usual pH and composition of the reaction medium. An optimization study of the reaction conditions has therefore been carried out with the aim to implement a method able to quantify low PLD activities in a reproducible way and without any interference.

3.1. Influence of Surfactant

Phospholipases are generally more active on liposomes, mixed micellar deriving from the dispersion of phospholipids substrate in detergents, or phospholipids solubilized in organic solvents [42]. Particularly, PLD from *Streptomyces Chromofuscus* is reported to exhibit higher activity towards monomers and mixed micelles than substrate present in lipid vesicles [43]. When assaying PLD activity it is then a common practice to employ mixtures of phospholipid and detergents. PLDs from different sources are activated by anionic or neutral detergents [44] whose influence on the enzymatic activity has been attributed mainly to the micellization of phospholipids substrates. Changes of the physicochemical properties of the phospholipid bilayers are not to be excluded. Practically, detergents exert their action by destroying liposome structure favoring the gradual formation of mixed micelles. Indeed, the formation of mixed micelles is the final goal achieved during the interaction of detergents with liposome. At the beginning of such an interaction, surfactant monomers are incorporated within the lipid bilayer and, as a consequence, vesicle dimensions increase. Then, as the surfactant concentration is increased, phospholipids solubilization gradually occurs and liposomes along with mixed micelles are both presents. Finally, only mixed micelles are present as phospholipids are completely solubilized.

Triton X-100 is one of the most common surfactants employed for PLD assay. It is a non-ionic detergent with an average length of approximately 9.5 oxyethylene units per detergent molecule. Its wide employment can be justified considering that it is a mild non-denaturing detergent: even at concentrations up to 0.5% it usually does not damage most enzymes. For lysing cells lower concentrations of about 0.1% are sufficient.

In the present work, the appropriate Triton X-100 concentration to employ for PLD assay has been evaluated. At this aim experiments have been carried out by varying the detergent concentration in the range 0.1–0.9% (v/v), while fixing the other experimental conditions close to the values adopted in the spectrophotometric assay (phosphatidylcholine 5 mM, PLD 0.22 IU/mL, calcium chloride 10 mM, Tris buffer pH 8.0, I 0.05 M). In Figure 1 the variation of the slopes computed from the current signals increments is illustrated as a function of the detergent concentration. As it can be seen, in the first part of the graph biosensor response increases as detergent concentration is increased in agreement with a more efficient phosphatidylcholine dissolution at higher contents of Triton X-100. The graph shows a maximum located at a concentration of about 0.75%. A further increase in detergent concentration probably causes a partial enzyme denaturation thus justifying the response decrease evidenced in the second part of the graph.

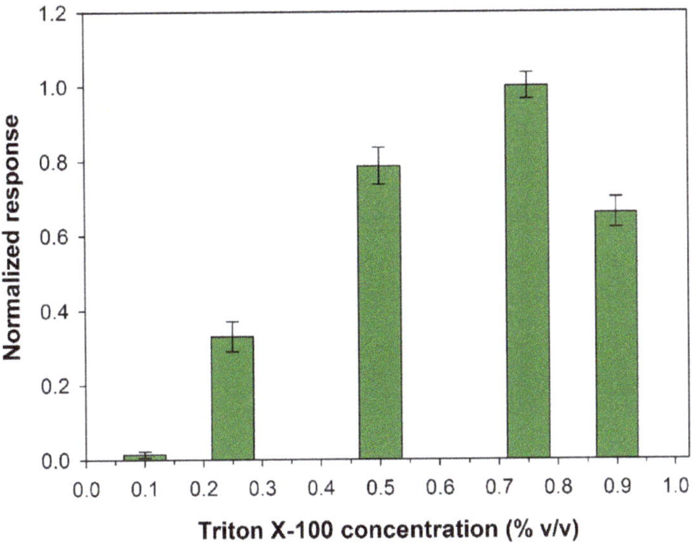

Figure 1. Normalized response (i.e., response/maximum response) of a typical Pt/PPyox/ChO biosensor to PLD 0.22 IU/mL as a function of Triton X-100 concentration. Supporting electrolyte: Tris buffer pH 8.0, I 0.05 M containing 10 mM calcium chloride, and 5 mM phosphatydylcholine. Electrode rotation rate: 1000 rpm.

Indeed, many enzymes remain active in the presence of Triton X-100 even at concentration of 1%, as it was reported for ChO [45]. As concern the effect of detergent on the activity of PLD, it was showed that when the activity was maximally increased by Ca^{2+} (10 mM) detergent was inhibitory at lower concentrations (completely at about 0.8 mM) and stimulating at higher concentrations (maximally at 4.5 mM) [38]. At the experimental conditions we adopted an inhibition effect on PLD from detergent is to be excluded. On the other hand, it was reported that the enzymatic reaction rate is optimal at a certain detergent to phospholipid ratio, a further increase in the molar ratio of Triton X-100 to phospholipid lead to progressive decrease in the enzymatic activities probably due to a sort of dilution of the micellar surface concentration of phosphatidylcholine [38]. In conclusion, the effect of the detergent on Phospholipase D activity would seem to be related essentially to the formation of mixed micelles that in the present experimental conditions is favored at a Triton concentration of 0.75%. This value was adopted from now on to assay PLD activity.

3.2. Influence of Calcium Concentration

The PLD from *S. Chromofuscus* is a water-soluble enzyme unlike its natural substrate which forms macromolecular associations. Since the PLD activity takes place in a heterogeneous medium, factors that could enhance enzyme removal from the soluble part of the medium (i.e. aqueous solution) to the particulate part (i.e. the lipid vesicles or monolayer) are of relevance. At this aim, ultrafiltration experiments showed that Ca^{2+} at micromolar concentration levels was essential for the PLD binding to PC vesicles suggesting that the binding of Ca^{2+} to PLD is followed by a conformational change of PLD and then the enzyme likely associates with PC vesicles [46]. Besides favoring the enzyme location at the water–lipid interface, calcium ion is reported to increase also the catalytic activity of PLD: it plays a key role in the phosphatidyl enzyme geometry for the nucleophilic attack of water on the P-O bond [47]. Considering then the PLD requirement for calcium, experiments were performed to determine the extent to which added Ca^{2+} affects its activity. Figure 2 shows the influence of calcium content on the biosensor response at fixed phosphatidylcholine and PLD concentrations. As it is possible to see, PLD activity increases as calcium concentration is increased up to a maximum at about 0.2 mM $(\log[Ca^{2+}] = -0.7)$ and then decreases to reach a quite constant value from calcium concentrations higher than 1 mM.

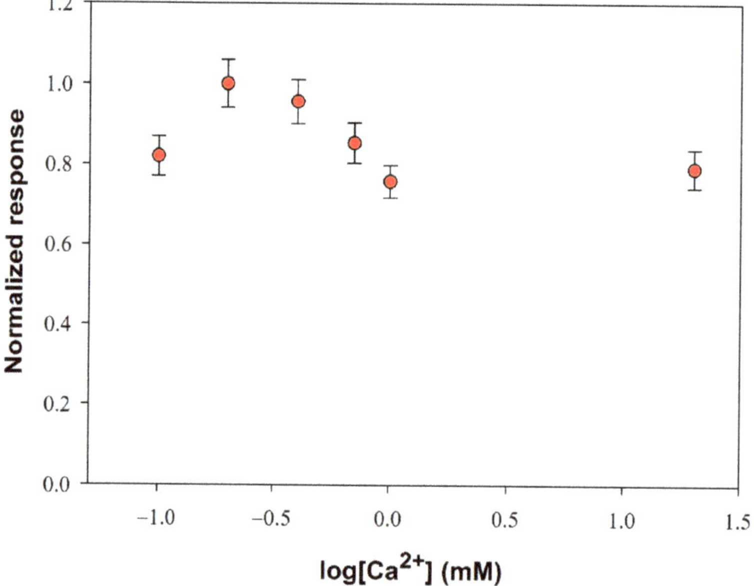

Figure 2. Normalized response (i.e., response/maximum response) of a typical Pt/PPyox/ChO biosensor to PLD 0.22 IU/mL as a function of Ca^{2+} concentration. Supporting electrolyte: Tris buffer pH 8.0, I 0.05 M containing 0.75% (v/v) Triton X-100 and 5 mM phosphatidylcholine. Electrode rotation rate: 1000 rpm. Each value represents the mean of triplicate measurements; the error bars represent the relevant standard deviations.

It is worth noting that, despite the small variations, an ANOVA analysis of data points (F-value 10.6, *p*-value 0.0004373 at a confidence level of 95%) evidenced that there's a significant difference between the mean response values of Figure 2. A more detailed treatment of the ANOVA analysis is reported in the Supplementary Materials. A calcium concentration value of 0.2 mM was therefore employed in the following experiments to optimize PLD assay.

3.3. Influence of Electrode Rotation Rate

The influence of substrate mass transport on the analytical performances of the biosensor was investigated. This study has been carried out in Tris-HCl buffer (pH 8.0, I 0.05 M) containing Triton X-100 and Ca^{2+} at the optimized concentrations respectively of 0.75% (v/v) and 0.2 mM. Under these experimental conditions, the variation of the slope of the signal current upon the enzymatic conversion of phosphatidylcholine 5 mM from PLD 0.22 IU/mL has been measured, by varying the rotation rate of the electrode from 50 to 3000 rpm. The upper limit of the explored range was limited by the well-known tendency of detergents to skim. Figure 3 shows the signal slopes variation as a function of the rotation rate at a Pt/PPyox/ChO biosensor. As expected in the case of a process limited by the mass transport in solution, the profile acquired shows in the first part an increment of the response as rotation rate is increased up to about 200 rpm. At higher rotation rates the slope of the signal decreases as rotation rate is increased, thus suggesting that the system is passing from an external diffusive control to an enzymatic kinetic control [48].

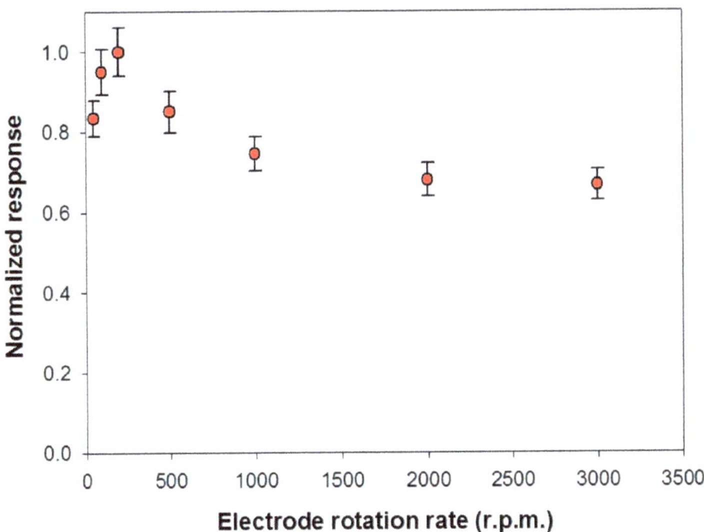

Figure 3. Normalized response (i.e., response/maximum response) of a typical Pt/PPyox/ChO biosensor to PLD 0.22 IU/mL as a function of electrode rotation rate. Supporting electrolyte: Tris buffer pH 8.0, I 0.05 M containing 0.75% (v/v) Triton X-100, 0.2 mM calcium chloride, and 5 mM phosphatidylcholine. Each value represents the mean of triplicate measurements; the error bars represent the relevant standard deviations.

Particularly, considering that solution stirring promotes the interaction PLD/phosphatidylcholine, the homogeneous catalytic process is unlikely to constitute the rate determining step. The whole bienzymatic process is reasonably controlled by the kinetic of the enzymatic reaction catalyzed by choline oxidase which occurs at the modified electrode. Indeed, the profile at higher rotation rates showed in Figure 3 is similar to that, here not reported, observed when using choline oxidase biosensor for choline sensing. On the other hand, hypothesizing the PLD catalysis in solution as the rate determining step, current signal would have been independent from the rotation rate. This behavior has already been noticed when the same choline biosensor was employed to assay ChE activity in solution following a similar reactions scheme [29].

In conclusion, to assure an appreciable sensitivity under controlled mass transport, a rotation rate of 200 rpm has been adopted in the following experiments.

3.4. Influence of pH

Being the proposed method based on the employment of an auxiliary enzyme, i.e., choline oxidase, before optimizing the pH for PLD assay, the effect of the buffer composition on the response of choline oxidase biosensor was examined using phosphate, glycine, tris, and borate buffers. The results, reported in Figure S2 of Supplementary Materials, showed that phosphate and borate buffer allowed to achieve the highest signals. One of the most employed biological buffers is surely Tris which however does not always assure the best analytical results [49]. As a confirmation, in the present study its employment resulted in the lowest response among all the studied buffers. Campanella et al. indeed evidenced a partial inhibition of choline oxidase in Tris buffer [50].

The reaction environment in which the enzymatic hydrolysis of phospholipids containing choline is carried out is usually composed of 50 mM Tris buffer at pH 8.0, calcium chloride and Triton X-100 [51]. Considering that the response of the choline biosensor was dramatically reduced in Tris buffer, as it was previously showed, the possibility of carrying out PLD assay in alternative buffers has been evaluated. At this aim phosphate, glycine, Tris, HEPES and borate buffers (0.05 M, pH 8.0) containing 0.75% (v/v) Triton X- 100- and 0.2-mM calcium chloride were employed. With the aim to discriminate which buffer allowed high sensitivity while assuring appreciable reproducibility, triplicate assay experiments have been carried out for each investigated system. The mean values of the current slopes with the relevant standard deviations are shown in Figure 4.

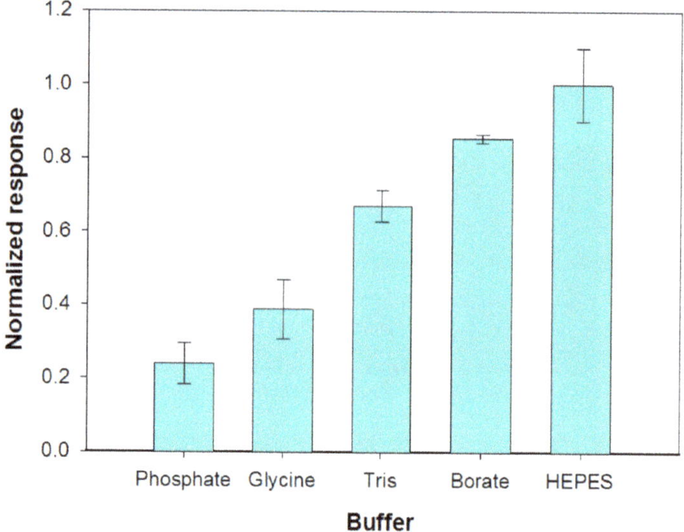

Figure 4. Effect of buffer composition (pH 8.0, I 0.05 M) on choline biosensor response to PLD 0.22 IU/mL. Triton X-100 concentration: 0.75% (v/v). Calcium chloride concentration: 0.2 mM. Phosphatidylcholine concentration: 5 mM. Electrode rotation rate: 200 rpm. Each value represents the mean of triplicate measurements; the error bars represent the relevant standard deviations.

The best results in term of sensitivity were obtained by employing HEPES and borate buffer. Nevertheless, HEPES buffer provided a poor repeatability of the analytical signal which progressively decreased after successive measurements, as it was already evidenced by Marazuela et al. [45]. Phosphate buffer showed poor reproducibility and sensitivity attributable to precipitation processes at the micromolar calcium concentrations required to maximize PLD activity. Such a buffer was therefore not used even if its employment assured the best performances for choline biosensor (Figure S2). Borate buffer, allowing the best response in terms of both precision (relative standard deviation of 1.3% (n = 3) on intraday measurements) and sensitivity was therefore employed for all further experiments.

The effect of the pH solution on the analytical signal was then tested within the range 7–9. At this aim the behavior of both the primary enzyme and the auxiliary enzyme upon varying pH was considered. Preliminary experiments, reported in Figure S3 of Supplementary Material, showed that the maximum sensitivity towards choline of the device Pt/PPyox/ChO falls in the pH range 8.5–9.5.

PLD is reported to show in solution a maximum activity at a pH close to 8.0 [38]. Figure 5 shows the signal slopes variation as a function of pH at a Pt/PPyox/ChO biosensor upon the addition of phosphatidylcholine (5 mM) and PLD (0.22 IU/mL). The curve evidences a maximum located at pH of about 8.0 and therefore in good agreement with the value reported in literature. Subsequent studies have been therefore carried out at pH 8.0 since at this value the activities of both PLD and ChO are almost at maximum.

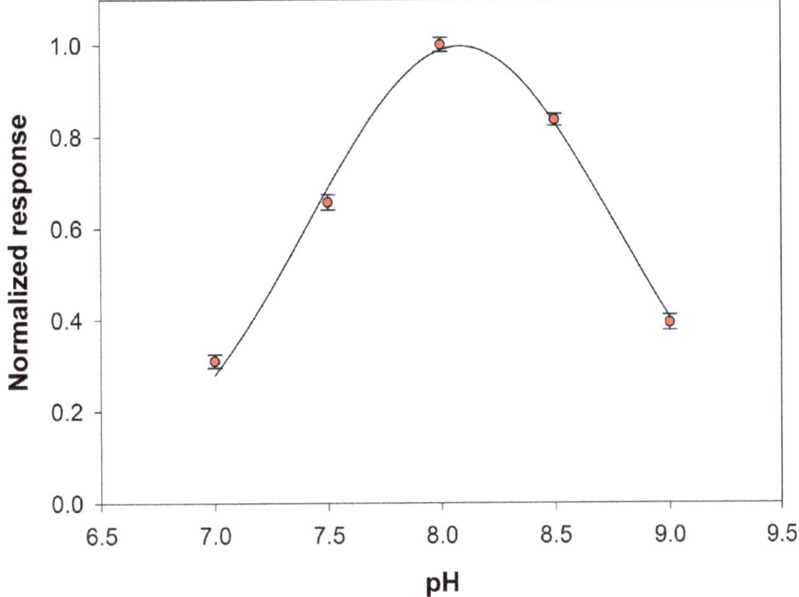

Figure 5. Normalized response (i.e., response/maximum response) of a typical Pt/PPyox/ChO biosensor to PLD 0.22 IU/mL as a function of pH. Supporting electrolyte: borate buffer pH 8.0, I 0.05 M containing 0.75% (v/v) Triton X-100, 0.2 mM calcium chloride, and 5 mM phosphatidylcholine. Electrode rotation rate: 200 rpm.

3.5. Optimization of Substrate Concentration

PLD from *Streptomyces Chromofuscus*, like bacterial PLDs in general, has broader substrate specificity than the eukaryotic enzymes. Such an enzyme is able to hydrolyze phosphatidylserine, phosphatidylethanolamine, phosphatidylglycerol, lysophosphatidylcholine, and sphingomyelin. Phosphatidylcholine is considered however the best substrate for hydrolysis. In the present work, L-α-phosphatidylcholine from egg yolk was employed with a mean molecular weight of about 768 Da. It is worth noting that complications arise when drawing kinetic data on phospholipases. These enzymes, indeed, acting on water-insoluble substrates, display an enhanced activity when using substrates organized in aggregates. This feature inevitably determines a dependence of the kinetic data on the particular phase in which phospholipids are organized in the reaction mixture. For this reason, we have experimentally verified at which substrate concentration, and then organization, PLD saturation was guaranteed at the experimental conditions already optimized. A calibration graph has been then constructed by varying the phosphatidylcholine concentration realized in the electrolyte solution (borate buffer pH 8.0, I 0.05 M containing 0.75% Triton X-100, and 0.2 mM Ca^{2+}) containing PLD at a concentration of 0.22 IU/mL, and holding the electrode under rotation at a speed of 200 rpm.

Figure 6 shows the behavior of the signal current slopes at substrate concentration values ranging from 0.5 to 8 mM. As it is possible to see, a phosphatidylcholine concentration of 5 mM assured substrate saturation condition for PLD and was therefore adopted to realize the calibration curve for enzyme assay.

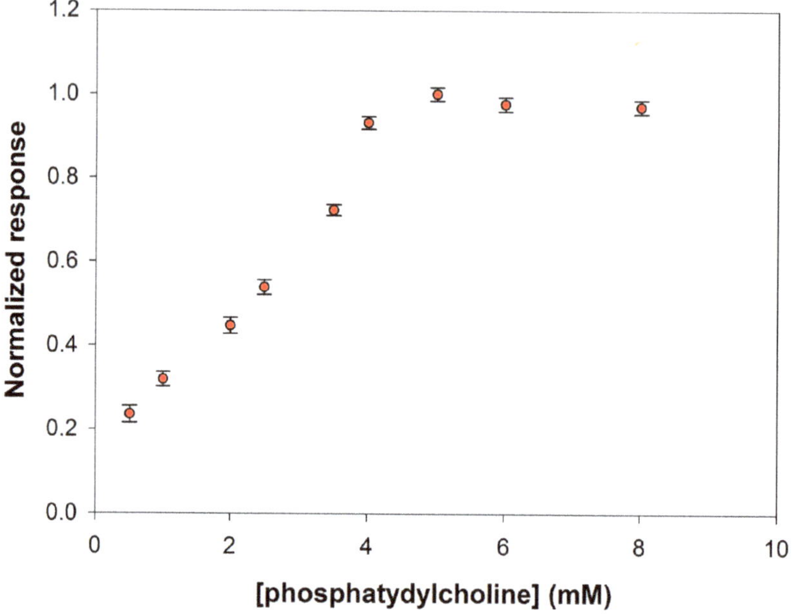

Figure 6. Normalized response (i.e., response/maximum response) of a typical Pt/PPyox/ChO biosensor to PLD 0.22 IU/mL as a function of phosphatidylcholine concentration. Supporting electrolyte: borate buffer pH 8.0, I 0.05 M containing 0.75% (v/v) Triton X-100, and 0.2 mM calcium chloride. Electrode rotation rate: 200 rpm.

3.6. Calibration Graphs for PLD Assay

The calibration curve for PLD assay has been derived by injecting in the electrochemical cell the substrate at the concentration value previously optimized and then by introducing an aliquot of an enzyme standard solution at a fixed temperature of 37 °C. The response has been evaluated, as usual, by the slope of the current signal acquired at the modified electrode. In Figure 7 the calibration graph obtained over all the PLD activities investigated is reported whereas in the inset the linear range of the calibration curve is also illustrated. Fitting of the linear part of the calibration graph gave a regression line with a slope of 23.9 ± 0.2 (nA/s)·(IU/mL)$^{-1}$ and an intercept of (−0.0148 ± 0.0382) nA/s (correlation coefficient better than 0.999). Due to the significant accumulation of the enzymatic products during the observation times, a deviation from linearity is observed from an enzyme activity of 0.33 IU/mL. The within-day coefficients of variation for replicate (n = 5) PLD assays were 4.15% and 5.18% at 0.22 and 0.04 IU/mL PLD levels, respectively; the between-days (n = 6) coefficient of variation was 9.4% at 0.14 IU/mL PLD levels. The operational stability of the assay was tested doing replicate PLD assays (at 0.22 IU/mL PLD level) during a day (n = 10): the relevant variation was comparable (confidence level of 95%) with the within-day coefficients of variation so the biosensor response could be considered almost stable. Finally, the long-term stability was proved by evaluating the sensitivity of several calibration curves (n = 15) for PLD assay (at 0.22 IU/mL PLD level) during an extended period (biosensor stored in buffer at 4 °C at the dark when not used): after 30 days, the sensitivity dropped to about 25% of its initial value.

The limit of detection estimated from the linear regression equation was 0.005 IU/mL, corresponding to an absolute amount of 0.03 IU (30 nmol of choline per minute). This amount is suitable for practical determination of PLD in real samples. As an example, in the clinical field PLD activity in normal mammary cell attests to levels detectable with our device. In mammary carcinoma cell lysate PLD activity was found to be considerable higher with respect to normal cell (2921.4 ± 216.7 U/L vs. 408.6 ± 28.7 U/L) [21].

A direct comparison with existing PLD assay methods is complicated by the different modalities in which enzyme activity is expressed. Chemiluminescence assays have been reported in which liberation of just 75 pmol of choline per minute was detectable [52]. Method precision was however affected by the presence of substances which react with hydrogen peroxide and proteins at concentrations above 0.2% (w/v) which can cause quenching phenomena of the luminol reaction therein employed. Fluorometric detection and radiometric assay are surely highly sensitive techniques [23,24] but need synthetically modified fluorogenic substrates or special equipment to be performed. As concerns the electrochemical method previously described to assay PLD, based on the employment of a choline amperometric biosensor [25], a limit of detection for choline of 0.05 mM was reported and then substantially higher with respect to the value of 0.1 µM at pH 8 that we evaluated in our previous work [29]. Moreover, as it was already stressed, the biosensor realization was quite laborious and time consuming [25]. Compared to existing analytical assays, the present method is therefore simple and easy to operate, representing a valid alternative in ascertaining PLD activity in various fields of interest.

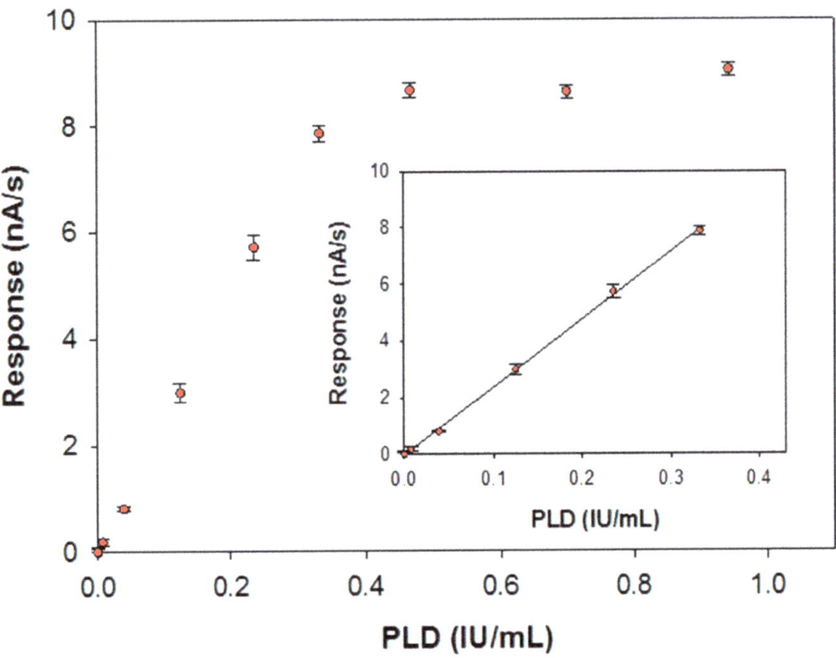

Figure 7. Calibration curve for phosphatidylcholine-PLD system carried out at a typical Pt/PPyox/ChO biosensor. Supporting electrolyte: borate buffer pH 8.0, I 0.05 M containing 0.75% (v/v) Triton X-100, 0.2 mM calcium chloride and 5 mM phosphatidylcholine. Electrode rotation rate: 200 rpm. Inset: linear interval of the calibration curve.

Preliminary experiments concerning the application of the proposed assay in real samples focused on the determination of PLD activity on crude extracts from pig lung. Lungs were chosen either for its high PLD activity [53] either because its crude extract is surely a complex biological matrix containing endogenous substances (e.g., ascorbate) which can interfere with the present electrochemical assay. Yet, the notable ability of PPyox polymer to reject interfering and fouling species in electrochemical analysis ensured proper measurements of PLD activity even in complex matrixes like this. In fact, analysis of pig lung crude extracts, after proper dilution, gave a specific activity of 1.50 ± 0.08 mIU/mg which agree with that obtained by the standard colorimetric method (see Materials and Methods section) used here for PLD assay (values from both assays not significantly different at a 95% confidence level); more importantly, the assay of pig lung crude extracts gave specificity activity values in agreement with those already reported elsewhere [31].

4. Conclusions

The novel electrochemical procedure for PLD assay proposed in this work is based on the employment of a bilayer choline biosensor composed of a co-crosslinked choline oxidase coating deposited on a permselective overoxidized polypyrrole film electrosynthesized on a platinum electrode. The optimization of the experimental parameters affecting PLD activity (e.g., pH and composition of the supporting electrolyte, hydrodynamic conditions, calcium chloride and surfactant concentration) allowed to obtain calibration curves characterized by appreciable sensitivity, wide linear range and adequately low limit of detection. Furthermore, the electrosynthesis of a polymer film with built-in permselectivity could realize a biosensor suitable for real matrices analysis being interferences from endogenous compound, electroactive at the applied potential, minimized. In conclusion, our device proved as an effective and novel method for the determination of PLD activity. Further experiments concerning its application to assay such an enzyme in biological samples are under way in our laboratory.

Supplementary Materials: The following are available online at http://www.mdpi.com/1424-8220/20/5/1304/s1, Section S1: Influence of calcium concentration: ANOVA analysis. Figure S1: Normalized response of a typical Pt/PPyox/ChO biosensor to PLD 0.22 IU/mL as a function of Ca^{2+} concentration. Experimental conditions as in Figure 2. Different labels indicate significantly different values. Figure S1: Dependence of the choline biosensor sensitivity on the buffer composition (pH 8.5, I 0.1 M). Electrode rotation rate: 1000 rpm, Figure S2: Influence of pH on the response of a typical Pt/PPyox/ChO biosensor. Supporting electrolyte: acetate/phosphate/borate buffer I 0.1 M. Electrode rotation rate: 1000 rpm.

Author Contributions: Conceptualization, A.G. and R.C.; Investigation, R.C.; Writing—original draft preparation, R.C.; Writing—review and editing, R.C.; Visualization, R.C.; Supervision, A.G.; Project administration, A.G. All authors have read and agreed to the published version of the manuscript.

Funding: The authors gratefully acknowledge financial support from Ministero dell'Istruzione, dell'Università e della Ricerca Italiana.

Acknowledgments: The authors thank Alberto Onzo and Giuliana Bianco for their helpful contribution to the ANOVA analysis. This work comes from Carmela Caivano's thesis.

Conflicts of Interest: The authors declare no conflict of interest.

References

1. Exton, J.H. Phosphatidylcholine breakdown and signal transduction. *Biochim. Biophys. Acta* **1994**, *1212*, 26–42. [CrossRef]

2. Eibl, H.; Kovatchev, S. Preparation of phospholipids and their analogs by phospholipase D. *Methods Enzymol.* **1981**, *72*, 632–639.

3. Selvy, P.E.; Lavieri, R.R.; Lindsley, C.W.; Brown, H.A. Phospholipase D: Enzymology, Functionality, and Chemical Modulation. *Chem. Rev.* **2011**, *111*, 6064–6119. [CrossRef] [PubMed]

4. Foster, D.A.; Xu, L. Phospholipase D in cell proliferation and cancer. *Mol. Cancer Res.* **2003**, *1*, 789–800. [PubMed]

5. Ryu, S.B.; Wang, X. Expression of Phospholipase D during Castor Bean Leaf Senescence. *Plant Physiol.* **1995**, *108*, 713–719. [CrossRef] [PubMed]

6. Ella, K.M.; Dolan, J.W.; Chen, Q.; Meier, K.E. Characterization of Saccharomyces Cerevisiae deficient in expression of phospholipase D. *Biochem. J.* **1996**, *314*, 15–19. [CrossRef]

7. McNamara, P.J.; Cuevas, W.A.; Songer, J.G. Toxic phospholipases D of Corynebacterium pseudotuberculosis, C. ulcerans and Arcanobacterium haemolyticum: Cloning and sequence homology. *Gene* **1995**, *156*, 113–118. [CrossRef]

8. Juneja, L.R.; Kazuoka, T.; Goto, N.; Yamane, T.; Shimizu, S. Conversion of phosphatidylcholine to phosphatidylserine by various phospholipases D in the presence of L- or D-serine. *Biochim. Biophys. Acta* **1989**, *1003*, 277–283. [CrossRef]

9. D'Arrigo, P.; Servi, S. Using phospholipases for phospholipid modification. *Trends Biotechnol.* **1997**, *15*, 90–96.

10. Kanaho, Y.; Nakai, Y.; Katoh, M.; Nozawa, Y. The phosphatase inhibitor 2,3-diphosphoglycerate interferes with phospholipase D activation in rabbit peritoneal neutrophils. *J. Biol. Chem.* **1993**, *268*, 12492–12497.

11. Morris, A.J.; Frohman, M.A.; Engebrecht, J. Measurement of Phospholipase D Activity. *Anal. Biochem.* **1997**, *252*, 1–9. [CrossRef] [PubMed]

12. Fezza, F.; Gasperi, V.; Mazzei, C.; Macarrone, M. Radiochromatographic assay of N-acyl-phosphatidylethanolamine-specific phospholipase D activity. *Anal. Biochem.* **2005**, *339*, 113–120. [CrossRef] [PubMed]

13. Lambrecht, R.; Ulbrich-Hofmann, R. A facile purification procedure of phospholipase D from cabbage and its characterization. *Biol. Chem. Hoppe-Seyler* **1992**, *373*, 81–88. [CrossRef]

14. Ikarashi, Y.; Maruyama, Y. Liquid chromatography with electrochemical detection for quantitation of bound choline liberated by phospholipase D hydrolysis from phospholipids containing choline in rat plasma. *J. Chromatogr. B* **1993**, *616*, 323–326. [CrossRef]

15. Majd, S.; Yusko, E.C.; MacBriar, A.D.; Yang, J.; Mayer, M. Gramicidin pores report the activity of membrane-active enzymes. *J. Am. Chem. Soc.* **2009**, *131*, 16119–16126. [CrossRef] [PubMed]

16. Do, L.D.; Buchet, R.; Pikula, S.; Abousalham, A.; Mebarek, S. Direct determination of Phospholipase D activity by infrared spectroscopy. *Anal. Biochem.* **2012**, *430*, 32–38. [CrossRef] [PubMed]

17. Thevenot, C.; Daussant, J.; Kas, J.; Valentova, O. Sensitive techniques for Phospholipase D determination in plants. *Trends Anal. Chem.* **1993**, *12*, 266–271. [CrossRef]

18. Hergenrother, P.J.; Haas, M.K.; Martin, S.F. Chromogenic assay for phospholipase D from Streptomyces chromofuscus: Application to the evaluation of substrate analogs. *Lipids* **1997**, *32*, 783–788. [CrossRef] [PubMed]

19. D'Arrigo, P.; Piergianni, V.; Scarcelli, D.; Servi, S. A spectrophotometric assay for phospholipase D. *Anal. Chim. Acta* **1995**, *304*, 249–254.

20. Liu, S.-J.; Wen, Q.; Tang, L.J.; Jiang, J.H. Phospholipid-graphene nanoassembly as a fluorescence biosensor for sensitive detection of Phospholipase D Activity. *Anal. Chem.* **2012**, *84*, 5944–5950. [CrossRef] [PubMed]

21. Cen, Y.; Wu, Y.-M.; Kong, X.-J.; Wu, S.; Yu, R.-Q.; Chu, X. Phospholipid-modified upconversion nanoprobe for ratiometric fluorescence detection and imaging of Phospholipase D in cell lysate and in living cells. *Anal. Chem.* **2014**, *86*, 7119–7127. [CrossRef] [PubMed]

22. Zhu, X.; Fan, L.; Wang, S.; Lei, C.; Huang, Y.; Nie, Z.; Yao, S. Phospholipid-tailored titanium carbide nanosheets as a novel fluorescent nanoprobe for activity assay and imaging of phospholipae D. *Anal. Chem.* **2018**, *90*, 6742–6748. [CrossRef] [PubMed]

23. Dippe, M.; Ulbrich-Hofmann, R. Spectrophotometric determination of phosphatidic acid via iron (III) complexation for assaying Phospholipase D activity. *Anal. Biochem.* **2009**, *392*, 169–173. [CrossRef] [PubMed]

24. Rahier, R.; Noiriel, A.; Abousalham, A. Development of a direct and continuous Phospholipase D assay based on the chelation-enhanced fluorescence property of 8-hydroxyquinoline. *Anal. Chem.* **2016**, *88*, 666–674. [CrossRef] [PubMed]

25. Vrbová, E.; Kroupová, I.; Valentová, O.; Novotná, Z.; Káš, J. Determination of Phospholipase D activity with a choline biosensor. *Anal. Chim. Acta* **1993**, *280*, 43–48.

26. Guerrieri, A.; Ciriello, R.; Cataldi, T.R.I. A novel amperometric biosensor based on a cocrosslinked l-lysine-α-oxidase/overoxidized polypyrrole bilayer for the highly selective determination of l-lysine. *Anal. Chim. Acta* **2013**, *795*, 52–59. [CrossRef] [PubMed]

27. Ciriello, R.; Cataldi, T.R.I.; Crispo, F.; Guerrieri, A. Quantification of l-lysine in cheese by a novel amperometric biosensor. *Food Chem.* **2015**, *169*, 13–19. [CrossRef]

28. Ciriello, R.; De Gennaro, F.; Frascaro, S.; Guerrieri, A. A novel approach for the selective analysis of L-lysine in untreated human serum by a co-crosslinked L-lysine–α-oxidase/overoxidized polypyrrole bilayer based amperometric biosensor. *Bioelectrochemistry* **2018**, *124*, 47–56. [CrossRef]

29. Ciriello, R.; Lo Magro, S.; Guerrieri, A. Assay of serum cholinesterase activity by an amperometric biosensor based on a co-crosslinked choline oxidase/overoxidized polypyrrole bilayer. *Analyst* **2018**, *143*, 920–929. [CrossRef]

30. Guerrieri, A.; De Benedetto, G.E.; Palmisano, F.; Zambonin, P.G. Amperometric sensor for choline and acetylcholine based on a platinum electrode modified by a co-crosslinked bienzymic system. *Analyst* **1995**, *120*, 2731–2736. [CrossRef]

31. Okamura, S.; Yamashita, S. Purification and Characterization of Phosphatidylcholine Phospholipase D from Pig Lung. *J. Biol. Chem.* **1994**, *269*, 31207–31213. [PubMed]

32. Centonze, D.; Guerrieri, A.; Malitesta, C.; Palmisano, F.; Zambonin, P.G. Interference-free glucose sensor based on glucose-oxidase immobilized in an overoxidized non-conducting polypyrrole film. *Fresen. J. Anal. Chem.* **1992**, *342*, 729–733. [CrossRef]

33. Guerrieri, A.; Palmisano, F. An Acetylcholinesterase/Choline Oxidase-Based Amperometric Biosensor as a Liquid Chromatography Detector for Acetylcholine and Choline Determination in Brain Tissue Homogenates. *Anal. Chem.* **2001**, *73*, 2875–2882. [CrossRef] [PubMed]

34. Guerrieri, A.; Lattanzio, V.; Palmisano, F.; Zambonin, P.G. Electrosynthesized poly(pyrrole)/poly(2-naphthol) bilayer membrane as an effective anti-interference layer for simultaneous determination of acethylcholine and choline by a dual electrode amperometric biosensor. *Biosens. Bioelectron.* **2005**, *21*, 1710–1718. [CrossRef]

35. Guerrieri, A.; Ciriello, R.; Crispo, F.; Bianco, G. Detection of choline in biological fluids from patients on haemodialysis by an amperometric biosensor based on a novel anti-interference bilayer. *Bioelectrochemistry* **2019**, *129*, 135–143. [CrossRef]

36. Shimbo, K.; Yano, H.; Miyamoto, Y. Two Streptomyces strains that produce phospholipase D with high transphosphatidylation activity. *Agric. Biol. Chem.* **1989**, *53*, 3083–3085. [CrossRef]

37. Carrera, G.; D'Arrigo, P.; Piergianni, V.; Roncaglio, S.; Secondo, F.; Servi, S. Purification and properties of two phospholipases D from Streptomyces species. *Biochim. Biophys. Acta* **1995**, *1255*, 273–279. [CrossRef]

38. Imamura, S.; Horiuti, Y. Purification of Streptomyces chromofuscus Phospholipase D by hydrophobic affinity chromatography on palmitoyl cellulose. *J. Biochem.* **1979**, *85*, 79–95. [CrossRef] [PubMed]

39. Roberts, M.F. Assays of phospholipases on short-chain phospholipids. *Methods Enzymol.* **1991**, *197*, 95–112. [PubMed]

40. Dennis, E.A. Kinetic dependence of phospholipase activity on the detergent Triton X-100. *J. Lipid Res.* **1973**, *14*, 152–159. [PubMed]

41. Holbrook, P.; Pannell, L.K.; Daly, J.W. Phospholipase D-catalyzed hydrolysis of phosphatidylcholine occurs with PO bond cleavage. *Biochim. Biophys. Acta* **1991**, *1084*, 155–158. [CrossRef]

42. Reynolds, L.J.; Washburn, W.N.; Deems, R.A.; Dennis, E.A. Assay strategies and methods for phospholipases. *Methods Enzymol.* **1991**, *197*, 3–23. [PubMed]

43. Yang, H.; Roberts, M.F. Cloning, overexpression, and characterization of a bacterial Ca^{2+} dependent phospholipase D. *Protein Sci.* **2002**, *11*, 2958–2968. [CrossRef] [PubMed]

44. Heller, M.; Greenzaid, P.; Lichtenberg, D. The activity of phospholipase D on aggregates of phosphatidylcholine, dodecylsulfate and Ca^{2+}. *Adv. Exp. Biol.* **1978**, *101*, 213–220.

45. Marazuela, M.D.; Moreno-Bondi, M.C. Determination of choline-containing phospholipids in serum with a fiber-optic biosensor. *Anal. Chim. Acta* **1998**, *374*, 19–29. [CrossRef]

46. Yamamoto, I.; Konto, A.; Handa, T.; Miyajima, K. Regulation of phospholipase D activity by neutral lipids in egg-yolk phosphatidylcholine small unilamellar vescicles and by calcium ion in aqueous medium. *Biochim. Biophys. Acta* **1995**, *1233*, 21–26. [CrossRef]

47. Kirat, K.E.; Besson, F.; Prigent, A.F.; Chauvet, J.P.; Roux, B. Role of calcium and membrane organization on phospholipase D localization and activity. Competition between a soluble and an insoluble substrate. *J. Biol. Chem.* **2002**, *277*, 21231–21236. [CrossRef]

48. Bartlett, P.N.; Whitaker, R.G. Electrochemical immobilisation of enzymes: Part I. Theory. *J. Electroanal. Chem.* **1987**, *224*, 27–35. [CrossRef]

49. Perrin, D.D.; Dempsey, B. *Buffers for pH and Metal Ion Control*; Chapman & Hall: London, UK, 1974.

50. Campanella, L.; Mascini, M.; Palleschi, G.; Tomassetti, M. Determination of choline-containing phospholipids in human bile and serum by a new enzyme sensor. *Clin. Chim. Acta* **1985**, *151*, 71–83. [CrossRef]

51. Takayama, T.; Itoh, S.; Nagasaki, T.; Tanimizu, I. A new enzymatic method for determination of serum choline-containing phospholipids. *Clin. Chim. Acta* **1977**, *79*, 93–98.

52. Becker, M.; Spohn, U.; Ulbrich-Hofman, R. Detection and Characterization of Phospholipase D by Flow Injection Analysis. *Anal. Biochem.* **1997**, *244*, 55–61. [CrossRef] [PubMed]
53. Wang, P.; Anthes, J.C.; Siegel, M.I.; Egan, R.W.; Billah, M.M. Existence of cytosolic phospholipase D. Identification and comparison with membrane-bound enzyme. *J. Biol. Chem.* **1991**, *266*, 14877–14880. [PubMed]

Communication

Application of a Glucose Dehydrogenase-Fused with Zinc Finger Protein to Label DNA Aptamers for the Electrochemical Detection of VEGF

Jinhee Lee [1,†], Atsuro Tatsumi [2,†], Kaori Tsukakoshi [2,†], Ellie D. Wilson [1], Koichi Abe [2], Koji Sode [1] and Kazunori Ikebukuro [2,*]

[1] Joint Department of Biomedical Engineering, The University of North Carolina at Chapel Hill and North Carolina State University, Chapel Hill, NC 27599, USA; jh.lee@unc.edu (J.L.); elliedw@email.unc.edu (E.D.W.); ksode@email.unc.edu (K.S.)

[2] Department of Biotechnology and Life Science, Graduate School of Engineering, Tokyo University of Agriculture and Technology, 2-24-16 Naka-cho, Koganei, Tokyo 184-8588, Japan; atsuro0916tatsumi@gmail.com (A.T.); k-tsuka@cc.tuat.ac.jp (K.T.); abe79kou@gmail.com (K.A.)

* Correspondence: ikebu@cc.tuat.ac.jp; Tel.: +81-42-388-7030

† These authors contributed equally to this work.

Received: 21 June 2020; Accepted: 9 July 2020; Published: 11 July 2020

Abstract: Aptamer-based electrochemical sensors have gained attention in the context of developing a diagnostic biomarker detection method because of their rapid response, miniaturization ability, stability, and design flexibility. In such detection systems, enzymes are often used as labels to amplify the electrochemical signal. We have focused on glucose dehydrogenase (GDH) as a labeling enzyme for electrochemical detection owing to its high enzymatic activity, availability, and well-established electrochemical principle and platform. However, it is difficult and laborious to obtain one to one labeling of a GDH-aptamer complex with conventional chemical conjugation methods. In this study, we used GDH that was genetically fused to a DNA binding protein, i.e., zinc finger protein (ZF). Fused GDH can be attached to an aptamer spontaneously and site specifically in a buffer by exploiting the sequence-specific binding ability of ZF. Using such a fusion protein, we labeled a vascular endothelial growth factor (VEGF)-binding aptamer with GDH and detected the target electrochemically. As a result, upon the addition of glucose, the GDH labeled on the aptamer generated an amperometric signal, and the current response increased dependent on the VEGF concentration. Eventually, the developed electrochemical sensor proved to detect VEGF levels as low as 105 pM, thereby successfully demonstrating the concept of using ZF-fused GDH to enzymatically label aptamers.

Keywords: aptamer; labeling; enzyme; zinc finger protein; glucose dehydrogenase; electrochemical sensor; vascular endothelial growth factor

---

1. Introduction

Aptamers are DNA or RNA ligands that recognize a specific target molecule with comparable affinity and specificity to antibodies [1,2]. The utilization of aptamers in biosensing has several advantages over antibodies owing to their small size, high thermal stability, and ease of chemical modification. Accordingly, aptamers have become a major target recognition element in a variety of biosensing principles, including fluorescence, electrochemical, and colorimetric methods. In particular, since our first report on aptamer-based electrochemical sensors [3], many electrochemical aptamer sensors have been described for point of care testing purposes, taking advantage of rapid response times, ease of miniaturization, and cost-effectiveness owing to their established platform over other electrochemical detection principles.

To construct such electrochemical sensors, enzymes are attractive and often used as labeling molecules for high-sensitivity detection owing to their signal amplification ability. In reported electrochemical aptamer sensors, alkaline phosphatase and horse radish peroxidase are the most common enzymatic labels [4]. To label aptamers, avidin is typically conjugated to these enzymes, and this avidin-enzyme conjugate interacts with a biotinylated aptamer to form a chimeric aptamer-enzyme molecule. Historically, alkaline phosphatase and horse radish peroxidase are both difficult enzymes to produce recombinantly via prokaryotic host cells, such as *Escherichia coli*. Therefore, the preparation of avidin-enzyme complexes synthesized via chemical conjugation is favorable. However, the procedure to chemically conjugate and purify avidin and enzymes of interest before their use is not only laborious, but it can result in heterogenous molecules due to random conjugation [5] as well as the loss of protein function [6].

Alternatively, we have focused on glucose dehydrogenase (GDH) as a labeling enzyme for electrochemical aptamer sensors [3,7,8]. As GDH is the enzyme employed in the representative miniaturized electrochemical biosensor, the glucose sensor for the self-monitoring of blood glucose, well-established electrochemical principles and platforms are available. Importantly, GDH can be produced via prokaryotes while retaining a high specific activity, allowing for easy engineering of the enzyme [9]. Therefore, GDH would serve as an attractive labeling enzyme to engineer and perform simple and site-specific labeling for electrochemical aptamer sensors.

With regard to site-specific modifications, several methods are reported utilizing unnatural amino acids [6], intein [10], enzyme-catalyzed tags [11], self-labeling enzymes [12], or DNA-templated conjugation [13,14]. However, additional materials other than the DNA and protein are required to perform these conjugation methods. Kobatake and Gordon's group have developed a unique strategy for DNA–protein crosslinking through a DNA-binding enzyme-mediated covalent bond formation [15–17]. However, the method still suffered from low bioactivity of the conjugated enzyme. Inspired by our previous study [18], we have focused on the zinc finger protein (ZF), a monomeric double-stranded DNA-binding protein, to create a GDH-labelled aptamer. ZF is a DNA-binding protein which is capable of recognizing and binding to a specific sequence of double-stranded DNA [19,20]. On account of its small molecular weight, various molecules, including enzymes and fluorescent proteins, have been genetically fused to ZF for the purpose of gene editing and DNA detection, and are efficiently produced using prokaryotic cells without compromising its original function [21–25]. Previously, we have reported the use of ZF-fused GDH for the detection of PCR products, and successfully detected target DNA quantitatively [25]. This study demonstrated that the binding of ZF toward DNA happens spontaneously in a neutral buffer solution in a site-specific manner in vitro. Therefore, we assumed the aptamer labeling of GDH would occur by simply mixing the aptamer with ZF-fused GDH.

To this end, we present a novel GDH labeling method using ZF-fused GDH for aptamers. As a proof of concept, we have utilized vascular endothelial growth factor (VEGF) as a model target. VEGF plays an important role in the formation of new blood vessels of tissues [26]. As anti-VEGF pharmaceuticals are being used in tumor and age-related macular degeneration therapies, the detection of VEGF is important to predict the effectiveness of these treatments [27–29]. Moreover, VEGF is suggested as a biomarker for the early diagnosis of several diseases, such as lymphangioleiomyomatosis [30], major depressive disorder [31], and diabetic nephropathy [32]. Therefore, the rapid and sensitive detection of VEGF levels is crucial for clinical purposes. Here, we present the construction and results of an electrochemical sensor based on a GDH-labeled aptamer for VEGF detection with a lower detection limit of 105 pM.

2. Materials and Methods

2.1. Chemicals and Materials

All oligonucleotides were purchased from Eurofins Genomics (Luxembourg, Luxembourg). The sequences are indicated in the Supplementary Materials Table S1.

Phenazine methosulfate (PMS), 2,6-dichlorophenolindophenol (DCIP), D(+)-glucose, sodium chloride, 2-amino-2-hydroxymethyl-propane-1,3-diol (Tris), glycerol, and Tween-20 were purchased from Kanto Chemical Co. Inc. (Tokyo, Japan). Kanamycin sulfate and d-desthiobiotin were purchased from Sigma-Aldrich Co. LLC (St. Louis, MO, USA). A self-assembled monolayer (SAM) forming thiol reagent (dithiobis(succinimidyl undecanoate)) was purchased from Dojindo Laboratories Co., Ltd. (Kumamoto, Japan). LB broth, the bacterial host strain *Escherichia coli* BL21(DE3), and the expression vector pET30c(+) were purchased form Merck KGaA (Darmstadt, Germany). Anti-VEGF antibodies (MAB293 and BAF293) were purchased from R&D Systems, Inc. (Minneapolis, MN, USA). Gold and platinum wires were purchased from TANAKA Kikinzoku (Tokyo, Japan). A silver/silver chloride (3M NaCl) reference electrode RE-1B (Ag/AgCl) was purchased from BAS Inc. (Tokyo, Japan).

2.2. Investigation of Aptamer and Antibody Combination for the Sandwich Assay

Either biotinylated VEGF aptamers (30 pmol/well) or anti-VEGF antibody; BAF293 (20 pg/well) was diluted by TBS (10 mM Tris-HCl and 100 mM NaCl; pH 7.0) and immobilized on a streptavidin-coated micro-titer plate (Nunc, Rochester, NY, USA) by incubating for 1 h at 25 °C. Subsequently, the wells were washed with TBS-T (10 mM Tris-HCl, 100 mM NaCl and 0.05% Tween-20; pH 7.0) and blocked by blocking buffer (TBS-T containing 4% skim milk). After blocking, 0 or 100 nM VEGF was added, incubated for 1 h, and washed again by TBS-T. Next, the 300 nM of the labeling VEGF aptamer harboring the Zif268 binding site was added and incubated. Finally, 100 nM ZF-GDH was added to each well and incubated. After a final washing by TBS-T, the residual GDH activity was measured using PMS and DCIP. For this measurement, 100 µL of assay buffer (TBS containing 0.06 mM DCIP, 0.6 mM PMS, and 100 mM glucose) was added to each well and the absorbance change at 595 nm was measured using a plate reader (Wallac 1420 ARVO MX, Perkin-Elmer, Waltham, MA, USA).

2.3. Investigation of Binding Specificity of ZF-GDH Towards Its Target Sequence

Biotinylated VEGF-binding aptamer (2G19) or Zif268 recognition sequence-inserted 2G19 (2G19-Z) was first heat-treated at 95 °C for 10 min then gradually cooled down to 25 °C in TBS to let them fold into a stable structure. Then, 100 µL of the folded oligonucleotide solution was added onto a streptavidin coated micro-titer plate (Nunc, Rochester, NY, USA) at a concentration of 1 µM. After immobilization, each well was washed with TBS-T three times and blocked with a blocking buffer. Finally, 100 µL of 100 nM ZF–GDH solution was added and incubated for 1 h at 25 °C. After washing with TBS-T, the binding of ZF–GDH to each oligonucleotide was detected by measuring the residual GDH activity using PMS and DCIP using a plate reader (Wallac 1420 ARVO MX, Perkin-Elmer).

2.4. Investigation of VEGF Concentration Dependency on Plate Using Antibody and Aptamer

To select the pair of ligands for the sandwich binding assay, either biotinylated aptamer (30 pmol/well) or the anti-VEGF antibody BAF293 (20 pg/well) was immobilized on a streptavidin coated clear micro-titer plate by incubating for 1 h at 25 °C. Subsequently, the wells were washed with TBS-T and blocked using a blocking buffer. After blocking, 0 or 100 nM VEGF was added and incubated for 1 h, followed again by washing with TBS-T. Next, 100 µL of 300 nM 2G19-Z or V7t1 aptamer was added and incubated. Finally, 100 nM ZF–GDH was added to each well and incubated. After a final wash by TBS-T, the residual GDH activity was measured as described in Section 2.2. For the investigation of VEGF concentration dependency, 0–100 nM VEGF solution was added on the antibody-immobilized micro-titer plate and incubated for 1 h at room temperature. After washing, the captured VEGF was detected by GDH labeled 2G19-Z.

2.5. Preparation of Antibody-Immobilized Gold Electrode

For the formation of the self-assembled monolayer (SAM) on a gold wire electrode, the wire was cleaned by piranha solution and immersed in ethanol containing 1 mM thiol reagent (dithiobis(succinimidy undecanoate)) overnight at 25 °C. Next, the SAM-modified electrode was immersed into HEPES buffer (pH 7.0) containing 0.15 mg/mL streptavidin and incubated overnight at 4 °C. Followingly, the electrode was incubated for 30 min in HEPES buffer containing 5 µg/mL biotinylated anti-VEGF antibody (BAF293). Finally, the N-hydroxysuccinimide (NHS) ester was blocked through incubation of 1 M Tris-HCl (pH 8.0) for 15 min.

2.6. Investigation of the Non-Specific Adsorption on a Gold Electrode

A bare gold wire electrode, SAM-modified electrode, or antibody immobilized electrode was incubated with 100 nM ZF-GDH in TBS for 30 min. After washing, the residual GDH activity was electrochemically measured using chronoamperometry. The prepared gold electrode (working electrode) was immersed in 10 mL TBS containing 1 mM electron mediator (m-PMS: 1-methoxy-5-methylphenazinium methylsulfate) with Ag/AgCl and Pt wire as reference and counter electrodes, respectively. After applying a potential of 0.1 V (vs. Ag/AgCl) in a VersaSTAT4 potentiostat (Princeton Applied Research, Princeton, NJ, USA), the background response current was measured as the background in the absence of glucose. The applied potential was determined by considering the standard redox potential of mPMS (+0.063 V) [33], and we confirmed the potential was able to effectively oxidize mPMS in our previously reported GDH-based amperometric aptamer sensor [3,7,8]. Then, 25 s after application of the potential, glucose was added at a final concentration of 100 mM and the response current was measured for 100 s. The delta current was determined by subtracting the background current before the addition of glucose from the plateau signals after the addition of glucose.

2.7. Electrochemical Detection of VEGF Using GDH-Labeled Aptamer

The BAF293-immobilized gold wire electrode was blocked with 4% skim milk in TBS-T and incubated in TBS containing 0.25–25 nM of VEGF for 30 min at room temperature. Following washing with TBS-T, the electrode was incubated with 300 nM of 2G10-Z for 30 min and was subsequently incubated with 100 nM of ZF–GDH for 30 min again. After a final wash, residual GDH activity was detected via the same amperometric method described in the Section 2.6 with 1 mM m-PMS and 100 mM glucose utilizing VersaSTAT4 potentiostat (Princeton Applied Research, Princeton, NJ, USA). The calibration of VEGF was obtained by plotting the delta current against VEGF concentration. As a negative control, bovine serum albumin (BSA) was incubated with the electrode instead of VEGF, and after incubation of the aptamer and ZF-GDH, the residual GDH activity was measured by the chronoamperometry with an applied potential of 0.1 V (vs. Ag/AgCl).

3. Results

3.1. GDH Labeling of VEGF-Binding Aptamer Using ZF-GDH and Characterization of the Labeled Aptamer

To construct the fusion of GDH and ZF, we chose GDH derived from *Aspergillus flavus* (AfGDH) [34] and a mouse-derived ZF (Zif268) [35]. AfGDH is a monomeric flavin adenine dinucleotide-dependent GDH which has high catalytic activity of several hundred Unit/mg protein. Zif268 is a very well-characterized C_2H_2-type ZF which binds to its target double-stranded DNA (forward strand: 5′-GCGTGGGCG-3′ and reverse strand: 5′-CGCCCACGC-3′) in a one-to-one manner with high sequence specificity and binding affinity [35]. We confirmed the fusion protein maintained both the target double-stranded DNA binding ability of Zif268 [25] and the enzymatic activity of AfGDH. Remarkably, the enzymatic activity was the almost same level as that of original AfGDH (Supplementary Materials Figure S1). This is due to the small molecular weight of ZF (<10 kDa), one of the advantages of ZF as a fusion partner.

For electrochemical detection of VEGF, we selected a sandwich manner of detection. Therefore, first, the best combination of ligands was investigated on a sandwich plate assay. For this comparison, we chose two types of VEGF binding aptamers (Supplementary Materials Table S1). Both aptamers have a single binding region against VEGF with a nano-molar order K_d value, but fold into different unique structures. Moreover, 2G19 aptamer is predicted to form a stem loop structure [36], whereas V7t1 aptamer was revealed to form a G-quadruplex (G4) structure [37,38]. To label them by ZF-GDH, 2G19 and V7t1 were fused to a ZF recognition site and referred to as 2G19-Z and V7t1-Z, respectively (Supplementary Materials Figure S2). As the capture ligand, we immobilized either biotinylated VEGF aptamer or anti-VEGF antibody on a streptavidin-coated micro-titer plate. After the addition of 100 nM VEGF, GDH-labeled 2G19-Z or V7t1-Z was further added and washed. Eventually, the residual GDH activity was detected with a colorimetric redox dye.

When we used the anti-VEGF antibody as the capture ligand, the GDH-labeled aptamers showed a signal increase in the presence of VEGF (Figure 1). This indicated that both aptamers were correctly labeled by ZF-GDH and can be used to detect VEGF regardless of folding structure. Whereas, all of the aptamer-aptamer pairings showed a higher background signal without the addition of VEGF compared to that of antibody-aptamer pairs (Figure 1). Considering the low background signal with the antibody–aptamer pairs, this higher background signal would be due to the direct interaction between the capture and labeling aptamers rather than the non-specific adsorption of ZF-GDH. Therefore, we chose the anti-VEGF antibody and 2G19 aptamer pair to construct the VEGF sensor, as this combination provided the highest signal with the lowest background noise.

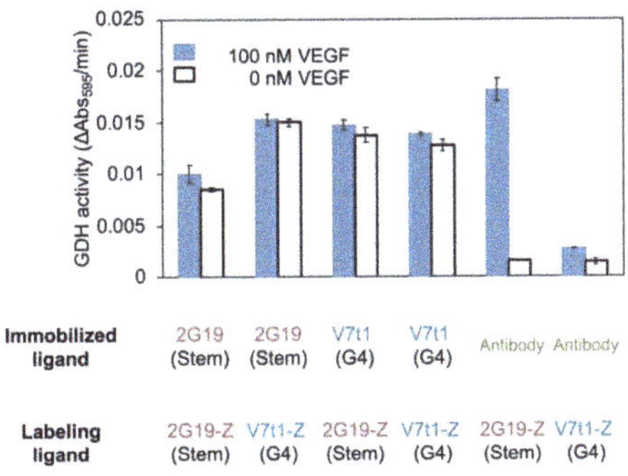

Figure 1. Investigation of aptamer and antibody combinations for the sandwich assay. To capture VEGF, aptamers (2G19 or V7t1) or anti-VEGF antibody (BAF293) was immobilized on a streptavidin-coated microtiter plate. As the detection ligand, ZF-GDH-labeled aptamers (2G19-Z or V7t1-Z) were used. Residual GDH activity on the plate was detected with colorimetric redox dye. Error bars indicate the standard deviation (n = 3).

The specific binding of ZF-GDH to the inserted Zif268-binding site in 2G19 was further investigated on a micro-titer plate. Here, we immobilized either the original 2G19 or Zif268 binding site inserted 2G19 (2G19-Z) using biotin-avidin interaction and allowed ZF-GDH to bind to the DNA. After washing, the residual GDH activity was measured with a colorimetric redox dye. We observed a significantly higher signal from the wells immobilized with 2G19-Z (Supplementary Materials Figure S3). This confirmed that ZF-GDH bound to the specific sequence inserted in the 2G19-Z aptamer. In other words, ZF-GDH labeled the aptamer with GDH.

Finally, we investigated VEGF concentration dependency using anti-VEGF antibody and GDH-labeled 2G19-Z on a micro-titer plate. Various concentrations of VEGF solution were added to the plate, and the VEGF captured by the antibody was detected using the GDH-labeled aptamer. When GDH activity was measured, signal increase was dependent on VEGF concentration, while there was no signal increase with the negative control protein, bovine serum albumin (BSA) (Supplementary

Materials Figure S4). Therefore, as we expected, ZF-GDH functioned to label the aptamer, which showed a specific binding against VEGF.

3.2. Construction of Electrochemical Detection System Using GDH-Labeled 2G19-Z to Detect VEGF

To construct the electrochemical VEGF detection system using the GDH-labeled aptamer, we modified the SAM on the gold wire electrode (approximate working surface area: 16 mm^2) using a decyl alkyl chain harboring a thiol group and *N*-hydroxysuccinimide (NHS) group to immobilize anti-VEGF antibody and to reduce the non-specific adsorption of ZF-GDH. The non-specific adsorption of ZF-GDH on the SAM-modified electrode was investigated by measuring the residual GDH activity after the electrode was immersed in the ZF-GDH solution. GDH activity was measured using the chronoamperometry in the presence of 1 mM m-PMS and 100 mM glucose with an applied potential of 0.1 V (vs. Ag/AgCl). When the bare gold electrode was incubated in the ZF-GDH solution, a high response current was observed even after washing with a buffer; however, the current was reduced with the SAM- and antibody-modified electrode (Supplementary Materials Figure S5). Therefore, the immobilization of this SAM on the electrode efficiently reduced non-specific adsorption.

To immobilize the antibody on the SAM-modified electrode, we used biotin-avidin interactions to control the orientation of the anti-VEGF antibody. First, streptavidin was immobilized on the SAM-modified electrode via amine coupling, and then biotinylated anti-VEGF antibody was immobilized on a streptavidin-modified electrode. After antibody immobilization, the surface was blocked by skim milk and used to detect VEGF. For this detection, we incubated the electrode with 0.25–25 nM of VEGF and after washing, the captured VEGF was labeled by 2G19-Z and ZF-GDH (Figure 2A). In this principle, the amount of GDH on the electrode would be dependent on the amount of the captured VEGF. Upon addition of glucose, GDH obtains an electron and the cofactor (flavin adenine dinucleotide) is reduced. Subsequently, the cofactor is then re-oxidized by the free electron mediator existing in solution, mPMS. Eventually, the reduced mPMS is oxidized by the electrode to generate the response current for the chronoamperometry measurement. Therefore, as the VEGF concentration increases, more reduced mediator is generated, and a higher response current is observed.

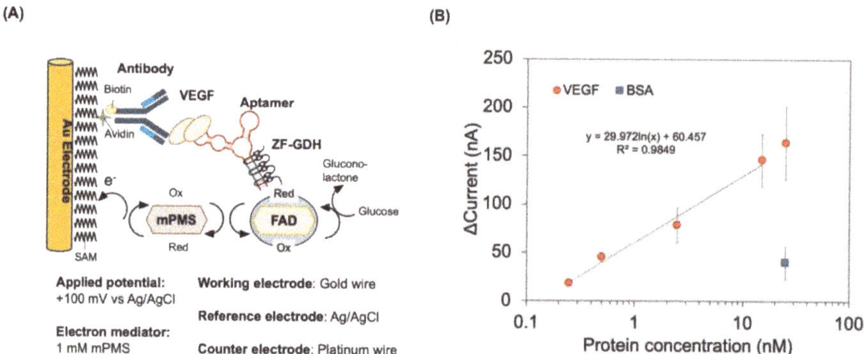

Figure 2. Electrochemical detection of VEGF. (**A**) Schematic diagram of the electrochemical detection of VEGF. (**B**) Electrochemical detection of VEGF by using ZF-GDH-labelled aptamer. The GDH activity was measured by chronoamperometry in the presence of 1 mM mPMS and 100 mM glucose with an applied potential of 0.1 V (vs. Ag/AgCl). The delta current is defined as the subtraction of background current before addition of glucose from the plateau signals after the addition of glucose. Error bars indicate standard deviations (*n* = 3).

As a result of detection, we observed a concentration-dependent signal increase with a linear relationship between the current responses and the logarithmic values of VEGF concentration ranging from 0.25–15 nM (Figure 2B). The limit of detection was calculated to be 105 pM according to the method of M_b + 3 SD (where M_b and SD are the mean value and standard deviation of the blank, respectively). However, the addition of the negative control, BSA instead of VEGF, gave a higher current value than the blank and was comparable to the current value obtained with 0.5 nM VEGF. This indicated that further optimization of the electrode surface modification is necessary.

Although the sensitivity can be improved as the estimated VEGF concentration in a sample would be sub-pico molar, these results showed that the aptamer was correctly labeled with GDH and generated an electrochemical signal. Most importantly, the spontaneous binding of the ZF to the binding sequence inserted into the aptamer enabled GDH labeling of the aptamer by merely mixing the solution.

4. Discussion

For the simple fabrication and construction of an enzyme-based amperometric aptamer sensor, we focused on GDH in terms of its facile recombinant production and engineering. Specifically, we circumvented GDH labeling via chemical conjugation by using ZF-fused GDH. As we expected, the ZF-GDH fusion protein was expressed as soluble protein in *E. coli*-based recombinant production system. In contrast to chemical conjugation, the fusion of ZF had little effect on GDH function, as ZF-fused GDH maintained more than 90% of original GDH activity (Supplementary Materials Figure S1). With chemical modification, surface lysine residues are commonly targeted, and the random modification of these residues can decrease the solubility and stability of protein. As we fused ZF genetically to the N-terminus of GDH, the surface lysine residues remained intact. Additionally, considering the fused position of ZF, which is separate from the active site, and the small molecular weight (<10 kDa) of ZF, the fusion of ZF did not alter protein folding of the GDH nor block the substrate access.

When we labeled two different VEGF-binding aptamers using ZF-GDH and performed sandwich detection of VEGF, both aptamers showed higher GDH activity in the presence of VEGF. Although their folding structures are different, both aptamers were labeled by GDH and served as a detection ligand for VEGF. Therefore, this suggests that this GDH labeling method can be generalized for different aptamers than those used in this study. In the sandwich assay, the G-quadruplex forming V7t1 aptamer showed lower signal intensity than 2G19 aptamer in spite of its better K_d value (1.4 nM vs. 52 nM). Therefore, the sensitivity depends on the paring of antibody and aptamer, not just on the aptamer's affinity.

We successfully detected VEGF electrochemically utilizing the aptamer labeled by ZF-GDH with a lower detection limit of 105 pM and a detection range of 250 pM to 15 nM. To date, several electrochemical aptamer sensors are reported for VEGF detection (Supplementary Materials Table S2). Some of these sensors utilize labeling enzymes, such as alkaline phosphatase or GDH, however they also require chemical conjugation of avidin. We were able to bypass this laborious process by using ZF-GDH to label the aptamer. Besides the simplicity of labeling, another advantage of the present system in comparison to other reported sensors is its detection principle. Namely, the system is an amperometric sensor based on GDH activity, and this principle is adopted in the commercially available glucose sensors. Therefore, we can construct this sensing system with already established platforms. On the other hand, the sensor showed a signal increase with BSA, a non-target molecule. This current increase can be attributed to the direct interaction of ZF-GDH with the non-specifically adsorbed BSA on the electrode surface. As seen in the result of the sandwich plate assay (Supplementary Materials Figure S4), once the surface was sufficiently blocked, the GDH-labeled 2G19-Z did not show a signal increase with the addition of BSA. Therefore, further optimization of the SAM composition and its density, as well as the conditions of blocking and washing would prove this signal increase with BSA.

Compared to previously reported GDH activity-based VEGF sensors, our sensor demonstrated a tenfold improvement in (Supplementary Materials Table S2). However, our sensor sensitivity was moderate in comparison to those reported for electrochemical sensors, as some demonstrated femto-molar levels of sensitivity. Indeed, the constructed sensor requires improvement considering the sub-pico molar concentration of VEGF in a clinical sample. To increase the signal, we can first consider fusing multiple GDH enzymes to ZF to obtain higher GDH activity. Secondly, we can increase the affinity of ZF to DNA by increasing the number of ZF motifs, as well as alter these motifs to recognize longer sequences. In this study, we used Zif268 which recognizes 9 bp and has a K_d with nanomolar affinity, but by fusing an additional ZF motif, we can improve the K_d to a picomolar level [39].

5. Conclusions

In this study, we employed GDH as a labeling enzyme for an amperometric aptamer sensor and developed a GDH labeling method for a VEGF-binding aptamer using ZF-fused GDH. Since GDH can be recombinantly produced and easily engineered, the connecter, ZF was genetically fused to GDH, eliminating the need for chemical conjugation methods to prepare ZF-fused GDH. The small molecular size of ZF (<10 kDa) contributed to the maintenance of GDH activity even after the fusion of ZF. Furthermore, the fusion protein enabled simple and easy labeling of the aptamer without laborious steps, such as purification and dialysis, to obtain the aptamer–GDH complex. Eventually, we could detect 105 pM VEGF in the GDH activity-based electrochemical detection system. Since the target specificity of ZF is designable, this GDH labeling method has the potential to be a universal labeling method to construct electrochemical aptamer sensors.

Supplementary Materials: The following are available online at http://www.mdpi.com/1424-8220/20/14/3878/s1, Supplemental Table S1: Sequences of oligonucleotides, Supplemental Table S2: Comparison of electrochemical VEGF aptamer sensors, Preparation of GDH-fused ZF, Figure S1: Kinetic parameters of purified AfGDH wild type and ZF-GDH, Figure S2: Predicted secondary structure of VEGF-binding aptamers after fusing ZF binding site, Figure S3: Binding specificity of ZF-GDH, Figure S4: VEGF concentration dependency on the microtiter plate, Figure S5: Effect of SAM modification on non-specific absorption of ZF-GDH.

Author Contributions: Conceptualization, J.L., A.T., K.A., K.S., and K.I.; methodology, J.L., and A.T.; data collection, J.L. and A.T.; data curation, J.L. and A.T.; writing—original draft preparation, J.L., A.T., and K.A.; writing—review and editing, K.T., E.D.W., and K.I.; supervision, K.I.; All authors have read and agreed to the published version of the manuscript.

Funding: This research received no external funding.

Acknowledgments: This work was supported by The Murata Science Foundation.

Conflicts of Interest: The authors declare no conflict of interest.

References

1. Tuerk, C.; Gold, L. Systematic evolution of ligands by exponential enrichment: RNA ligands to bacteriophage T4 DNA polymerase. *Science* **1990**, *249*, 505–510. [CrossRef] [PubMed]

2. Ellington, A.D.; Szostak, J.W. In vitro selection of RNA molecules that bind specific ligands. *Nature* **1990**, *346*, 818–822. [CrossRef] [PubMed]

3. Ikebukuro, K.; Kiyohara, C.; Sode, K. Electrochemical detection of protein using a double aptamer sandwich. *Anal. Lett.* **2004**, *37*, 2901–2909. [CrossRef]

4. Villalonga, A.; Pérez-Calabuig, A.M.; Villalonga, R. Electrochemical biosensors based on nucleic acid aptamers. *Anal. Bioanal. Chem.* **2020**, *412*, 55–72. [CrossRef]

5. Wang, L.; Amphlett, G.; Blattler, W.A.; Lambert, J.M.; Zhang, W. Structural characterization of the maytansinoid–monoclonal antibody immunoconjugate, huN901–DM1, by mass spectrometry. *Protein Sci.* **2005**, *14*, 2436–2446. [CrossRef]

6. Kazane, S.A.; Sok, D.; Cho, E.H.; Uson, M.L.; Kuhn, P.; Schultz, P.G.; Smider, V.V. Site-specific DNA-antibody conjugates for specific and sensitive immuno-PCR. *Proc. Natl. Acad. Sci. USA* **2012**, *109*, 3731–3736. [CrossRef]

7. Ikebukuro, K.; Kiyohara, C.; Sode, K. Novel electrochemical sensor system for protein using the aptamers in sandwich manner. *Biosens. Bioelectron.* **2005**, *20*, 2168–2172. [CrossRef]

8. Fukasawa, M.; Yoshida, W.; Yamazaki, H.; Sode, K.; Ikebukuro, K. An aptamer-based bound/free separation system for protein detection. *Electroanalysis* **2009**, *21*, 1297–1302. [CrossRef]

9. Ferri, S.; Kojima, K.; Sode, K. Review of glucose oxidases and glucose dehydrogenases: A bird's eye view of glucose sensing enzymes. *J. Diabetes Sci. Technol.* **2011**, *5*, 1068–1076. [CrossRef]

10. Barbuto, S.; Idoyaga, J.; Vila-Perello, M.; Longhi, M.P.; Breton, G.; Steinman, R.M.; Muir, T.W. Induction of innate and adaptive immunity by delivery of poly dA:dT to dendritic cells. *Nat. Chem. Biol.* **2013**, *9*, 250–256. [CrossRef]

11. Liang, S.I.; McFarland, J.M.; Rabuka, D.; Gartner, Z.J. A Modular approach for assembling aldehyde-tagged proteins on DNA scaffolds. *J. Am. Chem. Soc.* **2014**, *136*, 10850–10853. [CrossRef] [PubMed]

12. Duckworth, B.P.; Chen, Y.; Wollack, J.W.; Sham, Y.; Mueller, J.D.; Taton, T.A.; Distefano, M.D. A universal method for the preparation of covalent protein-DNA conjugates for use in creating protein nanostructures. *Angew. Chem. Int. Ed.* **2007**, *46*, 8819–8822. [CrossRef] [PubMed]

13. Yan, X.; Zhang, H.; Wang, Z.; Peng, H.; Tao, J.; Li, X.F.; Chris Le, X. Quantitative synthesis of protein–DNA conjugates with 1:1 stoichiometry. *Chem. Commun.* **2018**, *54*, 7491–7494. [CrossRef] [PubMed]

14. Rosen, C.B.; Kodal, A.L.; Nielsen, J.S.; Schaffert, D.H.; Scavenius, C.; Okholm, A.H.; Voigt, N.V.; Enghild, J.J.; Kjems, J.; Torring, T.; et al. Template-directed covalent conjugation of DNA to native antibodies, transferrin and other metal-binding proteins. *Nat. Chem.* **2014**, *6*, 804–809. [CrossRef] [PubMed]

15. Akter, F.; Mie, M.; Grimm, S.; Nygren, P.-Å.; Kobatake, E. Detection of Antigens Using a Protein–DNA Chimera Developed by Enzymatic Covalent Bonding with phiX Gene A*. *Anal. Chem.* **2012**, *84*, 5040–5046. [CrossRef] [PubMed]

16. Lovendahl, K.N.; Hayward, A.N.; Gordon, W.R. Sequence-directed covalent protein-DNA linkages in a single step using HUH-Tags. *J. Am. Chem. Soc.* **2017**, *139*, 7030–7035. [CrossRef] [PubMed]

17. Mie, M.; Niimi, T.; Mashimo, Y.; Kobatake, E. Construction of DNA-NanoLuc luciferase conjugates for DNA aptamer-based sandwich assay using Rep protein. *Biotechnol. Lett.* **2019**, *41*, 357–362. [CrossRef]

18. Abe, K.; Murakami, Y.; Tatsumi, A.; Sumida, K.; Kezuka, A.; Fukaya, T.; Kumagai, T.; Osawa, Y.; Sode, K.; Ikebukuro, K. Enzyme linking to DNA aptamers via a zinc finger as a bridge. *Chem. Commun.* **2015**, *51*, 11467–11469. [CrossRef]

19. Wolfe, S.A.; Nekludova, L.; Pabo, C.O. DNA recognition by Cys_2His_2 zinc finger proteins. *Annu. Rev. Biophys. Biomol. Struct.* **2000**, *29*, 183–212. [CrossRef]

20. Gaj, T.; Gersbach, C.A.; Barbas, C.F. ZFN, TALEN, and CRISPR/Cas-based methods for genome engineering. *Trends Biotechnol.* **2013**, *31*, 397–405. [CrossRef]

21. Urnov, F.D.; Rebar, E.J.; Holmes, M.C.; Zhang, H.S.; Gregory, P.D. Genome editing with engineered zinc finger nucleases. *Nat. Rev. Genet.* **2010**, *11*, 636–646. [CrossRef] [PubMed]

22. Stains, C.I.; Porter, J.R.; Ooi, A.T.; Segal, D.J.; Ghosh, I. DNA Sequence-Enabled Reassembly of the Green Fluorescent Protein. *J. Am. Chem. Soc.* **2005**, *127*, 10782–10783. [CrossRef] [PubMed]

23. Ooi, A.T.; Stains, C.I.; Ghosh, I.; Segal, D.J. Sequence-Enabled Reassembly of β-Lactamase (SEER-LAC): A Sensitive Method for the Detection of Double-Stranded DNA. *Biochemistry* **2006**, *45*, 3620–3625. [CrossRef] [PubMed]

24. Abe, K.; Kumagai, T.; Takahashi, C.; Kezuka, A.; Murakami, Y.; Osawa, Y.; Motoki, H.; Matsuo, T.; Horiuchi, M.; Sode, K.; et al. Detection of pathogenic bacteria by using zinc finger protein fused with firefly luciferase. *Anal. Chem.* **2012**, *84*, 8028–8032. [CrossRef]

25. Lee, J.; Tatsumi, A.; Abe, K.; Yoshida, W.; Sode, K.; Ikebukuro, K. Electrochemical detection of pathogenic bacteria by using a glucose dehydrogenase fused zinc finger protein. *Anal. Methods* **2014**, *6*, 4991–4994. [CrossRef]

26. Ferrara, N. Vascular endothelial growth factor: Basic science and clinical progress. *Endocr. Rev.* **2004**, *25*, 581–611. [CrossRef]

27. Burstein, H.J.; Chen, Y.-H.; Parker, L.M.; Savoie, J.; Younger, J.; Kuter, I.; Ryan, P.D.; Garber, J.E.; Chen, H.; Campos, S.M.; et al. VEGF as a marker for outcome among advanced breast cancer patients receiving anti-VEGF therapy with bevacizumab and vinorelbine chemotherapy. *Clin. Cancer Res.* **2008**, *14*, 7871–7877. [CrossRef]

28. Kim, K.J.; Li, B.; Winer, J.; Armanini, M.; Gillett, N.; Phillips, H.S.; Ferrara, N. Inhibition of vascular endothelial growth factor-induced angiogenesis suppresses tumour growth in vivo. *Nature* **1993**, *362*, 841–844. [CrossRef]

29. Girmens, J.-F.; Sahel, J.-A.; Marazova, K. Dry age-related macular degeneration: A currently unmet clinical need. *Intractable Rare Dis. Res.* **2012**, *1*, 103–114. [CrossRef]

30. Young, L.; Lee, H.-S.; Inoue, Y.; Moss, J.; Singer, L.G.; Strange, C.; Nakata, K.; Barker, A.F.; Chapman, J.T.; Brantly, M.L.; et al. MILES Trial Group Serum VEGF-D a concentration as a biomarker of lymphangioleiomyomatosis severity and treatment response: A prospective analysis of the Multicenter International Lymphangioleiomyomatosis Efficacy of Sirolimus (MILES) trial. *Lancet Respir. Med.* **2013**, *1*, 445–452. [CrossRef]

31. Lee, B.-H.; Kim, Y.-K. Increased plasma VEGF levels in major depressive or manic episodes in patients with mood disorders. *J. Affect. Disord.* **2012**, *136*, 181–184. [CrossRef] [PubMed]

32. Hanefeld, M.; Appelt, D.; Engelmann, K.; Sandner, D.; Bornstein, S.R.; Ganz, X.; Henkel, E.; Hasse, R.; Birkenfeld, A.L. Serum and plasma levels of vascular endothelial growth factors in relation to quality of glucose control, biomarkers of inflammation, and diabetic nephropathy. *Horm. Metab. Res.* **2016**, *48*, 529–534. [CrossRef] [PubMed]

33. Hisada, R.; Yagi, T. 1-Methoxy-5-methylphenazinium methyl Sulfate. A photochemically stable electron mediator between NADH and various electron acceptors. *J. Biochem.* **1977**, *82*, 1469–1473. [CrossRef] [PubMed]

34. Mori, K.; Nakajima, M.; Kojima, K.; Murakami, K.; Ferri, S.; Sode, K. Screening of *Aspergillus*-derived FAD-glucose dehydrogenases from fungal genome database. *Biotechnol. Lett.* **2011**, *33*, 2255–2263. [CrossRef]

35. Elrod-Erickson, M.; Rould, M.A.; Nekludova, L.; Pabo, C.O. Zif268 protein-DNA complex refined at 1.6 Å a model system for understanding zinc finger-DNA interactions. *Structure* **1996**, *4*, 1171–1180. [CrossRef]

36. Fukaya, T.; Abe, K.; Savory, N.; Tsukakoshi, K.; Yoshida, W.; Ferri, S.; Sode, K.; Ikebukuro, K. Improvement of the VEGF binding ability of DNA aptamers through *in silico* maturation and multimerization strategy. *J. Biotechnol.* **2015**, *212*, 99–105. [CrossRef]

37. Nonaka, Y.; Sode, K.; Ikebukuro, K. Screening and improvement of an anti-VEGF DNA aptamer. *Molecules* **2010**, *15*, 215–225. [CrossRef]

38. Marušič, M.; Veedu, R.; Wengel, J.; Plavec, J. G-rich VEGF aptamer with locked and unlocked nucleic acid modifications exhibits a unique G-quadruplex fold. *Nucleic Acids Res.* **2013**, *41*, 9524–9536. [CrossRef]

39. Moore, M.; Choo, Y.; Klug, A. Design of polyzinc finger peptides with structured linkers. *Proc. Natl. Acad. Sci. USA* **2001**, *98*, 1432–1436. [CrossRef]

sensors

MDPI

Article

Rapid and Sensitive Detection of miRNA Based on AC Electrokinetic Capacitive Sensing for Point-of-Care Applications

Nan Wan [1,2,3,†], Yu Jiang [2,†], Jiamei Huang [2], Rania Oueslati [2], Shigetoshi Eda [4], Jayne Wu [2,*] and Xiaogang Lin [1,*]

[1]   Key Laboratory of Optoelectronic Technology and System of the Education Ministry of China, Chongqing University, Chongqing 400044, China; wannan@sjtu.edu.cn
[2]   Department of Electrical Engineering and Computer Science, The University of Tennessee, Knoxville, TN 37996, USA; yjiang33@vols.utk.edu (Y.J.); jhuang37@vols.utk.edu (J.H.); oueslati.rania91@gmail.com (R.O.)
[3]   University of Michigan-Shanghai Jiao Tong University Joint Institute, Shanghai Jiao Tong University, Shanghai 200240, China
[4]   Department of Forestry, Wildlife and Fisheries, The University of Tennessee Institute of Agriculture, Knoxville, TN 37996, USA; seda@utk.edu
*   Correspondence: jwu10@tennessee.edu (J.W.); xglin@cqu.edu.cn (X.L.)
†   The authors contributed equally to the work.

Abstract: A sensitive and efficient method for microRNAs (miRNAs) detection is strongly desired by clinicians and, in recent years, the search for such a method has drawn much attention. There has been significant interest in using miRNA as biomarkers for multiple diseases and conditions in clinical diagnostics. Presently, most miRNA detection methods suffer from drawbacks, e.g., low sensitivity, long assay time, expensive equipment, trained personnel, or unsuitability for point-of-care. New methodologies are needed to overcome these limitations to allow rapid, sensitive, low-cost, easy-to-use, and portable methods for miRNA detection at the point of care. In this work, to overcome these shortcomings, we integrated capacitive sensing and alternating current electrokinetic effects to detect specific miRNA-16b molecules, as a model, with the limit of detection reaching 1.0 femto molar (fM) levels. The specificity of the sensor was verified by testing miRNA-25, which has the same length as miRNA-16b. The sensor we developed demonstrated significant improvements in sensitivity, response time and cost over other miRNA detection methods, and has application potential at point-of-care.

Keywords: capacitive sensing; alternating current electrokinetic effects; miRNA sensing; point-of-care diagnostics

Citation: Wan, N.; Jiang, Y.; Huang, J.; Oueslati, R.; Eda, S.; Wu, J.; Lin, X. Rapid and Sensitive Detection of miRNA Based on AC Electrokinetic Capacitive Sensing for Point-of-Care Applications. *Sensors* **2021**, *21*, 3985. https://doi.org/10.3390/s21123985

Academic Editor: Antonio Guerrieri

Received: 26 March 2021
Accepted: 3 June 2021
Published: 9 June 2021

Publisher's Note: MDPI stays neutral with regard to jurisdictional claims in published maps and institutional affiliations.

1. Introduction

MicroRNAs (miRNAs) are noncoding small RNAs of 18–25 nucleotides that regulate the expression of multiple genes. As the presence of distinct miRNAs indicates specific medical conditions—and because miRNAs can stably exist in various body fluids—miRNAs have been investigated as biomarkers for disease diagnosis [1]. The exosome-associated miRNA has been demonstrated to show potential diagnostic value in Alzheimer's disease, mild cognitive impairment and lung cancer [2,3]. Extracellular miRNAs not only circulate in the peripheral blood [4–7], but also widely exist in other body fluids such as saliva, urine, tears, amniotic fluid and breast milk [7–9]. For miRNA detection, Northern blotting (NB) is commonly used in molecular biology laboratories; however, it suffers the shortcomings of low sensitivity and long assay time. Reverse transcription quantitative polymerase chain reaction (RT-qPCR), as the gold standard for RNA quantification [10], requires unwieldy and expensive thermal cycling equipment for amplification and quantification, rendering it unsuitable for point-of-care (POC) detection of circulating miRNAs diagnostics. Furthermore, due to the short lengths of miRNAs, RT-PCR of miRNA is rather costly and technically demanding to perform, even compared with regular RT-PCR applications.

To overcome the drawbacks of existing methods for miRNA detection, we have presented herein the development of a rapid, high sensitivity, and easy-to-use biosensor for miRNA POC detection, based on the AC electrokinetic (ACEK)-enhanced capacitive sensing technique. ACEK effects, including the dielectrophoresis (DEP), AC electroosmosis (ACEO), and AC electrothermal (ACET) effects, have been widely studied and applied on biofluid handling and molecule manipulation [11,12]. DEP refers to the particle motion caused by the difference in polarizability between the particles and the fluid [13]. ACEO is induced by the moving charges in the electric double layer (EDL) and induces the motion of fluid [14]. The ACET effect arises from the induced temperature gradient in the fluids [15]. For relatively conductive biofluids, ACET effect dominates ACEO when generating microflows. In the ACEK capacitive sensing method utilized, AC capacitive sensing transduces biomolecular interactions to the change of interfacial capacitance. In the meantime, the AC electrical signal induces the ACET effect, which enriches biomolecules toward the electrode surface for sensing. Thus, rapid and sensitive detection for various biochemical molecules can be realized in a manner amenable to POC applications. In our previous work, we successfully detected progesterone, DNA, RNA, proteins, Bisphenol A and pathogens with the application of ACEK capacitive sensing [16–20].

In this work, miRNA-16b was adopted as a model target to perform a proof-of-concept demonstration. MiRNA-16b was derived from circulating extracellular vesicles and could potentially be used for early identification of the pregnancy status of cattle [21]. To be cost-effective for POC applications, interdigitated electrodes (based on low-cost gold-plated interdigitated printed circuit broad (PCB) electrodes) were adopted as sensors. A single-stranded DNA oligonucleotide probe was functionalized on the electrodes to specifically recognize miRNA-16b. The hybridization of the DNA probe with miRNA-16b resulted in a change at the electrode-electrolyte interface. Microfluidic experiments were conducted to show that the ACET effect dominated over other electrokinetic effects. It was used to accelerate the miRNA-16b target's travel to the sensor surface and speed up the binding of miRNA-16b with the capture DNA probe—shortening the testing time to 30s. With the combination of DNA probe recognition and ACET-enhanced capacitance sensing, we successfully detected miRNA-16b for bovine pregnancy diagnosis at the femtomolar (fM) level. To test the sensors' specificity, miRNA-25—which has the same length (21 bases) as miRNA-16b and coexists with miRNA-16b in pregnant cows—was tested as the interference, using the sensors with the same DNA probe. The results showed no observable response and confirmed the specificity of the miRNA sensor.

2. Materials and Methods

2.1. Interfacial Capacitance Sensing

The miRNA sensor in this study was based on ACEK capacitive sensing, and used to detect the binding of miRNA-16b molecules with the immobilized DNA probes. The single-stranded DNA probes were immobilized on the interdigitated microelectrodes, which captured and bound with the miRNA-16b molecules. The ssDNA (with the sequence of 5'- GCCAATATTTACGTGCTGCTGCTA-3') and the miRNA-16b (with the sequence of UAGCAGCACGÂUAAAUAUUGGC) were complementary strands. MiRNA-25 (with the sequence of CAUUGCACUUGUCUCGGUCUG), which has the same length (21 bases) as miRNA-16b and coexists with it in pregnant cows, was set as negative control. Open sites on the surface were then blocked with 6-mercaptohexanol (6-MH). During detection, the microelectrodes were stimulated with a low voltage (300 mVrms) AC signal to induce ACEK effects, such as the AC electrothermal (ACET) effect, to route the miRNA-16b molecules around the electrodes onto the electrode surface and hybridize with the immobilized DNA probes. The binding between miRNA-16b molecules and the immobilized DNA probes led to a change of the interfacial capacitance (Cint) at the microelectrode surface, which was detected electrically using the same AC signal, as shown in Figure 1.

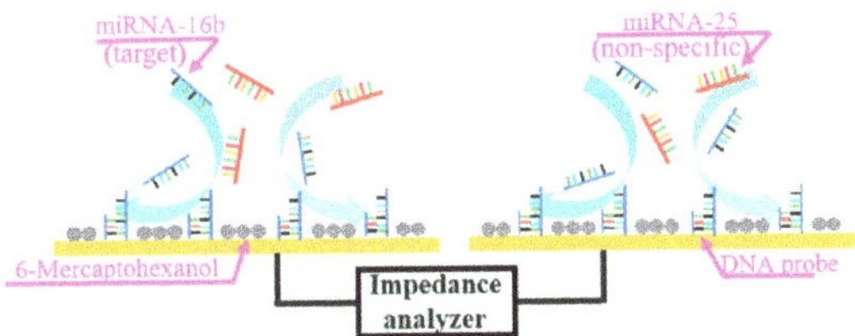

Figure 1. ACET microflow accelerates the binding of miRNA-16b and DNA probe.

As shown in Figure 2, before the DNA probe was immobilized on the electrode surface, the interfacial capacitance was determined by the formation of electrical double layer (EDL) (Figure 2a). A_0 represented the surface area of the interfacial capacitance before immobilization. After functionalization, the electrode surface was covered with the DNA probes and 6-MH at high density. Such a DNA/6-MH monolayer could be modeled as a dielectric capacitance that serially connects with the initial EDL capacitance (Figure 2b), while A_{b0} represents the surface area of the interfacial capacitance after immobilization of the probe DNA and 6-MH. After miRNA-16b targets bind with the probe DNA (Figure 2c), the total interfacial capacitance changes again, where A_b represents the surface area of the interfacial capacitance after binding. The capacitance change predominantly results from an increase in the surface area of the interfacial capacitance, which can be used for the detection of miRNA-16b.

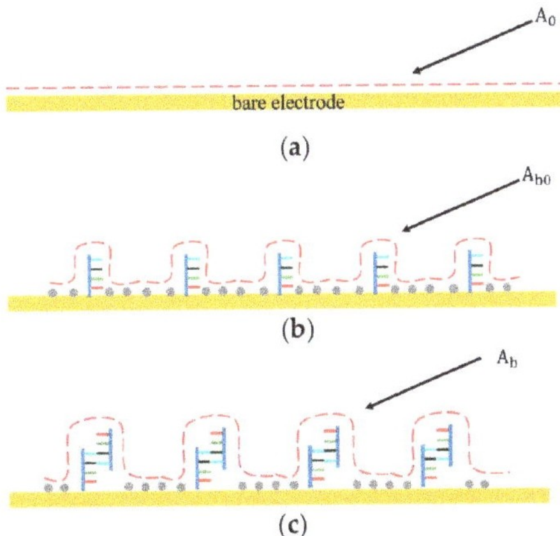

Figure 2. The interfacial capacitance changes at different stages of probe immobilization and miRNA-16 binding. (**a**) Before probe immobilization. (**b**) After probe immobilization and blocking. (**c**) After miRNA-16b binding with DNA probe.

As mentioned above, the sensor in this work could be represented by an equivalent electrical network. Specifically, when the interdigitated electrodes are immersed in an electrolytic solution, the electrode cell can be represented by an equivalent circuit network consisting of a solution resistance (R_{sol}), an interfacial capacitance (C_{int}) modeled by a constant phase element (CPE_{int}) and a resistive path (R_{leak}) in parallel with this capacitance for the non-faradaic sensors [22]. The complex impedance of a CPE can be expressed by $1/(j\omega A)^m$, where A is analogous to a capacitance, ω is the frequency in rad/s, and $0.5 < m < 1$ (m = 1 corresponds to a capacitor and m = 0.5 corresponds to a Warburg element) [19]. The bare sensor impedance was measured over a frequency range of 100 Hz to 1 MHz, with a buffer conductivity equivalent to that of $1\times$ Saline Sodium Citrate buffer (SSC) placed over the IDE. The equivalent circuit network and the impedance spectra are both shown in Figure 3.

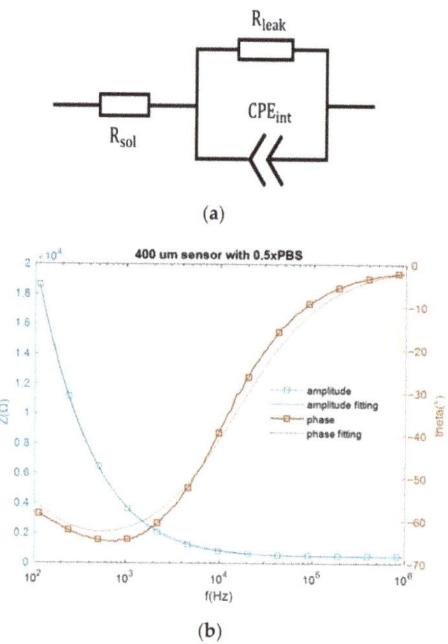

(a)

(b)

Figure 3. (**a**) The equivalent circuit network for an electrode biosensor, and (**b**) the measured impedance spectra and the fitting spectra.

The component values in the equivalent circuit network can be estimated from the fitting data. The solution resistance R_{sol} is 454.4 Ω and the resistive path R_{leak} equals 7.73 $\times 10^{-4}$ Ω. As for the constant phase element CPE_{int}, which can be expressed by $1/(j\omega A)^m$, $A = 4.465 \times 10^{-9}$ and m = 0.7842. At the assay frequencies of 10 kHz and 100 kHz, CPE_{int} equals 2578.8 Ω and 423.85 Ω, respectively, which are much smaller than R_{leak}. Since electricity takes the path of least resistance, the impedance network can be simplified as a series connection of the constant phase element (CPE_{int}) and the solution resistance (R_{sol}). In this work, the miRNA-16b detection was based on non-faradaic impedance measurement, and the interfacial capacitance C_{int}, which closely correlates with constant phase element (CPE_{int}), was monitored for target detection.

2.2. ACEK Enrichment Mechanism

In the process of detection, target molecules tended to diffuse to the electrode surface and bind with the immobilized probes. The random nature of passive diffusion is the underlying cause for low sensitivity and long testing times. To overcome these limitations for detection, ACEK effects were applied for molecule enrichment. Specifically, ACEK effects include dielectrophoresis (DEP) [23], AC electroosmosis (ACEO) [24] and AC electrothermal (ACET) effect [25]. For ACEO, the flow velocity diminished to close to zero when the conductivity of solution was increased to above 0.085 S/m [26]. Since the miRNA-16b sample solutions were based on $0.5\times$ SSC buffer and $1\times$ SSC buffer, with conductivities of 0.433 S/m and 0.865 S/m, ACEO flows were negligible in this work. In order to stimulate the fluids with high conductivity to convect to the electrode surface, the ACET effect was applied as we reported earlier [26]. ACET effects applied volume force on fluid and the microflows were generated to accelerate miRNA-16b molecules traveling to the surface of the electrode for binding, as shown in Figure 1. The high efficiency of binding shortened the testing time to 30 s and enhanced the sensitivity for detection [26]. The DEP force and ACET force can be expressed as

$$F_{DEP} = 2\pi a^3 \varepsilon_m \text{Re} \left[\frac{\varepsilon_p^* - \varepsilon_m^*}{\varepsilon_p^* + 2\varepsilon_m^*} \right] \nabla |E|^2 \tag{1}$$

$$F_{ACET} = -0.011 \nabla T \varepsilon_m |E|^2 \tag{2}$$

where a is the particle radius, ε_m is the medium permittivity, $|E|$ is electric field modulus, ε_p^* and ε_m^* are complex permittivities of particle and medium and ∇T is temperature gradient. As DEP force scales with particle size while the ACET is size-independent [25], the ACET effect was expected to be more effective than DEP for miRNA-16b enrichment due to miRNA-16b's small size [25,26]. The detailed analysis of the miRNA-16b enrichment mechanism is exhibited in Section 3.2, based on the experiment results.

2.3. Sensor Fabrication

Single-sided highly flexible clad laminate (PulsarProFX®) was used for the device prototyping. The laminate was a 0.005″ thick FR4 fiberglass with 1/2 oz copper. Interdigitated electrodes were patterned by PCB technique. The PCB fabrication process is simple and can be done in just a few minutes using benchtop equipment. The detailed sensor fabrication steps were as follows: (1) Draw the electrode pattern in Microsoft Visio and then print it onto a piece of toner transfer paper (PulsarProFX) using a regular laser printer. (2) Cover the copper laminate with the patterned toner transfer paper and roll it through a laminator (Micro-Mark, model VL110) when the 'HOT' indicator is on. Soak the copper laminate together with the toner transfer paper in water for several seconds and remove the toner transfer paper. The electrode pattern is then transferred to the copper laminate. (3) Cover the copper laminate with a piece of toner reactive foil (TRF) and roll it through the laminator. After removing the TRF from the copper laminate, only the electrode pattern on the copper laminate was covered by TRF, which is etch-resistant. (4) Heat 30 mL of ferric chloride solution (MG Chemicals) to 55 °C and put the copper laminate into the solution. Copper was etched within 10 min and the PCB showed the desired electrode pattern. Rinse the PCB with acetone and isopropyl alcohol to remove TRF from the electrode surface. Clean the PCB under running de-ionized water for 5 min. (5) Solder two lead wires to the electrode pads. (6) Smoothen the electrode surface with sandpaper (#5000 Grit) and clean it thoroughly. (7) Finally, coat the electrode with a thin layer of gold by electroplating. The plating solution used here was Caswell® 24CT Gold Plating Solution. An image of the fabricated interdigitated electrodes is provided in Figure S1 in Supplementary Materials. The characteristic length of the sensor was 400 μm (with widths of 400 μm and 400 μm gaps). The adhesive chamber in Figure S1b is Press-to-Seal® silicon isolator with a 2.5 mm diameter and a 2 mm depth.

2.4. Sample Preparation

Single-strained DNA probe for miRNA-16b target recognition was purchased from Integrated DNA Technologies, Inc. (IDT, Coralville, IA). Its sequence was 5′- GCCAATATTTA CGTGCTGCTGCTA-3′ while the 5′-end was dithiol-modified for self-assembly on the gold surface. Before experiments, the ssDNA probes were further reduced using TCEP (Sigma Aldrich, USA). The miRNA-16b was purchased from IDT and its sequence was rUrArG rCrArG rCrArC rGrUrA rArArU rArUrU rGrGrC [10]. For the sensor selectivity test, miRNA-25 [11], also from IDT, had the following sequence: rCrArU rUrGrC rArCrU rUr-GrU rCrUrC rGrGrU rCrUrG. After DNA probe incubation, the surface of the electrode was blocked with 1 mM 6-MH (Sigma-Aldrich, St. Louis, MO, USA). For probe solution, ultrapure water (Mili-Q) and 0.05× PBS (Thermo Fisher Scientific, Waltham, MA, USA) were used to prepare 20 μM DNA probe solution. For testing, miRNA-16 was diluted in

1× SSC buffer (AccuGENE™, Lonza, Rockland, ME) to make stock solutions. The final concentrations of miRNA-16b samples for testing were: 100 fM, 10 fM, 1 fM and 0.1 fM.

2.5. Preparation of Microelectrode Chips

Before immobilizing the DNA probe on the electrode, the electrodes needed to be cleaned rigorously. The steps for electrode cleaning are reported in our earlier work [27]. Briefly, the electrode chips were cleaned with acetone, isopropyl alcohol (IPA), ultrapure water, and then air-dried. After these steps, electrode chips were transferred into a UV ozone cleaner for 15 min. Then, the electrode chips were ready for probe functionalization.

The 10 µL DNA probe solution (100 µM in 0.05× PBS) was loaded on the electrode surface for incubation and the electrode chips were kept in humidor for 18 h. Thereafter, electrode chips needed to be blocked with 1 mM 6-MH for 3 h. Finally, the electrode chips were ready for detection.

2.6. Measurement and Analysis

As mentioned in Section 2.1, capacitance values were continuously measured to calculate the capacitance change ratio. In this work, the interfacial capacitance was measured for 30 s and sampled at 120 points with an AC signal of 10 kHz and 300 mV. The capacitance change ratio is expressed as normalized capacitance versus time (%/min), calculated by the least square linear fitting. The concentration of target samples from 0.1 fM to 100 fM were tested five times using a new functionalized sensor each time. An impedance analyzer was used for data acquisition (Aglient 4294A).

Before testing, the miRNA samples needed to be heated in a water bath at 95 °C for 5 min and then cooled on ice for 5 min. To obtain the best performance of the miRNA-16b detection sensor, the frequency was optimized and the sensors were prepared with different incubation buffers and testing buffers. All the target samples (miRNA-16b and miRNA-25) at different concentrations were tested five times and the average values were used for analysis.

3. Results

3.1. Treatment for miRNA-16b Samples

We used one single loop of heating and fast cooling to linearize miRNA for effective binding to the probe DNA. The first experiment was to confirm whether heat and fast annealing led to improvement, before other optimizations were employed. The melting temperature (Tm) of miRNAs are around 45 °C, so the samples were heated to 95 °C then put on ice to cool. We compared the responses of miRNA-16b samples that were not heated and cooled with the miRNA-16b samples that were heated in the water bath at 95 °C (5 min) and cooled on ice (1 min). The results are provided in Figure S2 in Supplementary Materials. The miRNA-16b samples were at a fixed concentration of 1 fM. The response of temperature cycled samples was 3.117%/min, while the response for the untreated sample was 2.677%/min. The 16.44% response improvement proved that temperature treatment for miRNA-16b samples is necessary. Therefore, in all the following experiments, all miRNA samples were heated and ice-cooled.

3.2. Frequency Optimization

Among ACEK effects, the ACET effect is not frequency dependent, while DEP forces may strongly depend on the applied frequency. Both of these ACEK forces may occur when an AC signal is applied. Thus, it was necessary to test different frequencies of the measuring signal, in order to determine the dominant ACEK mechanism. Therefore, two frequencies of 10 kHz and 100 kHz were used in the comparison experiments for the detection of miRNA-16b. The testing results are shown in Figure 4a. At the frequency of 10 kHz, the sensor responses of the background solution and 0.1 fM miRNA-16b were 0.788%/min and 0.943%/min, respectively, while the responses were 0.821%/min and 0.924%/min, respectively, when using 100 kHz. There were no significant differences between them. Also, under the frequency of 10 kHz, the 1 fM, 10 fM and 100 fM miRNA-16b showed responses of 3.117%/min, 4.366%/min and 5.428%/min, respectively. As for 100 kHz, the responses of 1 fM, 10 fM and 100 fM miRNA-16b decreased to 2.653%/min, 3.873%/min and 4.912%/min, respectively, and the differences in responses were about -14.89%, -11.29% and -9.51%, respectively, when compared with the cases at 10 kHz. These differences (~10%) indicated that the response of the sensor was relatively insensitive to the measuring frequency, pointing to the ACET effect's dominant sensing over DEP [28].

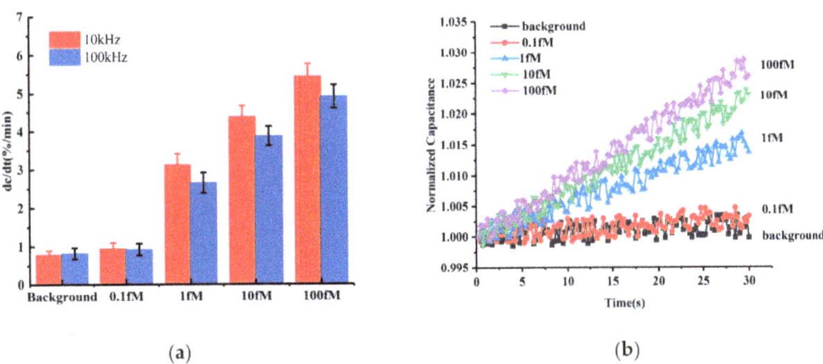

Figure 4. (**a**) Detection results at 100 kHz and 10 kHz (**b**) Normalized capacitance change as a function of time within 30 s for various levels of miRNA-16b.

The representative curves of normalized capacitances for different miRNA-16b concentrations are displayed in Figure 4b, corresponding to the 10 kHz data in Figure 3a. The capacitances were shown to linearly increase with time, and the capacitance change rates were commensurate with increasing concentrations of miRNA-16b, due to increasing levels of miRNA-16b binding.

Furthermore, according to the equivalent circuit network and the electrical impedance spectra discussed in Section 2.1, the impedance network can be simplified as a series connection of the constant phase element (CPE_{int}) and the solution resistance (R_{sol}). Compared to applying 100 kHz, when using 10 kHz, the CPE_{int} showed more capacitive components and resulted in a greater response of the capacitive sensor. Therefore, for frequency optimization, 10 kHz was selected to perform the tests.

Previous work demonstrated that the detection of lipopolysaccharides and genomic DNA was more pronounced at 100 kHz [29,30]. Because both were larger than the miRNA-16b tested here, the DEP effect was much more pronounced for enrichment. Thereby, it was reasonable to have a better response at 10 kHz for miRNA-16b detection due to the dominance of the ACET effect for miRNA enrichment [24,31].

To directly verify that the main effect of AC signal under our test conditions was ACET, we performed a microfluidic experiment based on fluorescent tracing particles. The solution of fluorescent particles (FluoSpheres® Carboxylate-Modified Microspheres, 1 μm, nile red fluorescent 535/575, 2% solids) was diluted 1000 times by 0.5× SSC. Even though both ACEO and ACET microflows were generated by nonuniform electric fields between electrodes, for electrodes with a thermally insulative substrate, flow patterns were fundamentally different [32,33]. While ACEO flows went from the electrode edge to its inside, ACET flows followed the direction of thermal gradients, i.e., going opposite to the ACEO directions, in this case. Thus, at the surface of side-by-side PCB electrodes, ACET flows went from electrode to gap, whereas ACEO flowed from gap to electrode. The particle movements at the surface of PCB electrodes under 10 kHz-7 Vpp signals are shown in Figure 5. Two moving particles are boxed out with a yellow box, indicating their distances at 0, 8 and 16 s. It shows that the two particles were moving away from the electrodes toward the gap, verifying that the particles were carried by ACET microflows.

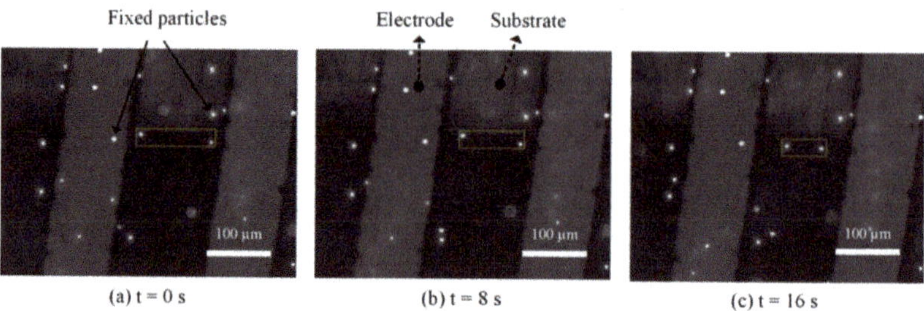

Figure 5. Image series showing particle movement in 0.5× SSC at 10 kHz.

3.3. Optimization of Sensor Preparation

Based on our earlier work, the ACEK-enhanced biosensor for biomolecule detection was expected to have different performance on responses, when different probe buffers or target buffers were used [18]. In this work, we compared the sensor responses using 0.05× PBS and ultrapure water for probe immobilization, respectively, and miRNA-16b samples with a fixed concentration of 1 fM were tested. The sensor responses are shown in Figure 6a. When applying 0.05× PBS solution to dissolve DNA probes, the response of miRNA-16b with the concentration of 1fM was 3.117%/min, which was up to 31.69% higher than the response (2.367%/min) with ultrapure water dissolving the DNA probe. This was because there were more ions in the 0.05× PBS solution than in ultrapure water, and the ions in the solution shielded the charge of nucleic acid. Thus, when the DNA probes were attached to the electrode surface, the coverage area increased and resulted in the improvement of binding efficiency.

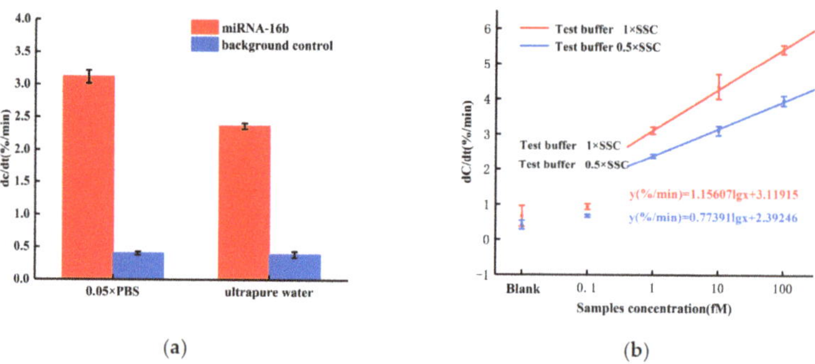

Figure 6. (**a**) Sensor response with probe diluting in 0.05× PBS and ultrapure water; (**b**) Dose responses with miRNA-16b diluting in 0.5× SSC and 1× SSC.

To optimize the concentration of the target buffer, we compared the results of diluting miRNA-16b molecules with 0.5× SSC solution and 1× SSC solution. The sensor responses are shown in Figure 6b.

According to Figure 6b, when the miRNA-16b molecule was diluted with 0.5× SSC solution, the responses of the miRNA-16b sample with concentrations of 1fM, 10fM, and 100 fM were 2.397%/min, 3.117%/min, and 3.967%/min, respectively. When the miRNA-16b molecule was diluted with 1× SSC solution, the responses of the miRNA-16b samples with concentrations of 1fM, 10fM, and 100fM were 3.117%/min, 4.366%/min and 5.428%/min, respectively. Compared with diluting targets using 0.5× SSC, miRNA-16b molecules diluted with 1× SSC solution increased the sensor response by 30.03%, 40.07% and 36.83%, respectively, with target concentrations of 1 fM, 10 fM, and 100 fM. In this work, the ACET effect accelerated the enrichment of miRNA-16b molecules at the electrode surface. Compared with the 0.5 × SSC solution, the conductivity of the 1 × SSC solution was higher and would generate stronger ACET microflows for target enrichment. Thus, the efficiency of binding between the DNA probe and miRNA-16b were improved, and the desired enhancement of the sensor response was achieved.

3.4. Dose Response of miRNA-16b Detection

To obtain peak performance of our sensor, we chose 0.05× PBS solution (5 mM phosphate buffer [pH 7] containing 7.5 mM sodium chloride) to dilute the DNA probe for incubation and 1× SSC solution as the background solution for the miRNA-16b molecules. The concentration of target samples from 0.1 fM to 100 fM were tested five times, and the average values were used to analyze specific responses to various concentrations of miRNA-16b.

The dose response of miRNA-16b detection is shown in Figure 7. As shown in Figure 7, the response of background solution 1× SSC was 0.788%/min, which was less than 1%/min that of the miRNA-16b samples. The significant response difference proved that the increase of interfacial capacitance resulting from the hybridization of miRNA-16b and DNA probe and the background solution of the samples did not bring any interference to the detection and did not cause the capacitance change. The response of the miRNA-16b sample with concentration of 0.1 fM was 0.943%/min. The response was less than 1%/min, indicating that the sensor detection limit was higher than 0.1 fM. The response of the miRNA-16b sample at 1 fM was 3.117%/min, that at 10 fM was 4.366%/min, and that at 100 fM was 5.428%/min. All the response values were greater than the background response, indicating that the electrode surface interface capacitance increased during the hybridization of the DNA probe and miRNA-16b.

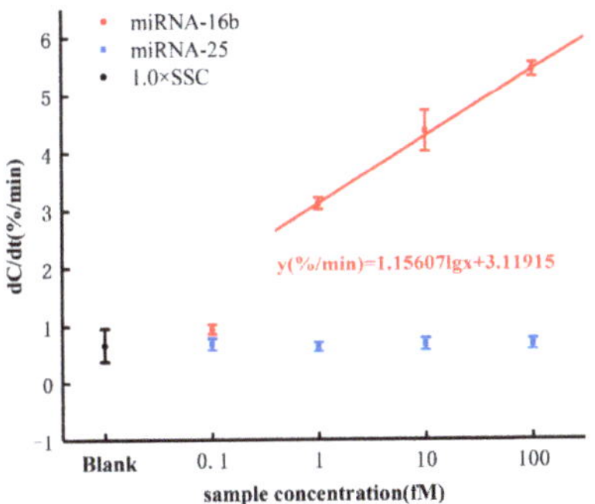

Figure 7. Capacitance change rates in response to different concentrations of miRNA-16b and miRNA-25 concentrations.

According to the equations introduced in Section 2.1, the increase of capacitance value indicated that additional topological structure was introduced when the hybridization occurred between miRNA-16b and DNA probe, resulting in increased surface area and leading to increased interface capacitance. More importantly, the response increased with the increase of the concentration of miRNA-16b samples following a semilogarithmic relationship. The dose response of the sensor can be expressed by the linear fitting equation $y(\%/min) = 1.15607 \lg(x) + 3.11915$, where x is the concentration of the miRNA-16b analytes with the unit of femtomolar. The cut-off response of the sensor was defined as three standard deviations from the response of the background solution. By inserting the cut-off response (1.676%/min) into the dose response equation $y(\%/min) = 1.15607 \lg(x) + 3.11915$, the detection limit of the sensor can be calculated to be 0.056 fM (theoretical value). However, this value (0.056 fM) was less than the lowest concentration tested (0.1 fM), from which we did not obtain a good response. Therefore, the detection limit of the sensor is theoretical.

3.5. Selectivity

MiRNA-25 was used as a negative control to evaluate the selectivity of the biosensor we designed. As shown in Figure 7, the responses of miRNA-25 at the concentrations of 0.1 fM, 1 fM, 10 fM, and 100 fM were less than 1%/min, and the responses were independent of the concentration. This showed that the selected DNA probe specifically hybridized with the target miRNA-16b but not with the irrelevant RNA, miRNA-25.

3.6. Spiked Serum Samples Detetction

To further validate whether this miRNA sensor could be applied to test complex matrix, miRNA-16b spiked bovine serum samples were also tested as a prove of concept. First, negative serum samples were mixed as negative serum pool, which was further 1:10 diluted using 0.5× PBS. Then, miRNA-16b samples (ranging from 0.1 fM to 100 fM) were spiked into diluted serum pool to simulate positive serum samples. To test the selectivity, the spiked miRNA-25 was also tested as a negative control.

The capacitance change rates are shown in Figure 8. The normalized capacitance changes for 0.1 fM to 100 fM were 1.536 ± 0.307%/min, 2.276 ± 0.806%/min, 3.323 ± 0.272%/min and 3.706 ± 0.698%/min, respectively. The response of control background (negative serum pool) was 1.327 ± 0.171%/min, which was slightly below the response of spiked 0.1 fM miRNA. The dose response can be represented by the linear fitting equation $y(\%/min) = 0.75 \lg(x) + 2.34$, where x is the concentration of the miRNA-16b analytes in fM. The cut-off response of sensor was defined as three standard deviations (1.840%/min) from the response of background solution, which was calculated to be 0.215 fM. Considering the dilution, the detection limit was 4.30 fM in neat serum, theoretically. Furthermore, the responses of the 1 fM samples and the control were statistically examined by *t*-test and the *p*-value was 0.03. This statistic result indicated that, with 5% serum samples, the signal-noise ratio of our sensor was sufficiently high for the detection of >1 fM miRNA from complex metrices. The results of spiked serum tests further supported that the sensor could be applied to real serum samples. As miRNA-16b at 1 fM or above in serum has already been proven to be a biomarker used to diagnose successful bovine pregnancy 8 days earlier than current tests [21], the miRNA detection method described in this study will help dairy farmers to avoid unnecessary animal management costs during that time period.

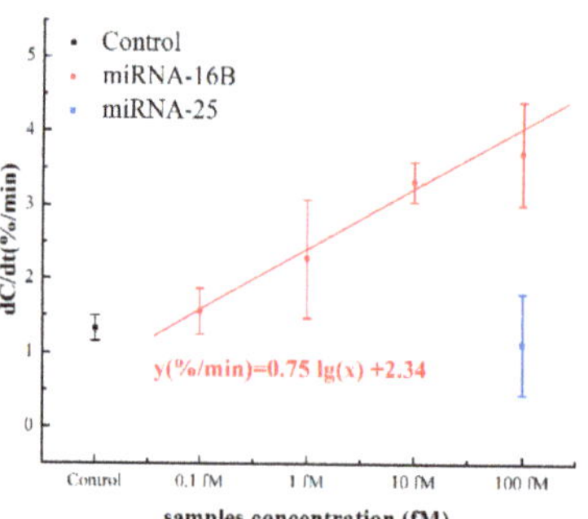

Figure 8. Capacitance change rates in response to different concentrations of spiked miRNA-16b and miRNA-25 serum samples.

4. Conclusions

This work presented a new method for detection of miRNA using ACEK-enhanced capacitive sensing. The ACET effect was utilized to accelerate the transport of miRNA and helped to accomplish the rapid and sensitive detection of miRNA within 30 s. The detection limit of the sensor was 1 fM in SSC or diluted serum. This work may form a basis for the development of point-of-care miRNA detection in general. Since the model miRNA used in this study, miRNA-16b, was shown to be a biomarker for early pregnancy in dairy cows, a specific application would be to realize a rapid, sensitive, simple, and inexpensive diagnostic for early bovine pregnancy detection. Further study would include testing more miRNA molecules for specificity—and clinical samples with qPCR confirmation.

Supplementary Materials: The following are available online at https://www.mdpi.com/article/10.3390/s21123985/s1, Figure S1: The interdigitated electrode. (a) bare electrode. (b) electrode with silicon chamber, Figure S2, Responses of miRNA-16b samples through water bath treatment and without water bath treatment.

Author Contributions: S.E., J.W. and X.L. conceived and designed the experiments; N.W., Y.J., J.H. and R.O. performed the experiments; N.W. analyzed the data; N.W., S.E. and J.W. wrote the paper. All authors have read and agreed to the published version of the manuscript.

Funding: This work acknowledges the support from the University of Tennessee (UT) Organized Research Unit-Initiative for PON/POC Nanobiosensing and UT Research Foundation Maturation Fund. The authors also acknowledge the support of USDA National Institute of Food and Agriculture (2017–67007-26150). X. Lin acknowledges the support of the Fundamental Research Funds for the Central Universities (Project no. 2019CDYGZD006), National Key R&D Program of China (No. 2020YFC1522900) and Venture & Innovation Support Program for Chongqing Overseas Returnees (Project No. cx2018017).

Institutional Review Board Statement: Not applicable.

Informed Consent Statement: Not applicable.

Data Availability Statement: The data used to support the findings of this study are available from the corresponding authors upon request.

Conflicts of Interest: The authors declare no conflict of interest.

References

1. Condrat, C.; Thompson, D.; Barbu, M.; Bugnar, O.; Boboc, A.; Cretoiu, D.; Suciu, N.; Cretoiu, S.; Voinea, S. miRNAs as biomarkers in disease: Latest findings regarding their role in diagnosis and prognosis. *Cells* **2020**, *9*, 276. [CrossRef] [PubMed]
2. Xing, W.; Gao, W.; Lv, X.; Xu, X.; Zhang, Z.; Yan, J.; Mao, G.; Bu, Z. The diagnostic value of exosome-derived biomarkers in alzheimer's disease and mild cognitive impairment: A meta-analysis. *Front. Aging Neurosci.* **2021**, *13*, 86. [CrossRef] [PubMed]
3. He, X.; Park, S.; Chen, Y.; Lee, H. Extracellular vesicle-associated miRNAs as a biomarker for lung cancer in liquid biopsy. *Front. Mol. Biosci.* **2021**, *8*, 19. [CrossRef]
4. Chen, X.; Ba, Y.; Ma, L.; Cai, X.; Yin, Y.; Wang, K.; Guo, J.; Zhang, Y.; Chen, J.; Guo, X.; et al. Characterization of microRNAs in serum: A novel class of biomarkers for diagnosis of cancer and other diseases. *Cell Res.* **2008**, *18*, 997–1006. [CrossRef] [PubMed]
5. Mitchell, P.S.; Parkin, R.K.; Kroh, E.M.; Fritz, B.R.; Wyman, S.K.; Pogosova-Agadjanyan, E.L.; Peterson, A.; Noteboom, J.; O'Briant, K.C.; Allen, A.; et al. Circulating microRNAs as stable blood-based markers for cancer detection. *Proc. Natl. Acad. Sci. USA* **2008**, *105*, 10513–10518. [CrossRef] [PubMed]
6. Tackett, M.; Doran, G.; Pregibon, D. Multiplex microRNA profiling from crude biofluids. *Genet. Eng. Biotechnol. News* **2015**, *35*, 14–15. [CrossRef]
7. Sempere, L. Tissue slide-based microRNA characterization of tumors: How detailed could diagnosis become for cancer medi-cine? *Expert Rev. Mol. Diagn.* **2014**, *14*, 853–869. [CrossRef] [PubMed]
8. Galimberti, D.; Villa, C.; Fenoglio, C.; Serpente, M.; Ghezzi, L.; Cioffi, S.; Arighi, A.; Fumagalli, G.; Scarpini, E. Circulating miRNAs as potential biomarkers in Alzheimer's disease. *J. Alzheimer Dis.* **2014**, *42*, 1261–1267. [CrossRef]
9. Weber, J.A.; Baxter, D.H.; Zhang, S.; Huang, D.Y.; Huang, K.H.; Lee, M.J.; Galas, D.J.; Wang, K. The microRNA spectrum in 12 body fluids. *Clin. Chem.* **2010**, *56*, 1733–1741. [CrossRef]
10. Tseng, H.H.; Tseng, Y.K.; You, J.J.; Kang, B.H.; Tsai, K.W. Next-generation sequencing for microRNA profiling: MicroRNA-21-3p promotes oral cancer metastasis. *Anticancer Res.* **2017**, *37*, 1059–1066.
11. Islam, N.; Lian, M.; Wu, J. Enhancing microcantilever capability with integrated AC electroosmotic trapping. *Microfluid. Nanofluidics* **2006**, *3*, 369–375. [CrossRef]
12. Wu, J.; Ben, Y.; Battigelli, D.; Chang, H.-C. Long-range AC electroosmotic trapping and detection of bioparticles. *Ind. Eng. Chem. Res.* **2005**, *44*, 2815–2822. [CrossRef]
13. Pethig, R. Review article—Dielectrophoresis: Status of the theory, technology, and applications. *Biomicrofluidics* **2010**, *4*, 022811. [CrossRef] [PubMed]
14. Chiou, P.-Y.; Ohta, A.; Jamshidi, A.; Hsu, H.-Y.; Wu, M. Light-actuated AC electroosmosis for nanoparticle manipulation. *J. Microelectromech. Syst.* **2008**, *17*, 525–531. [CrossRef]

15. Lian, M.; Islam, N.; Wu, J. AC electrothermal manipulation of conductive fluids and particles for lab-chip applications. *IET Nanobiotechnol.* **2007**, *1*, 36–43. [CrossRef]
16. Oueslati, R.; Jiang, Y.; Chen, J.; Wu, J. Rapid and sensitive point of care detection of MRSA genomic DNA by nanoelectro-kinetic sensors. *Chemosensors* **2021**, *9*, 97. [CrossRef]
17. Cheng, C.; Wu, J.; Fikrig, E.; Wang, P.; Chen, J.; Eda, S.; Terry, P. Unamplified RNA sensor for on-site screening of Zika virus disease in a limited resource setting. *ChemElectroChem* **2017**, *4*, 485–489. [CrossRef]
18. Cheng, C.; Cui, H.; Wu, J.; Eda, S. A PCR-free point-of-care capacitive immunoassay for influenza A virus. *Microchim. Acta* **2017**, *184*, 1649–1657. [CrossRef]
19. Lin, X.; Cheng, C.; Terry, P.; Chen, J.; Cui, H.; Wu, J. Rapid and sensitive detection of bisphenol a from serum matrix. *Biosens. Bioelectron.* **2017**, *91*, 104–109. [CrossRef] [PubMed]
20. Zhang, J.; Oueslati, R.; Cheng, C.; Zhao, L.; Chen, J.; Almeida, R.; Wu, J. Rapid, highly sensitive detection of Gram-negative bacteria with lipopolysaccharide based disposable aptasensor. *Biosens. Bioelectron.* **2018**, *112*, 48–53. [CrossRef] [PubMed]
21. Pohler, K.; Green, J.; Moley, L.; Gunewardena, S.; Hung, W.; Payton, R.; Hong, X.; Christenson, L.; Geary, T.; Smith, M. Cir-culating microRNA as candidates for early embryonic viability in cattle. *Mol. Reprod. Dev.* **2017**, *84*, 731–743. [CrossRef]
22. Oueslati, R.; Cheng, C.; Wu, J.; Chen, J. Highly sensitive and specific on-site detection of serum cocaine by a low cost aptasensor. *Biosens. Bioelectron.* **2018**, *108*, 103–108. [CrossRef] [PubMed]
23. Li, S.; Yuan, Q.; Morshed, B.I.; Ke, C.; Wu, J.; Jiang, H. Dielectrophoretic responses of DNA and fluorophore in physiological solution by impedimetric characterization. *Biosens. Bioelectron.* **2013**, *41*, 649–655. [CrossRef] [PubMed]
24. Meng, J.; Huang, J.; Oueslati, R.; Jiang, Y.; Chen, J.; Li, S.; Dai, S.; He, Q.; Wu, J. A single-step dnazyme sensor for ultra-sensitive and rapid detection of Pb2+ ions. *Electrochim. Acta* **2021**, *368*, 137551. [CrossRef]
25. Wu, J.; Lian, M.; Yang, K. Micropumping of biofluids by alternating current electrothermal effects. *Appl. Phys. Lett.* **2007**, *90*, 234103. [CrossRef]
26. Yuan, Q.; Wu, J. Thermally biased AC electrokinetic pumping effect for lab-on-a-chip based delivery of biofluids. *Biomed. Microdevices* **2012**, *15*, 125–133. [CrossRef]
27. Mirzajani, H.; Cheng, C.; Wu, J.; Chen, J.; Eda, S.; Aghdam, E.N.; Ghavifekr, H.B. A Highly sensitive and specific capacitive aptasensor for rapid and label-free trace analysis of Bisphenol A (BPA) in canned foods. *Biosens. Bioelectron.* **2017**, *89*, 1059–1067. [CrossRef]
28. Wu, J. Interactions of electrical fields with fluids: Laboratory-on-a-chip applications. *IET Nanobiotechnol.* **2008**, *2*, 14–27. [CrossRef]
29. Cheng, C.; Wu, J.; Chen, J. A highly sensitive aptasensor for on-site detection of lipopolysaccharides in food. *Electrophoresis* **2019**, *40*, 890–896. [CrossRef]
30. Cheng, C.; Oueslati, R.; Wu, J.; Chen, J.; Eda, S. Capacitive DNA sensor for rapid and sensitive detection of whole genome human herpesvirus-1 dsDNA in serum. *Electrophoresis* **2017**, *38*, 1617–1623. [CrossRef]
31. Zhang, J.; Fang, X.; Wu, J.; Hu, Z.; Jiang, Y.; Qi, H.; Zheng, L.; Xuan, X. An interdigitated microelectrode based aptasensor for real-time and ultratrace detection of four organophosphorus pesticides. *Biosens. Bioelectron.* **2020**, *150*, 111879. [CrossRef]
32. Lian, M.; Wu, J. Microfluidic Flow Reversal at Low Frequency by AC Electrothermal Effect. *Microfluid. Nanofluid.* **2009**, *7*, 757–765. [CrossRef]
33. Yang, K.; Wu, J. Investigation of Microflow Reversal by AC Electrokinetics in Orthogonal Electrodes for Micropump Design. *Biomicrofluidics* **2008**, *2*, 024101. [CrossRef]

Article

A Self-Calibrating IoT Portable Electrochemical Immunosensor for Serum Human Epididymis Protein 4 as a Tumor Biomarker for Ovarian Cancer

Valentina Bianchi [1], Monica Mattarozzi [2,*], Marco Giannetto [2,*], Andrea Boni [1], Ilaria De Munari [1] and Maria Careri [2]

[1] Department of Engineering and Architecture, University of Parma, 43124 Parma, Italy; valentina.bianchi@unipr.it (V.B.); andrea.boni@unipr.it (A.B.); ilaria.demunari@unipr.it (I.D.M.)

[2] Department of Chemistry, Life Sciences and Environmental Sustainability, University of Parma, 43124 Parma, Italy; maria.careri@unipr.it

* Correspondence: monica.mattarozzi@unipr.it (M.M.); marco.giannetto@unipr.it (M.G.)

Received: 9 March 2020; Accepted: 31 March 2020; Published: 3 April 2020

Abstract: Nowadays, analytical techniques are moving towards the development of smart biosensing strategies for the point-of-care accurate screening of disease biomarkers, such as human epididymis protein 4 (HE4), a recently discovered serum marker for early ovarian cancer diagnosis. In this context, the present work represents the first implementation of a competitive enzyme-labelled magneto-immunoassay exploiting a homemade IoT Wi-Fi cloud-based portable potentiostat for differential pulse voltammetry readout. The electrochemical device was specifically designed to be capable of autonomous calibration and data processing, switching between calibration, and measurement modes: in particular, firstly, a baseline estimation algorithm is applied for correct peak computation, then calibration function is built by interpolating data with a four-parameter logistic function. The calibration function parameters are stored on the cloud for inverse prediction to determine the concentration of unknown samples. Interpolation function calibration and concentration evaluation are performed directly on-board, thus reducing the power consumption. The analytical device was validated in human serum, demonstrating good sensing performance for analysis of HE4 with detection and quantitation limits in human serum of 3.5 and 29.2 pM, respectively, reaching the sensitivity that is required for diagnostic purposes, with high potential for applications as portable and smart diagnostic tool for point-of-care testing.

Keywords: human epididymis protein 4; competitive electrochemical immunosensor; WiFi portable potentiostat; on-board calibration; Internet of Things

1. Introduction

Ovarian carcinoma (OC) is the leading cause of death from gynecological malignancy worldwide: it is generally asymptomatic in the early stage, so the majority of women with OC are not diagnosed until the disease is in an advanced stage, with an overall five-year relative survival rate that is generally in the 30–40% range [1,2]. Therefore, it is crucial to detect OC as early as possible to correctly identify the cancer stage for effective treatment options and to early distinguish malignant from benign pelvic mass by means of annual routine gynecological and pelvic examinations, as well as dedicated screening programmes. In this context, noninvasive cancer detection at an early stage needs the identification of specific and sensitive biomarkers that are present at abnormal concentrations in body fluids, as well as the development of smart devices for analytical screening [3–6], in order to improve ovarian cancer survival rate [7]. Currently, carbohydrate antigen 125 (CA125) glycoprotein is the established biomarker for detecting OC occurrence and the monitoring therapeutic progress. However, the diagnostic potential of CA125 is mainly limited by its low specificity, since it can also be increased in a variety of other benign gynecological conditions affecting pre- and/or post-menopausal women, or in other malignant cancers [8]. In this context, over the last years, a number of alternative OC biomarkers, especially for the predominant OC subtype, i.e. epithelial ovarian cancer (EOC), have been identified and studied, either alone or combined, to improve the effectiveness of early diagnostic strategies. Among the recently proposed biomarkers, human epididymis protein 4 (HE4) resulted in being most promising in EOC differential diagnosis in detecting the disease at early stage, in monitoring the response to chemotherapy and in estimating the prognosis of ovarian cancer [1].

Electrochemical immunosensors are playing a growing role for improved clinical diagnosis and therapy monitoring by combining the specificity of antigen-antibody immunoreaction and the sensitivity of the electrochemical transduction [9]. As for HE4 determination, several sensing architectures that are based on "sandwich" approach have been recently exploited [10–13].

Electrochemical immunosensors can exploit enzyme labels for signal generation, such as the here presented study, but can also be designed according to label-free transduction and amplification mechanisms [14,15]. One of the main advantages associated to the enzyme-labelled electrochemical transduction is the possibility of method implementation on miniaturized electrodes as well as on portable and battery-operated devices [16]. Several instrumental approaches have been proposed in recent years and, some of these, such as 910 PSTAT and DropStat from Metrohm (Herisau, Switzerland), are already commercially available. DropStat, in particular, allows for making quantitative analyses and directly displaying the analyte concentration on the stand-alone device; the tool is custom configured with the electrochemical technique attending user's needs by simply inserting preprogrammed cards with calibration parameters available upon request from the producer. A PC connection is necessary to download acquired data for storage purposes. In addition, solutions by PalmSens (Houten, The Netherlands) and UNISCAN (Buxton, Derbyshire, UK) are battery-powered or USB-powered portable devices that are capable of performing voltammetric measurements, but they need to rely on external APP or software PC to determine the analyte concentration in a sample. Very recently, USB-based [17–19] and wireless [20–22] potentiostats were proposed, both involving data processing and visualization performed on an external device running a custom developed software. A Bluetooth communication for data collecting, processing, and sharing is exploited, even in the case of wireless potentiostats, thus requiring a PC that is close to the device. In 2015, Steinberg et al. proposed a device based on RFID or NFC data communication, where data are logged and then off-line analyzed in MS-Excel format [23]. However, all of the developed systems that allow for quantitative analyses to be performed involve off-line data processing based on a calibration function, even if using a Wi-Fi communication protocol. For example, Annamalai and coworkers recently described a Wi-Fi based solution for glucose determination, taking advantage from the Adafruit Cloud services to process and display data [24]. Chen et al. reported another recent example that involves Wi-Fi protocol that can locally visualize results on a TFT LCD display [25].

In this context, the possibility to perform data processing and calibration on-board, making the device completely autonomous, is very attractive. This approach allows, on the one hand, to reduce the device power consumption avoiding the raw data transmission and, on the other hand, to increase its portability, making it usable independently of connections or external devices at the time of measurement. However, the potentiostat portability must not affect the measurement accuracy in estimating the analyte concentration. Very recently, our research group proposed a new portable potentiostat having the Analog Front End (AFE) that was capable of very accurate analysis when compared to other portable solutions that are available in literature [26]. In this context, the present work represents a further improvement of that Wi-Fi portable potentiostat, proposing new additional functions through an enhanced firmware to introduce autonomous calibration and processing capabilities with the aim of applying it to the readout and processing of data from an innovative magneto-immunoassay for the determination of the ovarian cancer biomarker HE4 in human serum. This approach is consistent with the technological evolution of Analytics 4.0, representing the innovative integration of analytical chemistry into an Internet-of-Things (IoT) system, as recently defined by Mayer at al. [27].

2. Materials and Methods

2.1. Materials

Trizma® base, Tween-20, boric acid, sodium hydroxide, monosodium phosphate, disodium phosphate, magnesium chloride, human serum, albumin from bovine serum (BSA), Monoclonal Anti-WFDC2 clone 3F9 Antibody produced in mouse purified immunoglobulin (anti-HE4), PrEST Antigen WFDC2 (HE4) were purchased from Sigma Aldrich (Milan, Italy). Alkaline Phosphatase-conjugated secondary Rabbit Anti-Mouse IgG (RAM-AP), carcinoembryonic antigen (CEA), and Carbohydrate Antigen 125 (CA125) were purchased from Abcam (Cambridge, UK). DropSens® hydroquinone diphosphate (HQDP) was obtained from Metrohm Italiana (Origgio, Varese, Italy). The following buffers were prepared: 0.1 M borate buffer (BB) pH 9.5; 0.1 M Na-phosphate buffer (PB; 0.08 M Na_2HPO_4 and 0.02 M NaH_2PO_4) pH 7.4; 0.1 M TRIS Buffer (TB) containing 0.02 M of $MgCl_2$ pH 7.4; washing buffers PB-T and TB-T consisted, respectively, of PB and TB containing Tween-20 at 0.05 % (*v/v*). Reading buffer (RB) was prepared as TB, but adjusting the pH to 9.8.

Dynabeads®M-280 Tosylactivated (2.8 μm) MBs were purchased from Invitrogen (Thermo Fisher Scientific, Rodano, Milan, Italy). DropSens®Screen-Printed Carbon Electrodes (SPEs) for electrochemical readout were purchased from Metrohm Italiana; in particular, the working (WE; 4 mm diameter) and auxiliary (AE) electrodes are made of carbon, while pseudo-reference electrode (RE) is in silver.

2.2. Magnetic Immunoassay Set-Up

MBs were thoroughly suspended and a volume of 15 μL was transferred to a new tube, then, after triple-washing step with 200 μL of BB, the MBs were incubated with 400 μL of 10 μg mL^{-1} solution of HE4 antigen in BB (1 h; 37 °C; 1000 rpm). After the removal of unreacted HE4 solution by twice washing with BB, MBs were blocked with a 2% (*w/v*) solution of BSA dissolved in PB, and then washed with PB-T and PB. The HE4-modified MBs were used to carry out four independent competitive electrochemical immunoassays after their resuspension in 400 μL of TB and subdivision in four equivalent volume aliquots. After isolation, the MBs were resuspended with 100 μL of serum sample previously diluted ten-fold with TB and added with anti-HE4 monoclonal antibody to reach the final concentration of 1 μg mL^{-1}. The immunocompetition was left to undergo for 1 h at room temperature (RT) under agitation (1000 rpm), the reacted MBs was carefully washed with TB-T and TB and then incubated with 100 μL of 2 μgmL^{-1} RAM-AP secondary antibody (RT; 1 h; 1000 rpm). After washing with TB-T and TB, the beads were finally reacted with 100 μL of 1 mg mL^{-1} solution of HQDP enzyme substrate dissolved in RB for three minutes. Two 50 μL sub-aliquots of each sample were transferred on a SPE mounted on a DropSens®customized magnetic support that aimed at confining the MBs on the surface of WE and the electrochemical readout was carried out independently for each sub-aliquot of the sample suspension. Alkaline phosphatase (AP) dephosphorylates non-electroactive HQDP into hydroquinone (HQ), allowing for its electrochemical oxidation to quinone (Q) to achieve the signal related to the analyte.

Scheme 1 reports a schematic representation of the working principle of the electrochemical immunomagnetic assay.

The voltammetric readout of the assays were performed through our homemade IoT Wi-Fi portable potentiostat, while using the differential pulse voltammetry (DPV) scan. The potential was scanned between −0.4 and +0.2 V, setting on the IoT potentiostat a pulse amplitude of 0.05 V, a step potential of 0.005 V, and a pulse time of 100 ms. After the deposition of the MBs suspension on the SPE, a preconditioning step of 30 s at −0.4 V was applied to preconcentrate the reduced form of quinone.

The current responses observed for each concentration (S) were normalized with respect to the zero signal (S$_0$) obtained without HE4 in competition.

Scheme 1. Working principle of the electrochemical immunomagnetic assay on MBs supports.

2.3. IoT Architecture

The system architecture is depicted in the sketch that is shown in Scheme 2.

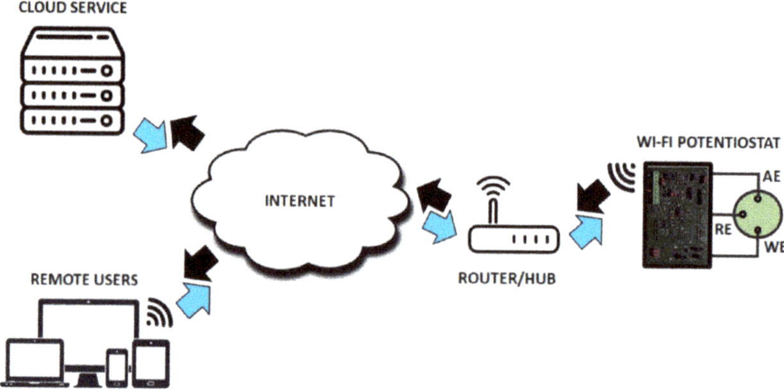

Scheme 2. Internet-of-Things (IoT) architecture for the proposed system.

The exploited AFE guarantees a sufficient measurement points to achieve higher accuracy with respect to other solutions presented in literature, as reported in our previous study [26]. In this work, the aforementioned AFE is connected to a CC3200 Microcontroller Unit (MCU) for signal processing with an improved firmware to develop a smart self-calibrating portable potentiostat, that was fully compliant with the IoT paradigm.

In particular, the proposed system takes a step forward with respect to current literature [17–25], since it is capable of on-board building a calibration function acquiring the signal from a given set of standards at known concentrations (*calibration mode*); this function is subsequently applied to calculate analyte concentration of unknown samples (*measurement mode*). Moreover, the proposed solution relies on a Wi-Fi connection to send the preprocessed data to a cloud service for storing and sharing exploiting the remote access feature. The on-board data elaboration allows for greatly reducing the device power consumption, enhancing battery lifetime. In fact, the main contribution to the device power consumption is the data transmission phase: sending preprocessed information allows to reduce the transmission task, with a benefit for the average absorbed current. In Bassoli et al. [28], a quantitative evaluation of this approach in the case of Wi-Fi protocol is reported. The data generated from the end device go directly, through the modem/router via Wi-Fi Protected Access (WPA) standard protocol, to the Internet cloud, where they are stored for subsequent sharing in a fast and efficient way. Thanks to the broad coverage guarantee by the Wi-Fi router, it is easy to ensure a stable sensor connection over a large area [29]. The scenario is fully compliant with an IoT paradigm, in which sensors "information flows rationally and orderly on the Internet, for being shared on a global scale" [30].

2.4. On-Board Data Management

The basedrift phenomenon has to be considered to correctly estimate the DPV peak current. In fact, the differential current baseline is often not parallel to the voltage axis, which prevents the computation of the peak by simply applying an algorithm for the search of the maximum. Hence, for each measurement point a baseline estimation algorithm is applied. Usually, for each curve, N minimum points have to be detected and the baseline is evaluated applying a polynomial fitting algorithm. Therefore, the baseline thus calculated is assumed as the reference for the correct evaluation of the peak current.

For calibration purposes, the peak current values have to be related to the HE4 concentration according to a suitable mathematical law. For this purpose, the normalized S/S_0 responses were plotted versus the HE4 concentration, so obtaining a dose-response inhibition curve, which was interpolated through a four-parameter logistic function (1) [31]:

$$S/S_0 = S_{min} + \frac{(S_{max} - S_{min})}{1 + ([C]/I_{50})^B} \, , \tag{1}$$

where S_{min} and S_{max} are the asymptotic minimum and maximum, I_{50} is the inflection point, namely the antigen concentration corresponding to 50% signal inhibition and B is the slope at I_{50}.

In the *calibration mode*, the device is able to compute on-board these four parameters when considering a set of measurements that were carried out on calibration standards. In particular, the parameters are computed, implementing the method of least squares according to the Levenberg–Marquardt algorithm [32,33]. When considering two vectors X and Y in Rn, containing, respectively, the n known concentration values and the n measurements of the corresponding peak current to be used for calibration, the vector of the residuals ε in Rn will be defined, as

$$\varepsilon = Y - F(X, \beta), \tag{2}$$

where F is the logistic function and β is the vector containing the four parameters to be computed. Hence, the aim of the proposed algorithm is to minimize the sum of the residual squares, SS, with an iterative approach. In doing so, the first step is to define the initial values for the parameters to be found, which are

$$S_{max\_0} = \max(Y), \tag{3}$$

$$S_{min\_0} = \min(Y), \tag{4}$$

$$B_0 = sign\left(\frac{y_n - y_1}{x_n - x_1}\right), \tag{5}$$

$$I_{50\_0} = X_k, \quad k \text{ is the position where } \min\left(\left\|Y - \frac{\max(Y) - \min(Y)}{2}\right\|\right), \tag{6}$$

Subsequently, during a generic step i the β vector is updated as

$$\beta_i = \beta_{i-1} - \Delta_{\beta_{i-1}} \, , \tag{7}$$

where Δ_{β_i} is defined as

$$\Delta_{\beta_i} = \left((J^T J)_{\beta_i} + \lambda (J^T J)_{\beta_i}\right)^{-1} J^T e_{\beta_i} \, , \tag{8}$$

λ is a parameter updated at each step and J is the Jacobian matrix defined as

$$J = \frac{\partial e}{\partial \beta} \, , \tag{9}$$

Finally, the sum of the residual square is computed as

$$SS_i = \sum_i (\varepsilon_i)^2 = \varepsilon_{\beta_i}^T \varepsilon_{\beta_i} \, , \tag{10}$$

If SS_i is equal to SS_{i-1} with a chosen tolerance of 1×10^{-6}, convergence is reached and the algorithm stops. Otherwise, the λ parameter is updated according the following law (11) and the algorithm continues.

$$\lambda_{i+1} = \begin{cases} 0.1\lambda_i \; if \; SS_i < SS_{i-1} \\ 10\lambda_i \; otherwise \end{cases} \, , \tag{11}$$

All of these operations were coded in C language for MCU programming and specific function to compute the derivative of residual with respect to each parameter in β vector was developed in order to compose the Jacobian matrix J.

Switching to the *measurement mode*, the inversion of the calibration function involves the computation of a B-th root. This type of calculation is trivial if it is performed on a platform with a high computing capacity, like the one made available by the main cloud services. However, processing the data in the cloud implies sending all of the collected raw data through the Wi-Fi energy-intensive connection, considerably increasing the power consumption and limiting the portability of the device. Processing data on board limits these issues, but, on the other hand, it entails fewer computational resources and the need to develop specific functions in C language. To implement the B-th root on the MCU platform, a generalization of the Heron's method was exploited [34]. Once the concentration has been computed, it can be made available on a cloud for remote monitoring or can be maintained in the device memory waiting a Wi-Fi connection.

2.5. Immunosensor Validation

The validation of sensing device was performed according to the Eurachem guidelines [35] on human serum as blank matrix. An assessment of the dose-response calibration curve was carried out using concentration values already corrected according to the serum dilution factor.

The detection (LOD) and quantitation (LOQ) limits were calculated from mean normalized signal (y_b) and standard deviation (s_b) recorded from ten independent replicates at HE4 level showing a not significant signal inhibition (35 fM, first point of dose-response curve, see 3.4 paragraph). More precisely, the corresponding signals were calculated, as $y_b - \left(3.3 \cdot \left(s_b / \sqrt{n}\right)\right)$ and $y_b - \left(10 \cdot \left(s_b / \sqrt{n}\right)\right)$, respectively, where n is the number of independent replicates for each concentration level. The linear dynamic range was determined as the concentration interval over which the slope (B parameter of the inhibition curve) varies within ±10%. Two concentration levels that were not explored for the calibration were used for the assessment of trueness.

3. Results and Discussion

3.1. MBs Functionalization and Immunocompetition Study

The competitive magneto-immunoassay was implemented on tosyl-activated magnetic microbeads as sensing substrate, involving the lysine residues of HE4 antigen for its covalent linking. The immobilization of the antigen is one of the most critical issues when developing immunoassays, as this procedure has to not compromise the immunoreactivity of the bioreceptor, i.e., not involving the antibody-binding regions.

Before carrying out the competition experiments, the maintenance of HE4 immunoreactivity was verified through direct binding titrations between the immobilized antigen and anti-HE4 antibody. These experiments were also aimed at investigating the dynamic range for the concentrations of (*i*) HE4 used for MBs functionalization and (*ii*) anti-HE4 used for the implementation of the immunocompetition. For this purpose, we fixed at 10 µg mL^{-1} the HE4 concentration for the MBs functionalization and, thus, we incubated the modified beads with anti-HE4 ranging from 0 to 2 µg mL^{-1}, and then with RAM-AP secondary antibody. Further experiments were carried out varying the antigen concentration from 0 to 20 µg mL^{-1}, keeping the anti-HE4 concentration at 2 µg mL^{-1} constant. The results of the immuno-titration experiments showed current responses increasing versus anti-HE4 and HE4 concentrations, with very low background signals recorded in the absence of antibody and antigen.

These findings can be rationalized as an evidence of the effective antigen immobilization, also being confirmed by the non-occurrence of unwanted non-specific binding on MBs surface.

On the basis of the above considerations, we performed a set of competition experiments keeping constant at 10 µg mL^{-1} the competing HE4 and varying the above discussed experimental parameters. The dataset of signal inhibition values, being calculated as 1-(S/S_0), was evaluated through a two-way Analysis of Variance (ANOVA), showing that the anti-HE4 factor and its interaction with HE4 factor significantly influence ($p < 0.01$) the immunosensor response, whereas the single HE4 factor resulted in being not significant ($p > 0.01$). Figure 1 reports the ANOVA interaction plot, showing the effect of the two studied factors. The observed trend clearly shows the higher signal inhibition rate being observed for the combination of the middle (1 µg mL^{-1}) anti-HE4 concentration level and the lowest (10 µg mL^{-1}) HE4 concentration explored for the functionalization of MBs. The best concentrations for the immunocompetition experiments were then chosen on the basis of these results.

As for functionalized MBs stability, intended as the maintenance of the reactivity of HE4 immobilized on the magnetic beads, it resulted in being good over at least two months stored in the dark at +4 °C.

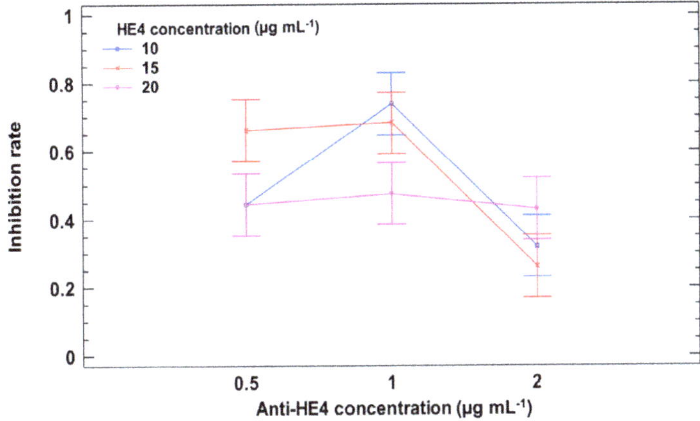

Figure 1. Interaction plot from two-way ANOVA. Factors: concentration of HE4 for functionalization and concentration of anti-HE4 antibody; Response: signal inhibition.

3.2. Data Acquisition and Processing

A *Vbias* ranging from −0.4 V to 0.2 V, which was suitable for acquisition of HQ oxidation current, was applied between WE and RE electrodes of the cell to measure the amperometric signal (i.e., the cell output current). Exploiting the high resolution of the AFE reported in our previous work [26], a total of 120 measurement points of differential output current were collected. Figure 2 shows the signals that were recorded over the 350 fM–350 nM HE4 concentration range.

In this application, a minimum number of point equal to 30 was set for baseline drift correction. Figure 3 illustrates an example of baseline estimation in the case of 350 fM HE4 concentration. The dots indicate the 30 minimum points that were exploited in the linear regression and the resulting baseline is shown. The actual peak value is calculated as the distance between the baseline and maximum current value.

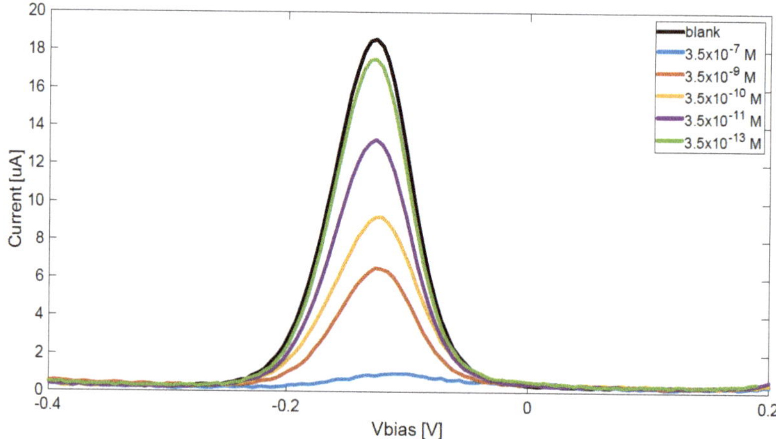

Figure 2. Differential pulse voltammetry (DPV) scans recorded carrying out immunocompetition assays over the 350 fM–350 nM HE4 concentration range in human serum samples.

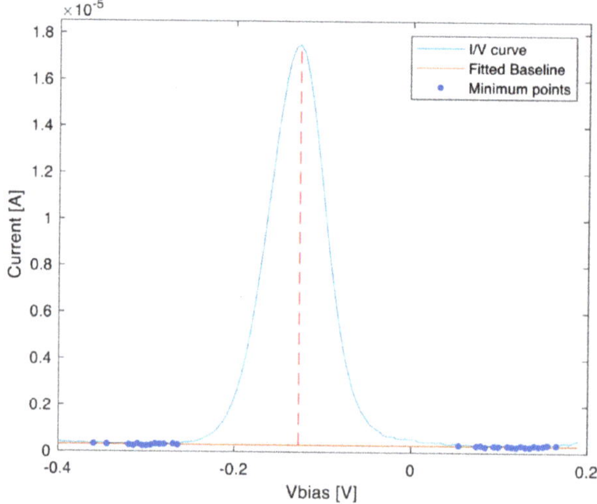

Figure 3. Example of baseline estimation from a voltammogram. Thirty minimum points were considered for the linear regression.

Figure 4 shows the dose-response inhibition curve that was obtained upon interpolation of the experimental dataset with the four-parameter logistic function (1). An increase of analyte concentration was found to reduce the amount of anti-HE4 that was bound to the antigen-modified MBs, thanks to the ability of HE4 to inhibit the antibody binding extent to the sensing surface that was functionalized with the antigen.

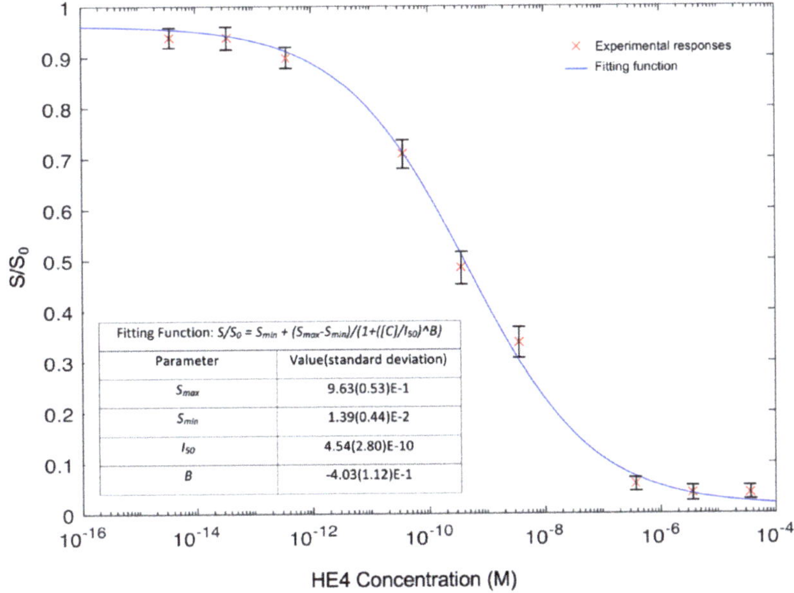

Parameter	Value(standard deviation)
S_{max}	9.63(0.53)E-1
S_{min}	1.39(0.44)E-2
I_{50}	4.54(2.80)E-10
B	-4.03(1.12)E-1

Figure 4. Dose-response inhibition curve for HE4-spiked human serum samples obtained interpolating the dataset from immunocompetition experiments carried out on HE4-modified MBs over the 3.5 fM–35 μM concentration range while using the four-parameter logistic function (1). Inset table: fitted curve parameters.

3.3. Electrochemical Immunoassay Analytical Performance

The developed IoT immunochemical device exhibited good analytical performance showing a linear dynamic range over three orders of magnitude. As for precision assessed on the basis of three independent assays, we observed relative standard deviations (RSD) always < 13%.

The analysis of HE4-spiked serum samples provided a LOD of 3.5 pM and a LOQ of 29.2 pM. A comparison of the calculated LOD with the threshold values set for premenopausal women (70 pM) and postmenopausal patients (140 pM) [36] highlights the suitability of the developed system in terms of quality criteria for routine analysis, while also considering the high automatization level in calibration and data processing and the assessment of all validation parameters in matrix, differently from previously published studies.

Trueness was assessed while using HE4-spiked serum samples with concentration levels that were not used for the calibration, i.e. LOQ and I_{50}, obtaining a mean recovery of 105 ± 12%. These values can be considered to be very satisfying while taking the developed on-board self-calibration innovative strategy, embedded in the portable automatized instrument, into account.

3.4. Selectivity Assessment

Protein markers potentially occurring in clinical samples of diagnostic interest for carcinomas of gynecologic concern were selected to assess the selectivity of the developed immunomagnetic assays. For this purpose, we evaluated the responses that were obtained with CA125 and CEA, both at I_{50} concentration levels, obtaining no significant signal inhibition ($p < 0.05$) with respect to blank reference samples only containing anti-HE4 antibody.

4. Conclusions

In the present work, we proposed an innovative approach to address disease diagnosis and management through the integration of IoT home-made portable potentiostat with electrochemical immunosensors for the determination of HE4 as ovarian cancer serological biomarker in the framework of the point-of-care (POC) testing. The choice of a mainstream protocol for the connection as Wi-Fi enhances the usability and portability of the system without requiring an additional device, such as PC or Smartphone. Thanks to the ability of the developed portable device to acquire a high number of measurement points (i.e. 350), an accurate quantitative evaluation of the analyte concentration can be carried out. The developed device is able to perform the interpolation of calibration function and the concentration evaluation directly on-board, enhancing the system portability, minimizing data transmission, and therefore the high power consumption that is generally required by the Wi-Fi connection. In the case of absence of connectivity, the device can store the data in a local memory waiting for an available wireless connection.

The analytical performance of the IoT-based competitive immunomagnetic electrochemical assay was assessed in human serum, obtaining LOD and LOQ values at the picomolar level, meeting the requirements for diagnostic applications in early diagnosis of OC. Finally, the sensing device was successfully applied in serum with adequate recovery values for the analysis of spiked samples as well as good selectivity toward potential protein interferents. This work shows great promise for the application of the developed self-calibrating IoT portable immunosensor for HE4 determination in serum samples, in accordance with Analytical 4.0 evolution towards smart platforms for decentralized POC testing.

Author Contributions: Conceptualization, V.B., M.M., I.D.M. and M.G.; methodology, V.B., M.M., M.G., A.B., I.D.M.; validation, M.M. and M.G.; investigation, V.B., M.M. and M.G.; writing—original draft preparation, V.B., M.M. and M.G.; writing—review and editing, V.B., M.M., M.G., A.B., I.D.M. and M.C.; funding acquisition, M.M., I.D.M. and M.C. All authors have read and agreed to the published version of the manuscript.

Funding: This work was financially supported by the Programme "FIL 2016-Quota Incentivante" of the University of Parma. It benefited also from the equipment and framework of the COMP-HUB Initiative, funded by the 'Departments of Excellence' program of the Italian Ministry for Education, University and Research (MIUR, 2018-2022).

Conflicts of Interest: The authors declare no conflict of interest.

References

1. Granato, T.; Porpora, M.G.; Longo, F.; Angeloni, A.; Manganaro, L.; Anastasi, E. HE4 in the differential diagnosis of ovarian masses. *Clin. Chim. Acta* **2015**, *446*, 147–155. [CrossRef]

2. Reid, B.M.; Permuth, J.B.; Sellers, T.A. Epidemiology of ovarian cancer: A review. *Cancer Biol. Med.* **2017**, *14*, 9–32. [PubMed]

3. Bianchi, F.; Giannetto, M.; Careri, M. Advances in molecular analysis of biomarkers for autoimmune and carcinogenic diseases. *Anal. Bioanal. Chem.* **2014**, *406*, 15–20. [CrossRef] [PubMed]

4. Sharma, S.; Raghav, R.; O'Kennedy, R.; Srivastava, S. Advances in ovarian cancer diagnosis: A journey from immunoassays to immunosensors. *Enzyme Microb. Technol.* **2016**, *89*, 15–30. [CrossRef] [PubMed]

5. Ranjan, R.; Esimbekova, E.N.; Kratasyuk, V.A. Rapid biosensing tools for cancer biomarkers. *Biosens. Bioelectron.* **2017**, *87*, 918–930. [CrossRef]

6. Giannetto, M.; Bianchi, M.V.; Mattarozzi, M.; Careri, M. Competitive amperometric immunosensor for determination of p53 protein in urine with carbon nanotubes/gold nanoparticles screenprinted electrodes: A potential rapid and noninvasive screening tool for early diagnosis of urinary tract carcinoma. *Anal. Chim. Acta* **2017**, *991*, 133–141. [CrossRef]

7. De La Franier, B.; Thompson, M. Early stage detection and screening of ovarian cancer: A research opportunity and significant challenge for biosensor technology. *Biosens. Bioelectron.* **2019**, *135*, 71–81. [CrossRef]

8. Su, Z.; Graybill, W.S.; Zhu, Y. Detection and monitoring of ovarian cancer. *Clin. Chim. Acta* **2013**, *415*, 341–345. [CrossRef]

9. Sadighbayan, D.; Sadighbayan, K.; Tohid-kia, M.R.; Khosroushahi, A.Y.; Hasanzadeh, M. Development of electrochemical biosensors for tumor marker determination towards cancer diagnosis: Recent progress. *Trends Anal. Chem.* **2019**, *118*, 73–88. [CrossRef]

10. Wang, J.; Song, J.; Zheng, H.; Zheng, X.; Dai, H.; Hong, Z.; Lin, Y. Application of $NiFe_2O_4$ nanotubes as catalytically promoted sensing platform for ratiometric electrochemiluminescence analysis of ovarian cancer marker. *Sensor. Actuators B Chem.* **2019**, *288*, 80–87. [CrossRef]

11. Fan, L.; Yan, Y.; Guo, B.; Zhao, M.; Li, J.; Bian, X.; Wu, H.; Cheng, W.; Ding, S. Trimetallic hybrid nanodendrites and magnetic nanocomposites-based electrochemical immunosensor for ultrasensitive detection of serum human epididymis protein 4. *Sensor. Actuators B Chem.* **2019**, *296*, 126697. [CrossRef]

12. Yan, Q.; Cao, L.; Dong, H.; Tan, Z.; Liu, Q.; Zhang, W.; Zhao, P.; Li, Y.; Liu, Y.; Dong, Y. Sensitive amperometric immunosensor with improved electrocatalytic Au@Pd urchin-shaped nanostructures for human epididymis specific protein 4 antigen detection. *Anal. Chim. Acta* **2019**, *1069*, 117–125. [CrossRef] [PubMed]

13. Klein, T.; Wang, W.; Yu, L.; Wu, K.; Boylan, K.L.M.; Isaksson Vogel, R.; Skubitz, A.P.N.; Wang, J.-P. Development of a multiplexed giant magnetoresistive biosensor array prototype to quantify ovarian cancer biomarkers. *Biosens. Bioelectron.* **2019**, *126*, 301–307. [CrossRef]

14. Ramanavicius, A.; Oztekin, Y.; Ramanaviciene, A. Electrochemical formation of polypyrrole-based layerfor immunosensor design. *Sensor. Actuators B* **2014**, *197*, 237–243. [CrossRef]

15. Tan, Y.; Li, M.; Ye, X.; Wang, Z.; Wang, Y.; Li, C. Ionic liquid auxiliary exfoliation of WS2 nanosheets and the enhanced effect of hollow gold nanospheres on their photoelectrochemical sensing towards human epididymis protein 4. *Sensor. Actuators B* **2018**, *262*, 982–990. [CrossRef]

16. Giannetto, M.; Bianchi, V.; Gentili, S.; Fortunati, S.; De Munari, I.; Careri, M. An integrated IoT-Wi-Fi board for remote data acquisition and sharing from innovative immunosensors. Case of study: Diagnosis of celiac disease. *Sensor. Actuators B Chem.* **2018**, *273*, 1395–1403. [CrossRef]

17. Loncaric, C.; Tang, Y.; Ho, C.; Parameswaran, M.A.; Yu, H.-Z. A USB-based electrochemical biosensor prototype for point-of-care diagnosis. *Sensor. Actuators B Chem.* **2012**, *161*, 908–913. [CrossRef]

18. Adams, S.D.; Doeven, E.H.; Quayle, K.; Kouzani, A.Z. MiniStat: Development and evaluation of a mini-potentiostat for electrochemical measurements. *IEEE Access* **2019**, *7*, 31903–31912. [CrossRef]

19. Pruna, R.; Palacio, F.; Baraket, A.; Zine, N.; Streklas, A.; Bausells, J.; Errachid, A.; López, M. A low-cost and miniaturized potentiostat for sensing of biomolecular species such as TNF-α by electrochemical impedance spectroscopy. *Biosens. Bioelectron.* **2018**, *100*, 533–540. [CrossRef]

20. Hanitra, I.N.; Criscuolo, F.; Pankratova, N.; Carrara, S.; De Micheli, G. Multi-Channel Front-End for Electrochemical Sensing of Metabolites, Drugs, and Electrolytes. *IEEE Sens. J.* **2019**, *20*, 3636–3645. [CrossRef]

21. De Campos da Costa, J.P.; Bastos, W.B.; da Costa, P.I.; Zaghete, M.A.; Longo, E.; Carmo, J.P. Portable laboratory platform with electrochemical biosensors for immunodiagnostic of hepatitis C virus. *IEEE Sens. J.* **2019**, *19*, 10701–10709. [CrossRef]

22. Ma, L.; Ju, F.; Tao, C.; Shen, X. Portable, Low Cost Smartphone-Based Potentiostat System for the Salivary α-Amylase Detection in Stress Paradigm. In Proceedings of the 41st Annual International Conference of the IEEE Engineering in Medicine and Biology Society (EMBC), Berlin, Germany, 23–27 July 2019; pp. 1334–1337.

23. Steinberg, M.D.; Kassal, P.; Kereković, I.; Murković Steinberg, I. A wireless potentiostat for mobile chemical sensing and biosensing. *Talanta* **2015**, *143*, 178–183. [CrossRef] [PubMed]

24. Annamalai, P.; Slaughter, G. Wireless Bipotentiostat Circuit for Glucose and H_2O_2 Interrogation. In Proceedings of the 41st Annual International Conference of the IEEE Engineering in Medicine and Biology Society (EMBC), Berlin, Germany, 23–27 July 2019; pp. 1567–1570.

25. Chen, G.; Xie, J.; Zhang, Z.; Meng, W.; Zhang, C.; Kang, K.; Wu, Y.; Guo, Z. A portable digital-control electrochemical system with automatic ohmic drop compensation for fast scan voltammetry and its application to ultrasensitive detection of chromium(III). *Sensor. Actuators B Chem.* **2019**, *301*, 127135. [CrossRef]

26. Bianchi, V.; Boni, A.; Fortunati, S.; Giannetto, M.; Careri, M.; De Munari, I. A Wi-Fi cloud-based portable potentiostat for electrochemical biosensors. *IEEE Trans. Instrum. Meas.* **2019**, in press. [CrossRef]

27. Mayer, M.; Baeumner, A.J. ABC spotlight on analytics 4.0. *Anal. Bioanal. Chem.* **2018**, *410*, 5095–5097. [CrossRef]

28. Bassoli, M.; Bianchi, V.; De Munari, I.; Ciampolini, P. An IoT approach for an AAL Wi-Fi-based monitoring system. *IEEE Trans. Instrum. Meas.* **2017**, *66*, 3200–3209. [CrossRef]

29. Bassoli, M.; Bianchi, V.; De Munari, I. A plug and play IoT Wi-Fi smart home system for human monitoring. *Electronics* **2018**, *7*, 200. [CrossRef]

30. Huang, Y.; Li, G. Descriptive Models for Internet of Things. In Proceedings of the International Conference on Intelligent Control and Information Processing, Dalian, China, 12–15 August 2010; pp. 483–486.

31. Manfredi, A.; Giannetto, M.; Mattarozzi, M.; Costantini, M.; Mucchino, C.; Careri, M. Competitive immunosensor based on gliadin immobilization on disposable carbon-nanogold screen-printed electrodes for rapid determination of celiotoxic prolamins. *Anal. Bioanal. Chem.* **2016**, *408*, 7289–7298. [CrossRef]

32. Levenberg, K. A method for the solution of certain nonlinear problems in least squares. *Quart. Appl. Math.* **1944**, *2*, 164–168. [CrossRef]

33. Marquaradt, D.W. An algorithm for least-squares estimation of nonlinear parameters. *J. Soc. Ind. Appl. Math.* **1963**, *11*, 431–441. [CrossRef]

34. Taisbak, C.M. Cube roots of integers: A conjecture about Heron's method in Metrika III. 20. *Hist. Math.* **2014**, *41*, 103–106. [CrossRef]

35. Eurachem. The Fitness for Purpose of Analytical Methods: A Laboratory Guide to Method Validation and Related Topics. Available online: http://www.eurachem.org (accessed on 3 March 2020).

36. Dochez, V.; Caillon, H.; Vaucel, E.; Dimet, J.; Winer, N.; Ducarme, G. Biomarkers and algorithms for diagnosis of ovarian cancer: CA125, HE4, RMI and ROMA: A review. *J. Ovarian Res.* **2019**, *12*, 28. [CrossRef] [PubMed]

Article

In Vivo Optogenetic Modulation with Simultaneous Neural Detection Using Microelectrode Array Integrated with Optical Fiber

Penghui Fan [1,2], Yilin Song [1,2], Shengwei Xu [1,2], Yuchuan Dai [1,2], Yiding Wang [1,2], Botao Lu [1,2], Jingyu Xie [1,2], Hao Wang [1,2] and Xinxia Cai [1,2,*]

[1] Aerospace Information Research Institute Chinese Academy of Sciences, Beijing 100190, China; fan_ph@163.com (P.F.); ylsong@mail.ie.ac.cn (Y.S.); swxu@mail.ie.ac.cn (S.X.); hongri1991@126.com (Y.D.); wangyiding18@mails.ucas.ac.cn (Y.W.); lslbt96@163.com (B.L.); xiejingyu16@mails.ucas.ac.cn (J.X.); wanghao161@mails.ucas.ac.cn (H.W.)

[2] University of the Chinese Academy of Sciences, Beijing 100049, China

* Correspondence: xxcai@mail.ie.ac.cn; Tel.: +86-10-5888-7193

Received: 16 July 2020; Accepted: 6 August 2020; Published: 13 August 2020

Abstract: The detection of neuroelectrophysiology while performing optogenetic modulation can provide more reliable and useful information for neural research. In this study, an optical fiber and a microelectrode array were integrated through hot-melt adhesive bonding, which combined optogenetics and electrophysiological detection technology to achieve neuromodulation and neuronal activity recording. We carried out the experiments on the activation and electrophysiological detection of infected neurons at the depth range of 900–1250 μm in the brain which covers hippocampal CA1 and a part of the upper cortical area, analyzed a possible local inhibition circuit by combining optogenetic modulation and electrophysiological characteristics and explored the effects of different optical patterns and light powers on the neuromodulation. It was found that optogenetics, combined with neural recording technology, could provide more information and ideas for neural circuit recognition. In this study, the optical stimulation with low frequency and large duty cycle induces more intense neuronal activity and larger light power induced more action potentials of neurons within a certain power range (1.032 mW–1.584 mW). The present study provided an efficient method for the detection and modulation of neurons in vivo and an effective tool to study neural circuit in the brain.

Keywords: optogenetics; micro-electrode array; in situ detection; electrophysiology; neural circuit recognition

---

1. Introduction

Optogenetic technology has been widely used in the research of behavior, diseases and other aspects of neural mechanism since it was put forward in 2005 [1], and not only on the central nervous system; it also had effects on the peripheral nervous system [2,3]. Optogenetic technology can modulate the activities of target neuron cells with high spatial-temporal resolution by combining photonics and genetic techniques and has been one of the most important neuroscience research tools [4–6]. The core of optogenetics is light-sensitive protein, also called opsins, which are mainly found in micro-organisms. The opsin gene is introduced into the subject's brain through viral vectors, and is eventually expressed in targeted cells after transcription [7,8]. The light-sensitive protein expressed in the cell membrane acts as an ion channel or pump controlled by light in a specific wavelength range. Channelrhodopsin-2 (ChR2) is one of the most commonly used opsin tools, which promotes cell depolarization when exposed to blue light.

When optogenetic technology was used to modulate neural activity, a large amount of neuronal activity needed to be recorded, mined and utilized to help us better understand the neurophysiology of the nervous system. Optogenetics combined neuronal recording can also help to identify neuronal subtypes and circuit [9]. Therefore, a tool that can simultaneously transmit light to targeted cells and detect electrophysiological information is particularly important. Microelectrode array (MEA), based on the Micro-Electro-Mechanical System (MEMS) method, is considered to be an ideal platform to detect the neural activity of a single neuron, which also has the potential of electrochemical detection. At present, the integrated device combining optical interface and MEA, also called optrode, which can be classified according to integrated optical fiber, waveguide and micro light emitting diode (μ-LED), has attracted attention [10,11]. The optrode integrating optical fiber is a kind of mature and effective optogenetic tool [12,13]. HyungDal Park formed the V-groove on the MEA by KOH wet etching process to assist the alignment of the optical fiber and electrode [14]. Wang and colleagues bonded a cannula to the open hole on the backside of the MEA to guide the fiber [15]. This method can be successfully applied to the field of optogenetics, but requires considerable time and effort to develop. The optrode based on waveguide is completed by directly depositing the waveguide material (usually SU-8) on the MEA through the MEMS technology. Since the optical waveguide only act as optical transmission medium, it is difficult to couple the light source and the optical waveguide, which usually generates a large light loss and increases cost [16–18]. The μ-LED itself is a low-power light source, and can directly stimulate tissues with minimal light loss, so it is often integrated on MEA [19,20]. Furthermore, the manufacturing technology is usually forming the μ-LED structure on silicon probe through metalorganic vapor phase epitaxy (MOVPE) or directly bonding the commercial μ-LED to the MEA. However, this method is complicated and even cause tissue damage due to overheating of the light source [21,22].

Considering the cost and operability, we produced an optrode that combines optical fiber and MEA through hot-melt adhesive. We used a homemade MEA that has been connected to Printed Circuit Board (PCB) which increased the operable space and provided a larger and harder contact area for fiber bonding. The device had the characteristics of low cost, easy operation and reusability for acute in vivo experiment. Here, we verified the performance of the integrated device, analyzed a possible circuit based on the performance of neurons under the optogenetic control and explored the effect of optical patterns and light powers on neuronal activity. This study provided an efficient tool for optogenetic neuromodulation and simultaneous in vivo electrophysiological detection.

2. Materials and Methods

2.1. Fabrication and Preparation of MEA Integrated with Optical Fiber

The Silicon-based MEA was fabricated based on MEMS technology as previously reported [23], which consisted of four 6 mm shanks. Each shank was 100 μm in width, 30 μm in thickness and 80 μm spacing. There were four electrode sites (16 μm in diameter) on the tip of every shank forming a 4×4 array. The fabrication process is shown as Figure 1a: (1) the SiO_2 layer was deposited on the silicon-on-insulator (SOI, 30 μm Si/2 μm SiO_2/450 μm Si) substrate to insulate the microelectrode from the substrate; (2) then, the Pt/Ti conductive layer pattern (including recording sites, bonding pads and conductive lines) was formed by photolithography, sputtering and lift-off process; (3) the SiO_2/Si_3N_4 insulating layer was deposited via plasma enhanced chemical vapor deposition (PECVD) and the recording sites and bonding pads were exposed by the second photolithography and CHF_3 reactive ion etching (RIE); (4) the shape of the MEA was established by the third photolithography and inductively coupled plasma deep reactive ion etching (ICP DRIE); (5) the backside silicon was wet etched and the silicon probes were released by self-stop etching in KOH solution. Then, the pads of individual probe were connected to the printed circuit board (PCB) through wire bonding and embedded in silicone rubber (Nanda 705#, Liyang Kangda Chemical Co. Ltd., Liyang, China) to isolate electrical connection. Finally, the assembled MEA was electroplated with platinum nanoparticles to improve recording performance.

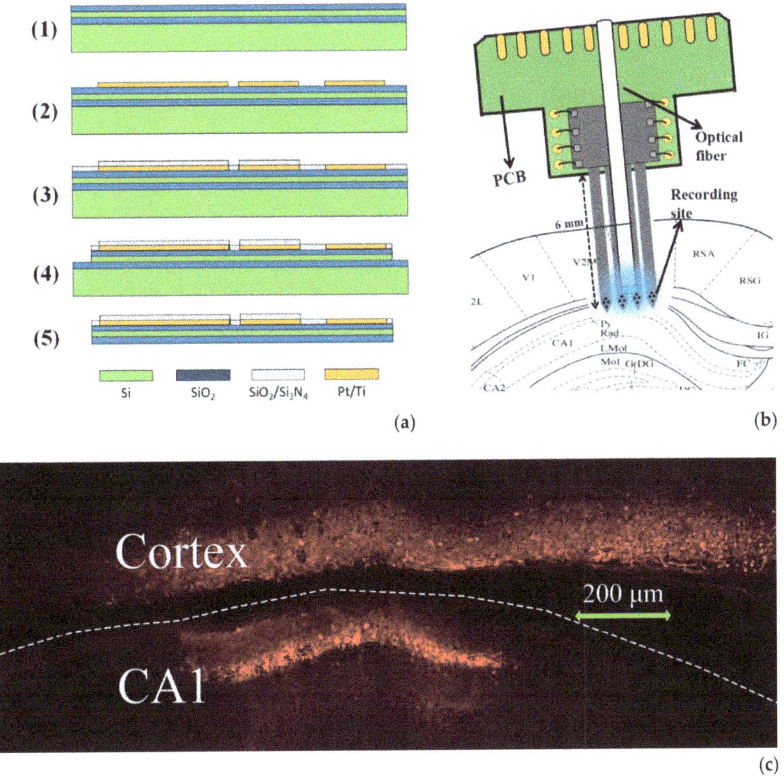

Figure 1. (**a**) Microelectrode array (MEA) manufacturing process: (1) Deposition of a layer of SiO_2 on silicon-on-insulator (SOI). (2) Photolithography, sputtering and lift-off to form conductive layer. (3) Deposition of the insulating layer and exposure of recording sites and bonding pads. (4) Formation of the shape of the MEA. (5) Release of the MEA from the substrate. (**b**) Schematic diagram of the integrated device inserted into hippocampal CA1 of brain. (**c**) The fluorescence image of brain area infected by rAAV.

In order to integrate with the optical interface, a 20 mm optical fiber (200 µm in diameter, NA = 0.39, CFMLC12L20, Thorlabs, USA) was closely attached to the PCB of the assembled MEA. Then the fiber was adjusted to be as parallel as possible with probes, and the fiber tip was about 200 µm above the recording sites. Finally, hot-melt adhesive was used to bond the optical fiber and MEA to ensure mechanical stability (Figure 1b). This method was simple in operation, short in time and low in light loss. The fiber and MEA could even be separated under heating, which undoubtedly improved their utilization rates. At the same time, the integrated device could be reused for acute experiments.

2.2. ChR2 Transduction and Optrode Implantation

The experimental procedures were conducted with the permission of Beijing Association on Laboratory Animal Care. The C57 mouse (7-week-old, 25 g) was anesthetized by the isoflurane anesthesia apparatus (RWD520, RWD Life Science, Shenzhen, China) and fixed in the stereotaxic frame (51, 600, Stenting, Wood Dale, IL, USA). The craniotomy was centered at AP = −2.3 mm and ML = 1.5 mm for targeting the mouse hippocampus. Then 0.5 µL adeno-associated viruses (rAAV-hSyn-hChR2(H134R) -mCherry-WPRE-hGH PolyA, BrainVTA, Wuhan, China) were injected into the mouse brain at a depth of 1 mm (Corresponding to the pyramidal cell layer of hippocampal CA1). After the injection, the scalp was sutured, and penicillin was applied to the wound to prevent infection. Then the mouse was placed on a soft cotton pad and returned to the rearing cage after being fully awake.

After 4 weeks, the integrated device was implanted into the target brain area at the same site for virus injection under anesthesia using a micropositioner (model 2662, David KOPF instrument, USA) for testing. The depth of electrophysiological signal recording is mainly concentrated from 800 µm to 1250 µm. The schematic diagram of device implantation is shown in Figure 1b.

2.3. Electrophysiological Recording and Optical Stimulation

For real-time electrophysiological recordings, the MEA was connected to the recordings system (USB-ME16-FAI-System, MultiChannel Systems, Reutlingen, Germany) via a 16-channel headstage. The electrophysiological signals were sampled at the rate of 25 kHz. A high pass filter (200 Hz) was used to obtain action potentials (spikes), and a low pass filter (200 Hz) was applied to obtain local field potentials (LFPs). During recording, the background noise was about ±13 µV, and the action potentials trigger threshold was set to −40 µV to ensure that the signal-to-noise ratio of the signal was greater than 3.

For optical stimulation, the light was applied by fiber-coupled LED (M450F, 450 nm, Throlabs, Newton, NJ, USA), and the optical fiber on the optrode was coupled to the light source through a ceramic mating sleeve. We used the Labview program to drive the LED light source, and change the output power of the LED by setting the driving voltage. The maximum driving voltage was 5000 mV, which corresponded to the maximum output power of the LED. The schematic diagram of the experiment is shown in Figure S1a.

2.4. Histology

Following completion of test, the mouse was injected with 10% chloral hydrate into deep anesthesia and perfused through the heart with 0.9% NaCl and 4% paraformaldehyde in sequence. After perfusion, the brain was removed and put into 20% sucrose solution for 24 h followed by 30% sucrose solution for 48 h for postfication. Then 40 µm thick sections were prepared with a freezing microtome and mounted on glass slides, and sections were viewed and photographed under the confocal fluorescence microscopy. The fluorescence image is shown as Figure 1c, due to the spread of the virus, the infection area covers the hippocampal CA1 and part of the cortex above CA1. The estimated spread of ChR2 was 400 µm in the vertical plane.

2.5. Statistics

We used a mouse in this article to verify the device performance and made repeated experiments to ensure the accuracy of the conclusion. There is enough blank time between different experiments to eliminate the interference between them. Therefore, all tests were administrated independently. Spike data were sent to Offline Sorter for sorting and clustering analyses. Spike waveforms were purified and extracted by principal component analysis (PCA). Then sorted spike data and LFP data were further analyzed by NeuroExplorer as the following:

(1) Spike firing rate analysis was defined as the mean number of spikes per second each bin. Here we took the average of multiple recording channels.
(2) Interspike interval (ISI) analysis was defined as the number of spikes each bin of the discharge interval (from 0–100 ms).
(3) Auto-correlation analysis was defined as the number of neuron firing time differences (from −50 to 50 ms) in each time window.

(4)   LFP power was defined as the integral of the mean squared LFP. Here we took the average of multiple recording channels.

Statistical analyses and graphs were administrated by Origin Software and all results were expressed as mean ± SD unless otherwise specified.

3. Results

3.1. Optical Stimulation Enhances Neuronal Activity

We conducted in vivo experiments with the integrated devices, first tested the response of neurons under optical stimulation and proved the performance of the MEA. In the region where the neurons successfully expressed ChR2, optical stimulation induced more neuronal activity. As shown in Figure 2a, the background noise of the electrode hardly changed before and during the illumination period. However, the spike firing rate significantly increased during the illumination period (Figure 2b). The LFP power also increased in the during-light period (Figure 2c). We also tested in non-viral transfection areas (as shown in Figure S2). It could be seen that the background noise had almost no change before and during the illumination, and the optical stimulation did not induce more spikes and increase lager LFP power, which proved that the application of light alone did not enhance neuronal activity. This indicated that the device successfully modulated the neuronal activity and measured the electrophysiology information, and these two experiments together illustrated the authenticity of optical stimulation-induced neuronal activity in the virus-infected area.

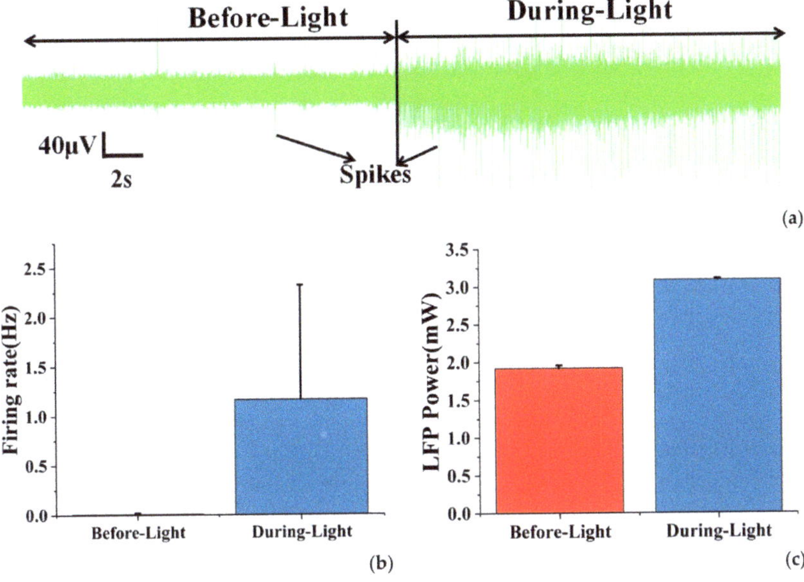

Figure 2. (**a**) The real-time recordings of electrophysiological signals before and during light illumination in viral transfection areas; (**b**) the average spike firing rate of neurons before and during optical stimulation; (**c**) the mean LFP power (0–30 Hz) of neurons before and during optical stimulation. Error bars indicate standard deviation of 3 channels.

The experiment verified that at least in the range of brain depth 900–1250 µm, optical stimulation enhanced neuronal activity, indicating the excellent performance of the device in modulation and detection. As shown in Figure 3a, spikes significantly increased during optical stimulation at different test depths. Figure 3b shows the spike waveform extracted from the recording channels, and the results indicated that the spike waveforms measured at different depths were different. Due to the different optical stimulation pattern and opsin expression level at different test depths, the neuronal activity during illumination is not the same, but their neuronal activity was all significantly enhanced compared to before light illumination.

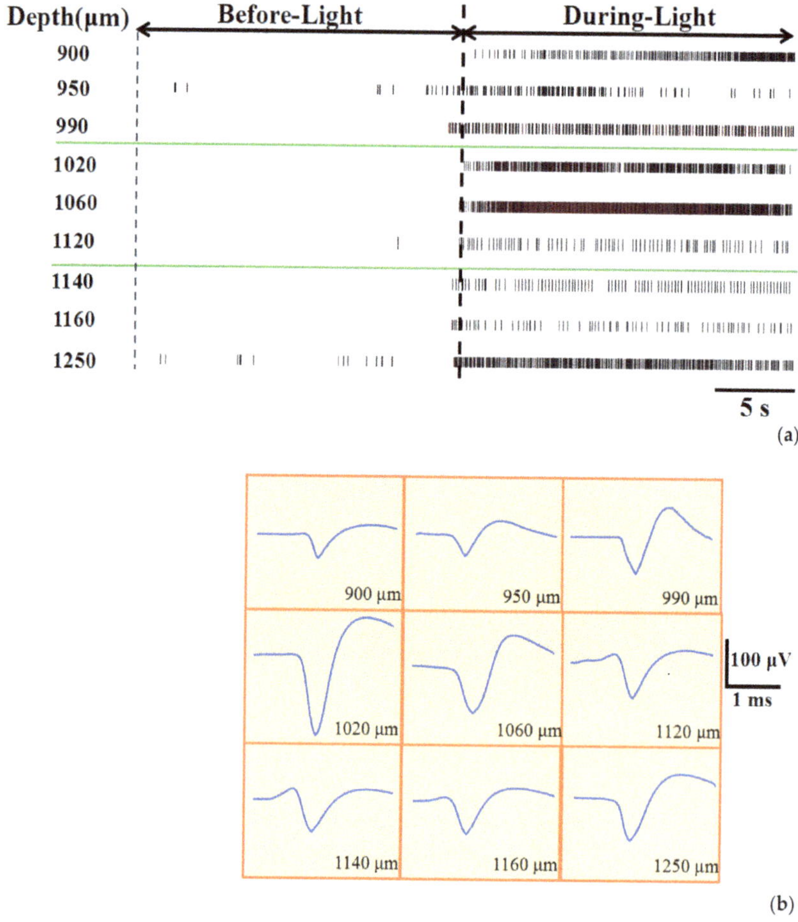

Figure 3. (a) The real-time recordings of spikes before and during blue light illumination at different depths (here we only selected one channel for each group of experiments as a display); (b) the spike waveform of neurons at different depths.

3.2. Optical Stimulation Activates a Possible Local Inhibitory Circuit

During the experiment, we attempted to analyze and study a local circuit combining optogenetics and electrophysiological recording. As shown in Figure 4a, the data was first recorded for 3 min without light, and then illumination (30 s light-on, 30 s light-off) lasting 6 min. Four neuron units were obtained (unit-a and unit-b were separated from the same recording channel) by principal component analysis and clustering from three recording channels. It can be seen from the figure that the spikes generated by unit-b,c,d increased significantly during optical stimulation. The spike firing rate of unit-b,c,d during optical stimulation was much greater than before optical stimulation (Figure 4b), which indicated that the neurons expressing ChR2 were successfully activated.

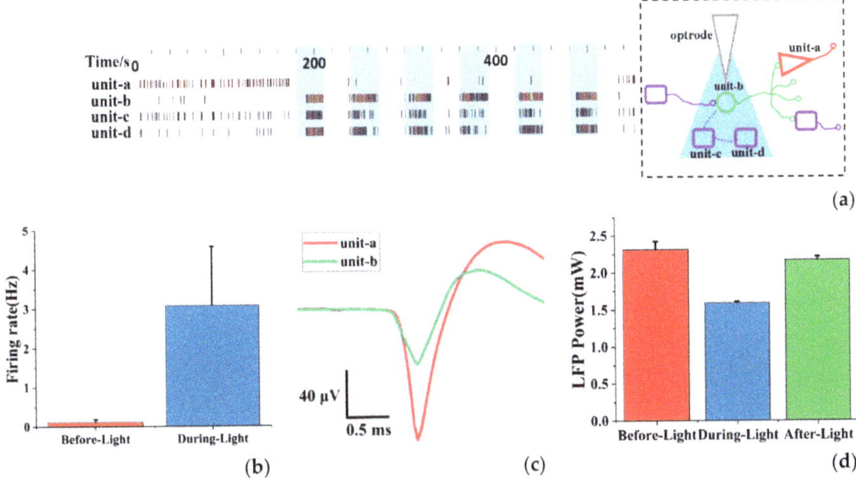

Figure 4. (**a**) The real-time recordings of spikes of 4 units separated from three recording channels (the shaded area is the period of during-light, 30sm) and inferred local circuit (the green circle represents the interneuron, the red triangle represents pyramidal neuron, the purple rectangles represent unknown type of neurons and dotted lines between neurons indicate uncertain connections); (**b**) the average spike firing rate of neurons before and during optical stimulation; (**c**) spike waveform of unit-a and unit-b; (**d**) the mean LFP power (0–8 Hz) of neurons before, during and after optical stimulation. Error bars indicate standard deviation of 3 channels and N = 6 (N is the number of experiments).

It was also found that the neuronal activity did not increase for unit-a, and even weakened during optical stimulation. In order to explore the reasons, we did a further analysis. Figure S3 shows the auto-correlograms and ISI histograms of unit-a (Figure S3a,c) and unit-b (Figure S3b,d). The ISI histogram of unit-a (yellow) showed a short and sharp peak, and its auto-correlation analysis showed a decreasing trend, indicating that it had obvious cluster-like spike characteristics. For the unit-b (green) auto-correlation analysis, several equally spaced peaks appeared, indicating that it has periodic spike characteristics under optical stimulation. In terms of spike waveform, the amplitude of unit-a was almost twice that of unit-b (Figure 4c). Combined with their respective auto-correlation analysis, we speculated that unit-a may be a putative pyramidal neuron, while unit-b may be a putative interneuron [24]. Furthermore, we speculated that they were in a local circuit (dotted box in the Figure 4a), and the direction of synaptic transmission was from interneuron to pyramidal neuron, i.e., unit-b was the upstream neuron. The types of unit-c and unit-d were unclassified and their connection with others was not clear. When optical stimulation activated unit-b interneurons, its neuronal activity increased and its inhibitory effect on unit-a also enhanced. Therefore, spikes of unit-a were reduced during light. The LFP power (0–8 Hz) during illumination was lower than before and after illumination (Figure 4d), which also seems to prove our speculation. Figure S3e showed the LFP power (0–30 Hz) before, during and after optical stimulation, which had the same trend as 0–8 Hz. Figure S3f showed the power of LFP in different frequency bands. It was found that LFP was mainly concentrated in 0–8 Hz, so the subsequent LFP analysis was mainly concentrated at 0–8 Hz.

3.3. Effects of Different Optical Stimulation Patterns on Optogenetic Neuromodulation

The spikes of the three channels in different optical stimulation patterns were recorded (see Figure S4a). According to the statistical average spike firing rate during optical stimulation, it was found that under the same duration, the spikes induced by s1 optical stimulation pattern were more than s2 optical stimulation pattern, the spikes induced by s2 optical stimulation pattern were more than s3 optical stimulation pattern (as shown in Figure 5a), the spike firing rate of the three channels had the same trend (Figure S4b). On the whole, within a certain range, the lower the light frequency and the greater the duty cycle, the stronger the neuronal activity were caused. At the same time, we calculated the LFP power (0–8 Hz) under different optical stimulation patterns. The power under s1 was greater than that of s2 and s3, but there was no significant difference between s2 and s3 (Figure 5b). This may be due to the fact that LFP depended on thousands of neurons in a wide range, and there may be inhibition or excitation circuits in this area, which had complex effects on neuronal activity. In short, we believed that different optical stimulation patterns had an effect on the intensity of induced neuronal activity.

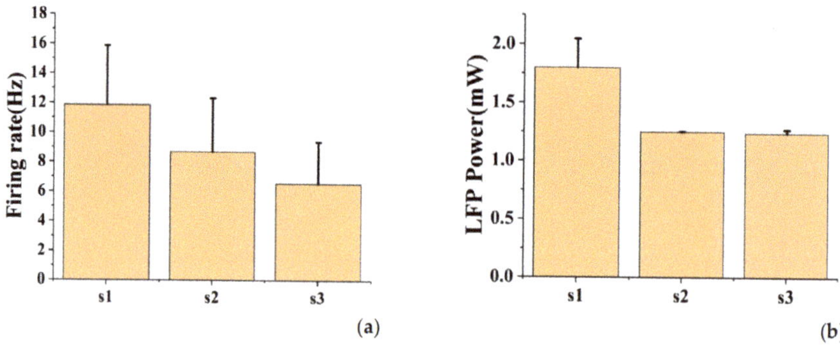

Figure 5. (**a**) The average spike firing rate of neurons under different optical stimulation patterns (s1: 10 Hz, duty ratio = 50%, 2 min; s2: 10 Hz, duty ratio = 25%, 2 min; s3: 16.6 Hz, duty ratio = 25%, 2 min); (**b**) The mean LFP power (0–8 Hz) of neurons under different optical stimulation patterns. Error bars indicate standard deviation of 3 channels.

3.4. Effects of Different Light Powers on Optogenetic Neuromodulation

The output power of LED light source (P_s) and the tip of optrode (P_t) under different driving voltages is shown in Figure S5a (unless otherwise specified, the output power refers to the output power of the optrode tip P_t). We could see the maximum tip output power was 1.7 mW. We also simulated the power density at a certain distance from the fiber tip of the integrated device [13,25,26], see supplementary materials and Figure S5b for detailed process. Since the distance between the optical fiber tip and the recording site was 200 μm, the neurons were exposed to less than 10 mW/mm^2 and the light would not cause great damage to the recorded neurons from the perspective of power density.

Then, the effect of different light power on neural activity was explored. We used 5000 mV, 4000 mV, 3000 mV, 2000 mV and 2500 mV to drive the laser one by one. The time of each stage was 2 min (after 40 s of illumination, the light was closed for 20 s, and the cycle was twice). According to the statistics of spike firing rate under different light power (Figure 6a), it was found that when the driving voltage of 2500 mV (corresponding to 1.032 mW output power) or above was used, the optical stimulation produced the effect of enhancing the activities of neurons; when the driving voltage of 2000 mV was used for optogenetic control, there was almost no obvious modulation effect. This showed that optogenetics required the light intensity to reach a certain threshold (1.032 mW in this study). At the same time, it was found that within a certain power range (1.032 mW–1.584 mW), the firing rate increased as the light power increased. When the driving voltage exceeds 4000 mV (corresponding to 1.584 mW), the firing rate decreased instead. On the one hand, it might be related to the switching property of photoreceptor itself, on the other hand, it might be due to the heat generated by light, resulting in the increase of neurons' temperature, which might have a inhibition effect on the activity of neurons. There was also no obvious relationship between LFP power and light power (Figure 6b).

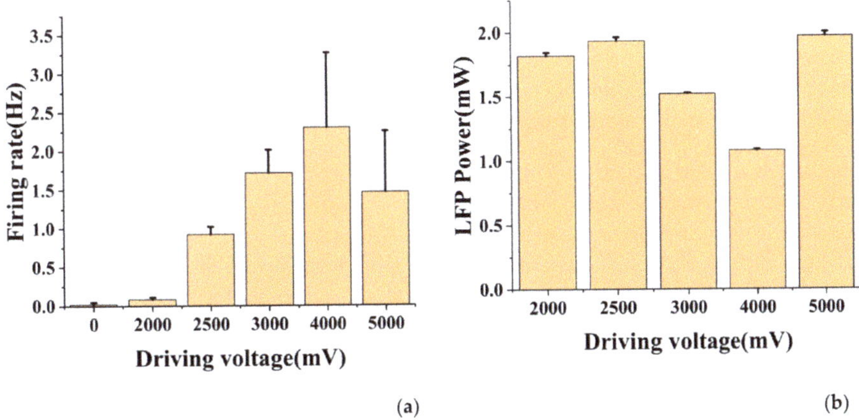

Figure 6. (**a**) The average spike firing rate of neurons under different driving voltages; (**b**) The mean LFP power (0–8 Hz) of neurons under different driving voltages. Error bars indicate standard deviation of 3 channels and N = 2 (N is the number of experiments).

4. Conclusions

In this paper, we employed a physical bonding method to integrate an optical fiber onto an MEA. This allowed us to characterize the activity of neural circuits in vivo, using optogenetic light delivery, and hence quantifying neural firing patterns. The changes of neuronal activity in a large range (depth from 900 to 1250 µm in the brain) under optogenetic control were detected and the infected neurons were successfully activated by optical stimulation, which proved the excellent performance of the integrated device. Furthermore, a possible circuit is analyzed by optogenetic modulation and electrophysiological detection, which provided a new method for circuit research. At the same time, we also studied the effect of different optical stimulation patterns and different light powers on the neuromodulation. We found that the optical stimulation patterns with lower frequency and larger duty cycle induced more intense neuronal activity. The light power needed to reach a certain response threshold (1.032 mW in this study), and greater optical power induced more intense neuronal activity within a certain power range.

In our work, the optical fibers are only used to transmit light. However, as far as we know, some new sensors based on optical fiber have great application potential in neuroscience research and attract many people's attention. For example, researchers integrate high-resolution twist/torsion optical fiber sensors to monitor eventual movements of the patients utilizing the electromagnetic interference independency (EMI) of optical fibers [27]. Furthermore, some optical fiber-based sensors also are studied for MRI (Magnetic Resonance Imaging) interventions [28]. The optical fiber force sensors can help to provide force sensing in the MRI interventions, which increases the safety or accuracy of interventions. If the optical fiber on our integrated device is combined with photo sensor and customized circuit, it will have the potential to become an optical fiber force sensor. However, there are many other issues to consider in order to make it MRI-compatible. First of all, small fiber diameter and small footprint should be used to meet the requirement of sensor miniaturization. Secondly, we should choose the materials carefully. Polymer optical fiber can be considered for better flexibility and compatibility and the electrode can be based on polyimide, which matches brain tissue well in magnetic susceptibility. Finally, due to the potential of the electric current inducing imaging artifacts, some measures should be taken to prevent the interference of electrophysiology detection on MRI. In summary, our devices have the potential to be compatible with MRI. At the same time, we can also add in vivo electrochemical analysis function of neurotransmitters to the MEA. Many events in the nervous system are accompanied by the release of neurotransmitters, so the detection of neurotransmitters is particularly important. Electrochemical analysis need to modify multilayer polymers and enzymes at the electrode sites and the measurement is carried out in the three electrodes system, which are compatible with our devices [29]. Furthermore, it is also possible to measure the concentration of different neurotransmitters on an MEA by using different polymers or enzymes. These ideas will make it possible to simultaneously detect electrophysiological signals and neurotransmitters under the neuromodulation of optogenetics, which will undoubtedly promote the development of neuroscience.

With the development of materials science, flexible electrodes have great advantages in the field of biosensors due to their close contact with brain tissue, many microelectrodes or optical devices based on flexible substrates have been studied [30,31]. The μ-LED array based on flexible substrate can accurately stimulate the brain at multiple points. Mccall and colleagues have made cellular-scale microscale, inorganic light-emitting diodes (μ-ILED, 6.45 μm thick, 50×50 μm^2), and the μ-ILED is transferred to Polyethylene terephthalate (PET, polyester film) substrate for long-term implantation [32]. The thickness of organic light-emitting diodes (OLED) can even be less than 1 μm, which is more suitable for implantation in the brain [33]. Therefore, we can combine flexible electrode and optical interface to explore long-term optogenetic modulation and multi-mode sensor integration in the future [34,35]. In a word, the method proposed in this paper was an effective and feasible way of optogenetic neuromodulation and simultaneous neural detection, which can be applied to the study of neuroscience mechanism.

Supplementary Materials: The following are available online at http://www.mdpi.com/1424-8220/20/16/4526/s1.

Author Contributions: P.F., Y.S. and X.C. conceived and designed the experiments; P.F., Y.D. and Y.W. performed the experiments; P.F., S.X., B.L., J.X. and H.W. analyzed the data; P.F., Y.S. and X.C. wrote the paper. All authors have read and agreed to the published version of the manuscript.

Funding: This research was funded by the National key Research and Development Program of nano science and technology of China [No. 2017YFA0205902], the National Natural Science Foundation of China [No. 61527815, No. 61960206012, No. 61975206, No. 61971400, No. 61973292, No. 61775216], and the Key Frontier Programs of Chinese Academy of Sciences [QYZDJ-SSW-SYS015].

Conflicts of Interest: The authors declare no conflict of interest.

References

1. Boyden, E.S.; Zhang, F.; Bamberg, E.; Nagel, G.; Deisseroth, K. Millisecond-timescale, genetically targeted optical control of neural activity. *Nat. Neurosci.* **2005**, *8*, 1263–1268. [CrossRef] [PubMed]
2. Hibberd, T.J.; Feng, J.; Luo, J.; Yang, P.; Samineni, V.K.; Gereau, R.W.t.; Kelley, N.; Hu, H.; Spencer, N.J. Optogenetic induction of colonic motility in mice. *Gastroenterology* **2018**, *155*, 514–528 e516. [CrossRef] [PubMed]
3. Spencer, N.J.; Hu, H. Enteric nervous system: Sensory transduction, neural circuits and gastrointestinal motility. *Nat. Rev. Gastroenterol. Hepatol.* **2020**, *17*, 338–351. [CrossRef]
4. Aston-Jones, G.; Deisseroth, K. Recent advances in optogenetics and pharmacogenetics. *Brain. Res.* **2013**, *1511*, 1–5. [CrossRef] [PubMed]
5. Chow, B.Y.; Boyden, E.S. Optogenetics and Translational Medicine. *Sci. Transl. Med.* **2013**, *5*, 177ps5. [CrossRef]
6. Yizhar, O.; Fenno, L.E.; Davidson, T.J.; Mogri, M.; Deisseroth, K. Optogenetics in neural systems. *Neuron* **2011**, *71*, 9–34. [CrossRef]
7. Zhang, F.; Vierock, J.; Yizhar, O.; Fenno, L.E.; Tsunoda, S.; Kianianmomeni, A.; Prigge, M.; Berndt, A.; Cushman, J.; Polle, J.; et al. The Microbial opsin family of optogenetic tools. *Cell* **2011**, *147*, 1446–1457. [CrossRef]
8. Bernstein, J.G.; Boyden, E.S. Optogenetic tools for analyzing the neural circuits of behavior. *Trends. Cogn. Sci.* **2011**, *15*, 592–600. [CrossRef]
9. Buzsaki, G.; Stark, E.; Berenyi, A.; Khodagholy, D.; Kipke, D.R.; Yoon, E.; Wise, K.D. Tools for probing local circuits: High-density silicon probes combined with optogenetics. *Neuron* **2015**, *86*, 92–105. [CrossRef]

10. Goncalves, S.B.; Ribeiro, J.F.; Silva, A.F.; Costa, R.M.; Correia, J.H. Design and manufacturing challenges of optogenetic neural interfaces: A review. *J. Neural Eng.* **2017**, *14*, 041001. [CrossRef]

11. Zhao, H. Recent progress of development of optogenetic implantable neural probes. *Int. J. Mol. Sci.* **2017**, *18*, 1751. [CrossRef]

12. Royer, S.; Zemelman, B.V.; Barbic, M.; Losonczy, A.; Buzsaki, G.; Magee, J.C. Multi-array silicon probes with integrated optical fibers: Light-assisted perturbation and recording of local neural circuits in the behaving animal. *Eur. J. Neurosci.* **2010**, *31*, 2279–2291. [CrossRef] [PubMed]

13. Stark, E.; Koos, T.; Buzsaki, G. Diode probes for spatiotemporal optical control of multiple neurons in freely moving animals. *J. Neurophysiol.* **2012**, *108*, 349–363. [CrossRef] [PubMed]

14. Park, H.; Shin, H.-J.; Cho, I.-J.; Yoon, E.-S.; Suh, J.-K.F.; Im, M.; Yoon, E.; Kim, Y.-J.; Kim, J.; IEEE. The First Neural Probe Integrated with Light Source (Blue Laser Diode) for Optical Stimulation and Electrical Recording. In Proceedings of the 2011 Annual International Conference of the IEEE Engineering in Medicine and Biology Society, Boston, MA, USA, 30 August–3 September 2011; pp. 2961–2964.

15. Wang, J.; Wagner, F.; Borton, D.A.; Zhang, J.; Ozden, I.; Burwell, R.D.; Nurmikko, A.V.; van Wagenen, R.; Diester, I.; Deisseroth, K. Integrated device for combined optical neuromodulation and electrical recording for chronic in vivo applications. *J. Neural Eng.* **2012**, *9*, 016001. [CrossRef] [PubMed]

16. Schwaerzle, M.; Seidl, K.; Schwarz, U.T.; Paul, O.; Ruther, P. Ultracompact optrode with integrated laser diode chips and su-8 waveguides for optogenetic applications. In Proceedings of the 26th IEEE International Conference on Micro Electro Mechanical Systems, Taipei, China, 20–24 January 2013; pp. 1029–1032.

17. Son, Y.; Lee, H.J.; Kim, J.; Shin, H.; Choi, N.; Lee, C.J.; Yoon, E.S.; Yoon, E.; Wise, K.D.; Kim, T.G.; et al. In vivo optical modulation of neural signals using monolithically integrated two-dimensional neural probe arrays. *Sci. Rep.* **2015**, *5*, 15466. [CrossRef]

18. Huang, W.C.; Chi, H.S.; Lee, Y.C.; Lo, Y.C.; Liu, T.C.; Chiang, M.Y.; Chen, H.Y.; Li, S.J.; Chen, Y.Y.; Chen, S.Y. Gene-embedded nanostructural biotic-abiotic optoelectrode arrays applied for synchronous brain optogenetics and neural signal recording. *ACS Appl. Mater. Interfaces* **2019**, *11*, 11270–11282. [CrossRef]

19. Cao, H.; Gu, L.; Mohanty, S.K.; Chiao, J.C. An integrated mu LED optrode for optogenetic stimulation and electrical recording. *IEEE Trans. Biomed. Eng.* **2013**, *60*, 225–229. [CrossRef]

20. Wu, F.; Stark, E.; Ku, P.C.; Wise, K.D.; Buzsaki, G.; Yoon, E. Monolithically integrated muLEDs on silicon neural probes for high-resolution optogenetic studies in behaving animals. *Neuron* **2015**, *88*, 1136–1148. [CrossRef]

21. Scharf, R.; Tsunematsu, T.; McAlinden, N.; Dawson, M.D.; Sakata, S.; Mathieson, K. Depth-specific optogenetic control in vivo with a scalable, high-density muLED neural probe. *Sci. Rep.* **2016**, *6*, 28381. [CrossRef]

22. Reddy, J.W.; Kimukin, I.; Stewart, L.T.; Ahmed, Z.; Barth, A.L.; Towe, E.; Chamanzar, M. High density, double-sided, flexible optoelectronic neural probes with Embedded muLEDs. *Front. Neurosci.* **2019**, *13*, 745. [CrossRef]

23. Fan, X.; Song, Y.; Ma, Y.; Zhang, S.; Xiao, G.; Yang, L.; Xu, H.; Zhang, D.; Cai, X. In situ real-time monitoring of glutamate and electrophysiology from cortex to hippocampus in mice based on a microelectrode array. *Sensors* **2016**, *17*, 61. [CrossRef] [PubMed]

24. Kuang, H.; Tsien, J.Z. Large-scale neural ensembles in mice: Methods for recording and data analysis. In *Electrophysiological Recording Techniques*; Vertes, R.P., Stackman, R.W., Eds.; Humana Press: Totowa, NJ, USA, 2011; Volume 54, p. 103126.

25. Aravanis, A.M.; Wang, L.-P.; Zhang, F.; Meltzer, L.A.; Mogri, M.Z.; Schneider, M.B.; Deisseroth, K. An optical neural interface:in vivocontrol of rodent motor cortex with integrated fiberoptic and optogenetic technology. *J. Neural Eng.* **2007**, *4*, S143–S156. [CrossRef] [PubMed]

26. Binding, J.; Ben Arous, J.; Leger, J.F.; Gigan, S.; Boccara, C.; Bourdieu, L. Brain refractive index measured in vivo with high-NA defocus-corrected full-field OCT and consequences for two-photon microscopy. *Opt. Express.* **2011**, *19*, 4833–4847. [CrossRef] [PubMed]

27. Zhang, X.; Chen, J.; González-Vila, Á.; Liu, F.; Liu, Y.; Li, K.; Guo, T. Twist sensor based on surface plasmon resonance excitation using two spectral combs in one tilted fiber Bragg grating. *J. Opt. Soc. Am. B* **2019**, *36*, 1176–1182. [CrossRef]

28. Su, H.; Iordachita, I.I.; Tokuda, J.; Hata, N.; Liu, X.; Seifabadi, R.; Xu, S.; Wood, B.; Fischer, G.S. Fiber optic force sensors for mri-guided interventions and rehabilitation: A Review. *IEEE Sens. J.* **2017**, *17*, 1952–1963. [CrossRef]

29. Xiao, G.; Xu, S.; Song, Y.; Zhang, Y.; Li, Z.; Gao, F.; Xie, J.; Sha, L.; Xu, Q.; Shen, Y.; et al. In situ detection of neurotransmitters and epileptiform electrophysiology activity in awake mice brains using a nanocomposites modified microelectrode array. *Sens. Actuators B Chem.* **2019**, *288*, 601–610. [CrossRef]

30. Noh, K.N.; Park, S.I.; Qazi, R.; Zou, Z.; Mickle, A.D.; Grajales-Reyes, J.G.; Jang, K.I.; Gereau, R.W.t.; Xiao, J.; Rogers, J.A.; et al. Miniaturized, battery-free optofluidic systems with potential for wireless pharmacology and optogenetics. *Small* **2018**, *14*, 1702479. [CrossRef]

31. Park, S.I.; Shin, G.; McCall, J.G.; Al-Hasani, R.; Norris, A.; Xia, L.; Brenner, D.S.; Noh, K.N.; Bang, S.Y.; Bhatti, D.L.; et al. Stretchable multichannel antennas in soft wireless optoelectronic implants for optogenetics. *Proc. Natl. Acad. Sci. USA* **2016**, *113*, E8169–E8177. [CrossRef]

32. Kim, T.I.; McCall, J.G.; Jung, Y.H.; Huang, X.; Siuda, E.R.; Li, Y.; Song, J.; Song, Y.M.; Pao, H.A.; Kim, R.H.; et al. Injectable, cellular-scale optoelectronics with applications for wireless optogenetics. *Science* **2013**, *340*, 211–216. [CrossRef]

33. Matarese, B.F.E.; Feyen, P.L.C.; de Mello, J.C.; Benfenati, F. Sub-millisecond Control of Neuronal Firing by Organic Light-Emitting Diodes. *Front. Bioeng. Biotechnol.* **2019**, *7*, 278. [CrossRef]

34. McCall, J.G.; Kim, T.I.; Shin, G.; Huang, X.; Jung, Y.H.; Al-Hasani, R.; Omenetto, F.G.; Bruchas, M.R.; Rogers, J.A. Fabrication and application of flexible, multimodal light-emitting devices for wireless optogenetics. *Nat. Protoc.* **2013**, *8*, 2413–2428. [CrossRef] [PubMed]

35. Schwaerzle, M.; Pothof, F.; Paul, O.; Ruther, P. High-Resolution Neural Depth Probe with Integrated 460 Nm Light Emitting Diode For Optogenetic Applications. In Proceedings of the 18th International Conference on Solid-State Sensors, Actuators and Microsystems, Anchorage, AK, USA, 21–25 January 2015; pp. 1774–1777.

Letter

Assessment of Gold Bio-Functionalization for Wide-Interface Biosensing Platforms

Lucia Sarcina [1], Luisa Torsi [1,2,3], Rosaria Anna Picca [1,2] **, Kyriaki Manoli [1,2,*] and Eleonora Macchia [3]**

[1]   Dipartimento di Chimica, Università degli Studi di Bari Aldo Moro, 70125 Bari, Italy; lucia.sarcina@uniba.it (L.S.); luisa.torsi@uniba.it (L.T.); rosaria.picca@uniba.it (R.A.P.)
[2]   CSGI (Centre for Colloid and Surface Science), Department of Chemistry, 70125 Bari, Italy
[3]   The Faculty of Science and Engineering, Åbo Akademi University, FI-20500 Turku, Finland; eleonora.macchia@abo.fi
*   Correspondence: kyriaki.manoli@uniba.it

Received: 28 May 2020; Accepted: 28 June 2020; Published: 30 June 2020

Abstract: The continuous improvement of the technical potential of bioelectronic devices for biosensing applications will provide clinicians with a reliable tool for biomarker quantification down to the single molecule. Eventually, physicians will be able to identify the very moment at which the illness state begins, with a terrific impact on the quality of life along with a reduction of health care expenses. However, in clinical practice, to gather enough information to formulate a diagnosis, multiple biomarkers are normally quantified from the same biological sample simultaneously. Therefore, it is critically important to translate lab-based bioelectronic devices based on electrolyte gated thin-film transistor technology into a cost-effective portable multiplexing array prototype. In this perspective, the assessment of cost-effective manufacturability represents a crucial step, with specific regard to the optimization of the bio-functionalization protocol of the transistor gate module. Hence, we have assessed, using surface plasmon resonance technique, a sustainable and reliable cost-effective process to successfully bio-functionalize a gold surface, suitable as gate electrode for wide-field bioelectronic sensors. The bio-functionalization process herein investigated allows to reduce the biorecognition element concentration to one-tenth, drastically impacting the manufacturing costs while retaining high analytical performance.

Keywords: surface plasmon resonance; biosensors; bio-functionalization optimization; cost-effective biosensors; lab-on-a-chip

1. Introduction

Single-molecule detection is a crucial task to accomplish [1,2], which could in fact allows to gather digital tracking of a biomarker from its physiologic to its pathogenic level. Such an analytical tool will enable to define the onset from a healthy to diseased patient. Early diagnostics in progressive diseases would, hence, become possible well before any symptom appears. Single-molecule biomarker detection would further allow marker quantification non-invasively, in readily available biofluids such as saliva, sweat, or even tears where they can be present at much lower concentrations. Along the same line, it would make possible ultrasensitive liquid biopsy, i.e., the assay of peripheral biofluids such as plasma, serum, or even saliva, a feasible medical procedure replacing the invasive inspection of diseased tissues. Among the single-molecule detection methods proposed so far, only a few are exploitable for real clinical sensing. This approach, addressed as wide-field sensing [2], involves the assay of a biomarker at the attomolar (aM, 10^{-18} M) or even zeptomolar (zM, 10^{-21} M) limit-of-detection with a large-area interface that is functionalized with a huge number of biorecognition elements (10^{11}–10^{12} cm^{-2}). Large-area organic-bioelectronic devices endowed with an electronic-interface capable to detect recognition element/biomarker complexes intrinsic properties, such as the electrostatic or dielectric ones, are emerging as a powerful tool capable of selective, label-free, and fast biomarker detection at the physical limit in real biofluids. Bioelectronic thin-film transistors (TFTs) [3–5], gated via an ionically-conducting and electronically-insulating electrolyte [6–8], are generating a lot of interest as they can potentially be produced by scalable large-area low-cost approaches. Such sensors [9,10] are capable of high selectivity via the bio-functionalization of the organic semiconductors (OSCs) [10] or the gate metal surface [11]. Electrolyte-gated TFTs (EG-TFTs) have been successfully engaged, lately, as wide-field bioelectronic sensors exhibiting limits of detection at zM - aM level also in real bio-fluids [12,13]. In the Single-Molecule with a large-Transistor (SiMoT) platform, based on an EG-TFT device [12], the gate is bio-functionalized [14] with 10^{12} cm^{-2} recognition elements covalently attached to a 0.5 cm^2 gold gate. The SiMoT platform has been demonstrated to successfully perform label-free detection at the physical limit of immunoglobulin-M [13], C-reactive protein in saliva [15], and HIV-p24 [16,17], as well as genomic biomarkers [18]. However, in clinical practice, to gather enough information to formulate a diagnosis, multiple biomarkers are normally quantified from the same biological sample. Therefore, it is necessary to develop a technology capable to perform multiplexing [19] and thus realized with an array of 96 or more transducing elements, so that the standard solutions, the negative-controls, and the sample can be assayed, with all the replicates for each biomarker, at the same time. All this calls for translating the lab-based SiMoT device into a cost-effective portable multiplexing array prototype that integrates, with a modular approach, novel materials, and standard components/interfaces. In this perspective, the assessment of cost-effective manufacturability represents a crucial step, with specific regard to the optimization of the bio-functionalization protocol of the gate modules. Therefore, in this study, we have developed, with the aid of surface plasmon resonance (SPR) technique, a novel and reliable cost-effective process to successfully bio-functionalize a gold surface suitable as gate electrode in a SiMoT-based platform. An uncommon approach has been chosen in the SPR settings by using a method that can be easily translated to bioelectronic device surface modifications. In particular, the bio-functionalization process herein proposed allows to reduce to one-tenth the concentration of the biorecognition elements. This can drastically reduce the manufacturing costs without sacrificing the sensing performance in terms of sensitivity. In general, for a biomolecular reaction to occur, the two reagents need to be confined in an adequately small volume for a sufficiently long time. The recognition element can be attached on a surface that serves as detecting interface. Regardless, the interaction cross-section of the two reagents has to be reasonably high. In this respect, a volume of 1 μm^3 (1 femtoliter - fL) has been proven sufficiently small for a single enzyme to interact with its substrate (present in excess concentration though) on the minute time-scale [20,21]. Indeed, a solution comprising n = 1 ± 1 ($\sqrt{n}$ = Poisson error) molecules in each 1fL sub-volumes has a concentration of ~1 × 10^{-9} mol × l^{-1} (nM). Smaller volumes (attoliter, aL, or zeptoliter, zL), each occupied by a single molecule, entails even larger concentrations. Since the number of molecules in a volume V = 100 μL of a solution of molar concentration [c] is n = [c]·V·N$_A$ (N$_A$ = Avogadro's number), 1 nM equals ~10^{11} molecules or, equivalently, ~10^{11} 1 fL sub-volumes. As two molecules need to be confined in a volume of 1 fL or smaller to rapidly interact, at least one of them is to be present at a concentration of 1 nM. Every 1 fL (or lower) statistically contains one reagent, so wherever the other single reagent is, there is always one fL sub-volume comprising both reagents. A single-molecule interaction can, therefore, occur when the recognition-elements are present at nM concentration (or higher) along with a single biomarker or the opposite way around. In clinical assays, the former is to be preferred. Therefore, the aim of the study herein presented is to determine which is the minimum biorecognition element concentration requested to achieve a sufficiently high surface coverage. Interestingly, it has been demonstrated that it is possible to reduce to one tenth the biorecognition elements concentration without impacting on the analytical performances, thus optimizing the biofunctionalization protocol of the SiMBiT platform as well as of other bioelectronic devices [22–24]. This study paves the way toward a multiplexing single molecule technology that will open to a massive use of high-throughput array-based assay not only in clinical laboratory analysis but also in point-of-care and low resources settings.

2. Materials and Methods

2.1. Materials

3-mercaptopropionic acid (3-MPA), 11-mercaptoundecanoic acid (11-MUA), 1-ethyl-3-(3-dimethylaminopropyl)-carbodiimide (EDC), N-hydroxysulfosuccinimide sodium salt (NHSS), ethanolamine hydrochloride (EA), and 66 kDa molecular weight bovine serum albumin (BSA) were purchased from Sigma–Aldrich and used with no further purification. HPLC-grade water, ethanol grade puriss. p.a. assay, $\geq$99.8%, ammonium hydroxide solution (NH_4OH) 28.0–30.0%, hydrogen peroxide (H_2O_2) 30% (w/w) in H_2O were purchased from Sigma–Aldrich and used with no further purification. Anti-human immunoglobulin M (anti-IgM) produced in goat polyclonal antibodies was purchased from Sigma–Aldrich and used with no further purification. Human IgM (~950 kDa) affinity ligand, purchased from Sigma–Aldrich, was isolated from pooled normal human serum and used with no further purification. Phosphate buffered saline (PBS pH 7.4, Sigma-Aldrich- Merck KGaA, Darmstadt, Germany) solution was prepared according to previous works [12]. 2-(N-morpholino)ethane-sulfonic acid (MES) buffer (Sigma-Aldrich- Merck KGaA, Darmstadt, Germany) 0.1 M was adjusted with sodium hydroxide solution (NaOH 1 M) at pH 4.8–4.9.

2.2. Preparation of Mixed Self-Assembled Monolayers

The sensor slides (SPR Navi-200), comprising a 50 nm gold on glass, were cleaned in a freshly prepared "basic piranha" $NH_4OH/H_2O_2/H_2O$ solution (1:1:5 v/v) at c.a. 80–90 °C for 10 min. The slides were rinsed with HPLC water, dried with N_2, and then treated for 10 min in ozone cleaner. For the assembly of the chemical SAM (chem-SAM) on the gold surface, the slides were immediately immersed in a 10 mM thiol solution of 11-MUA: 3-MPA (1:10 molar ratio) in degassed ethanol. The slides were kept in contact with the mixed thiol solution for 18 h at 22 °C under a nitrogen atmosphere in the dark. Afterward, the samples were rinsed with ethanol and mounted in the SPR apparatus, drying the glass back-surface of the chip with N_2.

2.3. Surface Plasmon Resonance Real-Time Functionalization

A BioNavis Multi-parameter Surface plasmon resonance (MP-SPR) Navi^{TM} instrument, in the Kretschmann configuration, was used. The SPR instrument, equipped with two laser sources (670 and 785 nm wavelengths) was used to study the gold surface bio-functionalization in situ, by using the 670 nm source for both sampling areas. A wide angular range (50–78 °C) was measured, with real angular resolution of 0.001 °C. The variation of the plasmon peak angular response was monitored over time. All the experiments were performed at 22 °C in a one-channel cell in which the solutions were manually injected and kept in static conditions. This configuration allows the exposure of a gold area of c.a. 0.42 cm^2 to be functionalized. The strong gold-sulfur interaction results in the exposure of the carboxylic groups of thiols anchored to the sensor surface. For the immobilization of the biorecognition element, the COOH activation can be performed by the well-known EDC/NHSS chemistry [25,26]. Two sensor slides modified with the chem-SAM were tested by using two different concentrations for the functionalizing antibody. In the first protocol, addressed as protocol A, the modified slide was exposed first to HPLC water to acquire a stable SPR response as baseline. Then, 1 mL of an EDC (200 mM) and NHSS (50 mM) aqueous solution was injected through the cell (internal volume 100 μL, plus 100 μL capillary tubing) and left in contact for 2 h. The surface was subsequently rinsed first with H_2O and then PBS to inject 500 μL of a 100 μg/mL anti-IgM solution in PBS. The antibody was left in contact until a complete bio-conjugation was achieved, i.e., a plateau response was observed in the sensogram (SPR angle vs time). Then, the sensor slide was rinsed thoroughly with PBS to remove unbound antibodies. This preconcentration step was followed by the injection of 1 M ethanolamine

PBS solution (EA) for 45 min. The bio-functionalization was completed with the injection of 0.1 mg/mL bovine serum albumin (BSA) solution in PBS, to gain a more compact SAM, less prone to the nonspecific adsorption [27]. The modified sensor surface was then exposed to an IgM solution in PBS (50 nM) for testing the antibody binding efficacy. Protocol A has been described in detail elsewhere [14]. In the second protocol, addressed as protocol B and adapted from Reference [28] (Figure 2c), the baseline was set by injecting, instead of water, MES buffer through the chem-SAM modified surface. The solution of EDC/NHSS, prepared also in MES, was then injected and left in the SPR cell for 15 min. After the surface was washed with MES, PBS was injected to acquire a new baseline. At this stage, the activated chem-SAM gold surface was exposed to 500 μL of 10 μg/mL anti-IgM solution, and the conjugation was monitored in the sensogram, until a plateau was observed. The succeeding steps involving EA and BSA were performed as in protocol A. Finally, the IgM solution in PBS was injected through the functionalized sensor surface, recording the immunoglobulin binding response.

3. Results and Discussion

The amount of anti-IgM capturing proteins immobilized on the gate with the two bio-functionalization protocols A and B was estimated by measuring the SPR shift Δθ occurring when the anti-IgM capturing proteins are conjugated to the activated chem-SAM. The final capturing SAM segregated on the SPR slide is schematically depicted in Figure 1. It is well known that SPR technique ensures the direct correlation of plasmon peak shift with the thickness and optical properties of the medium that contacts the metal surface [29]. The activated carboxylic groups of SAM thiols on the modified gold surface can stably bind antibody primary amines. Thus, when antibodies approach the sensor slide, a new layer can be created at the surface and an increase of the detected angle in the sensogram is observed [30]. The efficacy of each step in the functionalization process can be verified by the SPR real-time monitoring for both the investigated protocols.

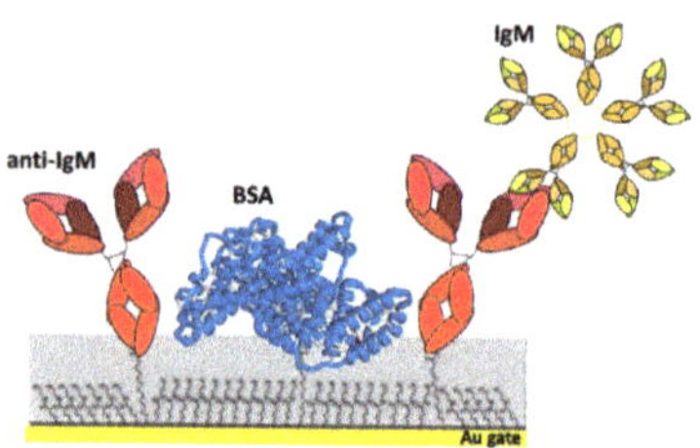

Figure 1. Capturing SAM, comprising both a chem-SAM of activated-and-blocked 3-mercaptopropionic acid (3-MPA) and 11-mercaptoundecanoic acid (11-MUA) and a bio-SAM of capturing antibodies.

The experimental trend for the immobilization of anti-IgM loaded at different concentrations is shown in Figure 2. In protocol A, shown in Figure 2a,b, a saturating trend for the preconcentration of anti-IgM is observed and completed soon after 60 min from the exposure to the solution. The high provision of ligand to the surface entails the typical fast increase of SPR angle within the first minutes of incubation [31]. Then, a slower rate of binding occurs until all the available sites on the surface are covered and an equilibrium state is reached. The injection of the PBS buffer (crossed arrow) removes excess of anti-IgM not anchored to the chem-SAM. A successful binding can be confirmed, as no significant decrease of the signal is registered after the PBS injection. The receptor conjugation is completed by injecting the EA solution, after which the unreacted carboxylic groups are deactivated and the electrostatically bound antibodies are washed over [32].

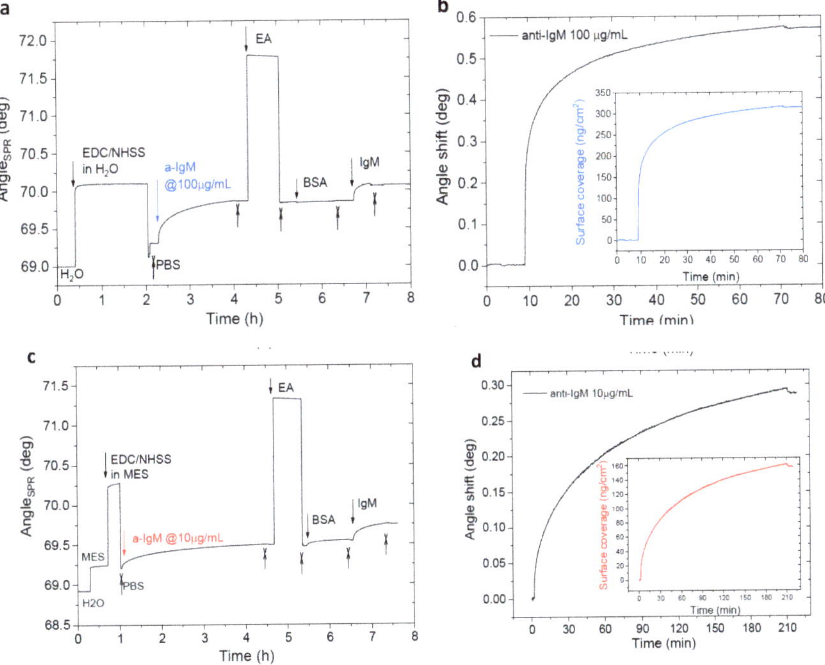

Figure 2. SPR real-time functionalization in which top-down arrows refer to injected solutions and reverse crossed-arrows to phosphate buffered saline (PBS) rinsing steps. (**a**) Sensogram for the immobilization of anti-IgM on the gold surface pre-modified with mixed SAM. (**b**) Zoom of the angular response (angle shift vs time) for anti-IgM exposure at 100 µg/mL and inset showing the corresponding surface coverage (ng/cm^2). (**c**) Sensogram for anti-IgM immobilization performed with reduced antibody concentration and (**d**) zoom in the 10 µg/mL anti-IgM preconcentration step with surface coverage (ng/cm^2) in the inset.

The SPR signal after the rinsing of EA will be related to the amount of effectively bound receptors as reported in Table 1. This angular shift does not differ significantly from the one recorded in the preconcentration. Moreover, the final step involving BSA does not produce significant changes in the angular response, thus a partial insertion of this blocking agent in the well-packed biolayer can be assumed [12,27].

Table 1. Surface plasmon resonance (SPR) response as angle shift ($\Delta\theta$) recorded for the anchored anti-human immunoglobulin M (IgM) and IgM exposure in nM range for both protocols. Calculated surface coverage expressed in ng/cm^2 and number of immobilized molecules for cm^2 surface area.

		SPR $\Delta\theta$ (°)	*SC* Γ (ng/cm^2)	*SC (particles/cm^2)*
Protocol A	*Anti-IgM* [#] *100 µg/mL*	0.53	294	1.2×10^{12}
	IgM [##]	0.23	127	8.0×10^{10}
Protocol B	*Anti-IgM* [#] *10 µg/mL*	0.27	146	5.9×10^{11}
	IgM [##]	0.21	116	7.3×10^{10}

*SC: Surface coverage; # anti IgM MW 150 kDa; ## IgM MW 950 kDa.

The sensogram recorded for immobilization protocol B is reported in Figure 2c, with a zoom on the injection of anti-IgM 10 µg/mL in Figure 2d. Independently of the used solvent, the optimization of the procedure can be settled first with a decrease in the reaction time of EDC/NHSS solution, without substantial consequences on the activation efficacy. Indeed, the EDC action is completed within 20 min, so that an extended reaction will only result in a longer functionalization time with no enhancement in the subsequent binding rate of the receptor [33]. Moreover, the control over the pH obtained by means of MES buffer for the activating solution (pH ~5) leads to a more efficient reaction [34]. In a two-step reaction, to gain best results, the first activation step (i.e., EDC/NHSS) should be performed in a MES buffer at pH 5–6, then the pH should be raised to 7.2–7.5 with a phosphate buffer for a more effective reaction with the amine-containing groups of the antibody (anti-IgM) [35]. The pH plays an important role during the immobilization since the ligand is uncharged or positively charged in the preconcentration process to promote an electrostatic attraction between the amino group of the antibody and the negatively charged SAM surface. A pH 0.5–1 units below the isoelectric point of the ligand (typically ~ 8) is needed while preserving the negative charge on the sensor surface keeping the pH above 4 [36–38]. Optimal reaction conditions have been chosen in the present work according to the results already reported for other receptors [28].

As soon as the activation solution is washed away by the buffer, the ligand solution in PBS is injected into the SPR cell.

The main observed difference, related to the usage of a more diluted solution, relies on the time required to reach the steady state. Indeed, the first 30 min give a slower increase of the signal (Figure 2d) if compared to protocol A (Figure 2b). In static conditions, the molecule replenishment to the surface will be mostly related to their concentration in the bulk, since there is no flow that opposes the depletion of ligands near the surface [39,40]. Hence, for reaching the equilibrium state at a lower concentration, the time needed for the incubation of anti-IgM is at least 3.5 h. Once a plateau in the antibody binding signal is reached, the buffer is slotted over the cell. After rinsing, the signal does not drop-down thus the binding can be considered effective and the functionalization can be completed through the injection of EA and BSA afterward. A slight increase of the SPR angle is observed after BSA step for the second protocol (c.a. 0.07 °C shift from anti-IgM level). For this different behavior, major adsorption of BSA on the SAM can be assumed considering that the amount of the anti-IgM immobilized is half of that obtained in the first protocol. This means that the anchored receptors are more spaced and BSA can stabilize them by prominent steric hindrance, producing an appreciable signal onto the sensor surface.

The assessment of sensor surface modification is fundamental when developing a biolayer-based device since the properties of the biorecognition element regarding its orientation, surface density, and activity toward the binding analyte can dramatically influence the assay analytical performances [41,42]. In SPR direct assays, the typical explored analyte concentration returns limit of detections in the 1–10 nM range [43,44], with further improvement only by using nanostructures or sandwich assays [45]. Hence, for testing the response of the functionalized surface at saturating concentrations, in this study, the sensor slides for both protocols were exposed to a standard solution of IgM at nominal concentration of 50 nM (10^{-9} M), recording the corresponding SPR angle variation. The response obtained as angle shift and the equivalent surface density is compared in Table 1. This evaluation is done according to the literature [46], which states the relation between surface coverage (Γ, in ng/cm^2) and the plasmon resonance shift by means of the Feijter's equation: $\Gamma = (n_a - n_b)d_a \cdot (dn/dC)^{-1}$ [47]. Here, the coverage is related to the difference in refractive index between the antibody layer (n_a) and the bulk solution (n_b), the average thickness (d_a) of bounded species, and the refractive index increment (dn/dC). The difference in average refractive index corresponds to $(n_a - n_b) = \Delta\theta \cdot k$, where $\Delta\theta$ is the measured angular shift and k is the wavelength dependent sensitivity coefficient. For thin layer beyond the evanescent field depth (less than 200 nm) at a source wavelength of 670 nm, the equation can be simplified: the product ($k \cdot d_a$) is approximated to 1.0×10^{-7} cm/deg and dn/dc to 0.182 cm^3/g. Thus, the equation becomes $\Gamma = \Delta\theta \cdot 550$ (ng/cm^2) [12,14]. The conversion into a surface coverage expressed in number of molecules per cm^2 can be performed by considering the molecular weight of the species under investigation. At a fixed analyte concentration, by comparing the total bound IgM obtained with the two protocols, an analogous result is achieved (Table 1). Knowing the average response at this saturating analyte concentration for the same sensing platform used in protocol A [13,14], the reduction to one-tenth of biorecognition elements concentration still produces comparable analytical performances, as can be stated considering the response obtained with protocol B. Although the number of anti-IgM available on the surface is slightly lower (5.9×10^{11} particles/cm^2) with this latter protocol, the available binding sites are indeed enough to measure the same response of protocol A. By fitting the exponential growth relative to the analyte binding, a plateau was obtained correspondent to: $\Delta\theta = (0.238 \pm 0.005)$ °C for the exposure in the anti-IgM (100 µg/mL) modified SAM of protocol A, and $\Delta\theta = (0.227 \pm 0.002)$ °C for the anti-IgM (10 µg/mL) modified SAM of protocol B. Although the response is comparable in both protocols, one should notice that the 90% of the signal can be reached for protocol A soon after 16 min, meanwhile, 43 min are necessary for protocol B. This highlights the importance of selecting the correct time of analyte exposure within the assay for getting results not affected by time-dependent signals. Interestingly, protocol B leads to an improvement in the functionalization process, drastically reducing the concentration of capturing antibodies in the preconcentration solution, without affecting the assay readout. Moreover, the observed angular shifts for the antibody immobilization is consistent with values already reported in literature [48–50]. The main goal in a biosensing platform based on a wide-field approach is that of having a high density of receptor on the sensor surface, for the investigation of extremely low analyte concentrations [2]. These working conditions are hardly tied with low consumption of costly reagents, especially if compared with other miniaturized sensing techniques [42]. As SPR has been largely employed as surface-sensitive technique, many standardized protocols are reported in literature [51–53]. The usage of a SAM modified surface allows a controlled antibody binding, and the consequent usage of EA and BSA is well-established [26,54]. However, they were mostly focused on the flow-system facility, which enables the consumption of low volume of reagents while continuous replenishment of molecules to the surface. However, the applicability of these protocols to other sensing methods, such as wide-interface bioelectronics, is not straightforward. The assessment of a robust protocol by means of the SPR apparatus in non-conventional ways is presented here. The authors suggest an improvement in the fabrication of the sensor bio-active surface already tested for an immunoassay application, in the SiMoT device described elsewhere [55,56]. The SPR experimental conditions (i.e., solution volume loaded, the gold area exposed, and the manual injections) have been set to be feasible also for the SiMoT, as well as further bioelectronic platforms. After testing the proposed protocol for the bio-functionalization with a real-time technique like SPR, the modification of the gold electrode used in the electronic sensor can be optimized accordingly.

4. Conclusions

In conclusion, a modified bio-functionalization protocol of a gold surface is proposed and compared with the one previously adopted for gold gate electrodes in the ultrasensitive SiMoT biosensing platform. The two protocols are compared using SPR technique. The amount of anti-IgM capturing proteins immobilized on the mm^2 gate area with the two protocols was estimated by measuring the SPR angle shift $\Delta\theta$ when the anti-IgM capturing proteins are conjugated in real-time to the activated chem-SAM. It was shown that the amount of the capturing antibodies can be reduced at least ten times, up to 10 µg/mL, without affecting the assay analytical performance. The new protocol allows better control of the pH during the different steps of bio-functionalization and reduces the cost of the process, therefore, the cost of the SiMoT platform production. Indeed, this study sets the ground for the assessment of scalable manufacturability of biosensing platform based on EG-TFT devices such as multiplexing single-molecule technologies.

Author Contributions: Conceptualization, E.M., L.T. and K.M.; methodology, R.A.P, K.M. and E.M.; formal analysis, E.M., L.S.; investigation, E.M., L.S., R.A.P.; data curation, L.S., K.M. and E.M.; writing—original draft preparation, E.M. and L.S.; writing—review and editing, L.S., L.T., R.A.P. and K.M.; supervision, E.M. and K.M.; resources and funding acquisition, L.T. All authors have read and agreed to the published version of the manuscript.

Funding: This research was partially funded by H2020—Electronic Smart Systems—SiMBiT: Single molecule bio-electronic smart system array for clinical testing (Grant agreement ID: 824946), Italian MIUR PON project "PMGB -Sviluppo di piattaforme meccatroniche, genomiche e bioinformatiche per l'oncologia di precisione"—ARS01_01195—Dottorati innovativi con caratterizzazione industriale—PON R&I 2014–2020 "Sensore bio-elettronico usa-e-getta per l'HIV autoalimentato da una cella a combustibile biologica" (BioElSens&Fuel), and CSGI.

Acknowledgments: The authors would like to thank Ronald Österbacka and Davide Blasi for useful discussions.

Conflicts of Interest: The authors declare no conflict of interest.

References

1. Jain, A.; Liu, R.; Ramani, B.; Arauz, E.; Ishitsuka, Y.; Ragunathan, K.; Park, J.; Chen, J.; Xiang, Y.K.; Ha, T. Probing cellular protein complexes using single-molecule pull-down. *Nature* **2011**, *473*, 484–488. [CrossRef]
2. Gooding, J.J.; Gaus, K. Single-molecule sensors: Challenges and opportunities for quantitative analysis. *Angew. Chem. Int. Ed.* **2016**, *55*, 11354–11366. [CrossRef]
3. Torsi, L.; Farinola, G.M.; Marinelli, F.; Tanese, M.C.; Omar, O.H.; Valli, L.; Babudri, F.; Palmisano, F.; Zambonin, P.G.; Naso, F. A sensitivity-enhanced field-effect chiral sensor. *Nat. Mater.* **2008**, *7*, 412–417. [CrossRef] [PubMed]
4. Gualandi, I.; Marzocchi, M.; Achilli, A.; Cavedale, D.; Bonfiglio, A.; Fraboni, B. Textile organic electrochemical transistors as a platform for wearable biosensors. *Sci. Rep.* **2016**, *6*, 33637. [CrossRef] [PubMed]
5. Piro, B.; Wang, D.; Benaoudia, D.; Tibaldi, A.; Anquetin, G.; Noël, V.; Reisberg, S.; Mattana, G.; Jackson, B. Versatile transduction scheme based on electrolyte-gated organic field-effect transistor used as immunoassay readout system. *Biosens. Bioelectron.* **2017**, *92*, 215–220. [CrossRef] [PubMed]
6. Berggren, M.; Richter-Dahlfors, A. Organic bioelectronics. *Adv. Mater.* **2007**, *19*, 3201–3213. [CrossRef]
7. Kergoat, L.; Herlogsson, L.; Braga, D.; Piro, B.; Pham, M.-C.; Crispin, X.; Berggren, M.; Horowitz, G. A water-gate organic field-effect transistor. *Adv. Mater.* **2010**, *22*, 2565–2569. [CrossRef] [PubMed]
8. Pettersson, F.; Remonen, T.; Adekanye, D.; Zhang, Y.; Wilén, C.-E.; Österbacka, R. Environmentally friendly transistors and circuits on paper. *Chem. Phys. Chem.* **2015**, *16*, 1286–1294. [CrossRef] [PubMed]
9. Mulla, M.Y.; Tuccori, E.; Magliulo, M.; Lattanzi, G.; Palazzo, G.; Persaud, K.; Torsi, L. Capacitance-modulated transistor detects odorant binding protein chiral interactions. *Nat. Commun.* **2015**, *6*, 6010. [CrossRef]
10. Manoli, K.; Magliulo, M.; Mulla, M.Y.; Singh, M.; Sabbatini, L.; Palazzo, G.; Torsi, L. Printable bioelectronics to investigate functional biological interfaces. *Angew. Chem. Int. Ed.* **2015**, *54*, 12562–12576. [CrossRef]
11. Palazzo, G.; De Tullio, D.; Magliulo, M.; Mallardi, A.; Intranuovo, F.; Mulla, M.Y.; Favia, P.; Vikholm-Lundin, I.; Torsi, L. Detection beyond Debye's length with an electrolyte-gated organic field-effect transistor. *Adv. Mater.* **2015**, *27*, 911–916. [CrossRef] [PubMed]
12. Macchia, E.; Manoli, K.; Holzer, B.; Di Franco, C.; Ghittorelli, M.; Torricelli, F.; Alberga, D.; Mangiatordi, G.F.; Palazzo, G.; Scamarcio, G.; et al. Single-molecule detection with a millimetre-sized transistor. *Nat. Commun.* **2018**, *9*, 3223. [CrossRef] [PubMed]
13. Macchia, E.; Tiwari, A.; Manoli, K.; Holzer, B.; Ditaranto, N.; Picca, R.A.; Cioffi, N.; Di Franco, C.; Scamarcio, G.; Palazzo, G.; et al. Label-Free and Selective Single-Molecule Bioelectronic Sensing with a Millimeter-Wide Self-Assembled Monolayer of Anti-Immunoglobulins. *Chem. Mater.* **2019**, *31*, 6476–6483. [CrossRef]
14. Holzer, B.; Manoli, K.; Ditaranto, N.; Macchia, E.; Tiwari, A.; Di Franco, C.; Scamarcio, G.; Palazzo, G.; Torsi, L. Characterization of covalently bound anti-human immunoglobulins on self-assembled monolayer modified gold electrodes. *Adv. Biosyst.* **2017**, *1*, 1700055. [CrossRef]
15. Macchia, E.; Manoli, K.; Holzer, B.; Di Franco, C.; Picca, R.A.; Cioffi, N.; Scamarcio, G.; Palazzo, G.; Torsi, L. Selective single-molecule analytical detection of C-reactive protein in saliva with an organic transistor. *Anal. Bioanal. Chem.* **2019**, *411*, 4899–4908. [CrossRef] [PubMed]
16. Sailapu, S.K.; Macchia, E.; Merino-Jimenez, I.; Esquivel, J.P.; Sarcina, L.; Scamarcio, G.; Minteer, S.D.; Torsi, L.; Sabaté, N. Standalone operation of an EGOFET for ultra-sensitive detection of HIV. *Biosens. Bioelectron.* **2020**, *156*, 112103. [CrossRef]
17. Macchia, E.; Sarcina, L.; Picca, R.A.; Manoli, K.; Di Franco, C.; Scamarcio, G.; Torsi, L. Ultra-low HIV-1 p24 detection limits with a bioelectronic sensor. *Anal. Bioanal. Chem.* **2020**, *412*, 811–818. [CrossRef]

18. Macchia, E.; Manoli, K.; Di Franco, C.; Picca, R.; Osterbacka, R.; Palazzo, G.; Torricelli, F.; Scamarcio, G.; Torsi, L. Organic field-effect transistor platform for label-free single-molecule detection of genomic biomarkers. *ACS Sens.* **2020**, *5*, 1822–1830. [CrossRef]

19. He, X.P.; Hu, X.L.; James, T.D.; Yoon, J.; Tian, H. Multiplexed photoluminescent sensors: Towards improved disease diagnostics. *Chem. Soc. Rev.* **2017**, *46*, 6687–6696. [CrossRef]

20. Iino, R.; Lam, L.; Tabata, K.V.; Rondelez, Y.; Noji, H. Single-molecule assay of biological reaction in femtoliter chamber array. *Jpn. J. Appl. Phys.* **2009**, *48*, 6–11. [CrossRef]

21. Engvall, E.; Jonsson, K.; Perlmann, P. Antiserum, immunoglobulin fraction, and specific antibodies Iodinated proteins. *Biochim. Biophys. Acta.* **1971**, *251*, 427–434. [CrossRef]

22. Casalini, S.; Leonardi, F.; Cramer, T.; Biscarini, F. Organic field-effect transistor for label-free dopamine sensing. *Org. Electron.* **2013**, *14*, 156–163. [CrossRef]

23. Berto, M.; Diacci, C.; Theuer, L.; Di Lauro, M.; Simon, D.T.; Berggren, M.; Biscarini, F.; Beni, V.; Bortolotti, C.A. Label free urea biosensor based on organic electrochemical transistors. *Flex. Print. Electron.* **2018**, *3*, 24001. [CrossRef]

24. Yang, S.Y.; DeFranco, J.A.; Sylvester, Y.A.; Gobert, T.J.; Macaya, D.J.; Owens, R.M.; Malliaras, G.G. Integration of a surface-directed microfluidic system with an organic electrochemical transistor array for multi-analyte biosensors. *Lab Chip* **2009**, *9*, 704–708. [CrossRef] [PubMed]

25. Patel, N.; Davies, M.C.; Hartshorne, M.; Heaton, R.J.; Roberts, C.J.; Tendler, S.J.B.; Williams, P.M. Immobilization of protein molecules onto homogeneous and mixed carboxylate-terminated self-assembled monolayers. *Langmuir* **1997**, *13*, 6485–6490. [CrossRef]

26. Lee, J.W.; Sim, S.J.; Cho, S.M.; Lee, J. Characterization of a self-assembled monolayer of thiol on a gold surface and the fabrication of a biosensor chip based on surface plasmon resonance for detecting anti-GAD antibody. *Biosens. Bioelectron.* **2005**, *20*, 1422–1427. [CrossRef] [PubMed]

27. Hideshima, S.; Sato, R.; Inoue, S.; Kuroiwa, S.; Osaka, T. Detection of tumor marker in blood serum using antibody-modified field effect transistor with optimized BSA blocking. *Sens. Actuators B Chem.* **2012**, *161*, 146–150. [CrossRef]

28. Blasi, D.; Sarcina, L.; Tricase, A.; Stefanachi, A.; Leonetti, F.; Alberga, D.; Mangiatordi, G.F.; Manoli, K.; Scamarcio, G.; Picca, R.A.; et al. Enhancing the sensitivity of biotinylated surfaces by tailoring the design of the mixed self-assembled monolayers synthesis. *ACS Omega* **2020**, in press. [CrossRef]

29. Homola, J. Present and future of surface plasmon resonance biosensors. *Anal. Bioanal. Chem.* **2003**, *377*, 528–539. [CrossRef]

30. Biacore Sensor Surface Handbook. Available online: https://timothyspringer.org/files/tas/files/biacore3000-sensorsurface.pdf. (accessed on 29 June 2020).

31. Miura, N.; Sasaki, M.; Gobi, K.V.; Kataoka, C. Highly sensitive and selective surface plasmon resonance sensor for detection of sub-ppb levels of benzo [a] pyrene by indirect competitive immunoreaction method. *Biosens. Bioelectron.* **2003**, *18*, 953–959. [CrossRef]

32. Disley, D.M.; Cullen, D.C.; You, H.X.; Lowe, C.R. Covalent coupling of immunoglobulin G to self-assembled monolayers as a method for immobilizing the interfacial-recognition layer of a surface plasmon resonance immunosensor. *Biosens. Bioelectron.* **1998**, *13*, 1213–1225. [CrossRef]

33. Liu, L.; Deng, D.; Xing, Y.; Li, S.; Yuan, B.; Chen, J.; Xia, N. Activity analysis of the carbodiimide-mediated amine coupling reaction on self-assembled monolayers by cyclic voltammetry. *Electrochim. Acta* **2013**, *89*, 616–622. [CrossRef]

34. Thermofisher INSTRUCTIONS NHS and Sulfo-NHS. Available online: https://www.thermofisher.com/document-connect/document-connect.html?url=https%3A%2F%2Fassets.thermofisher.com%2FTFS-Assets%2FLSG%2Fmanuals%2FMAN0011309_NHS_SulfoNHS_UG.pdf&title=VXNlciBHdWlkZToglE5IUyBhbmQgU3VsZm8tTkhT (accessed on 29 June 2020).

35. Grabarek, Z.; Gergely, J. Zero-length crosslinking procedure with the use of active esters. *Anal. Biochem.* **1990**, *185*, 131–135. [CrossRef]

36. de Mol, N.J.; Fischer, M.J.E. Surface plasmon resonance: Methods and protocols. *Life Sci.* **2010**, *627*, 255. [CrossRef]

37. Tsai, T.C.; Liu, C.W.; Wu, Y.C.; Ondevilla, N.A.P.; Osawa, M.; Chang, H.C. In situ study of EDC/NHS immobilization on gold surface based on attenuated total reflection surface-enhanced infrared absorption spectroscopy (ATR-SEIRAS). *Colloids Surf. B Biointerfaces* **2019**, *175*, 300–305. [CrossRef]

38. Goyon, A.; Excoffier, M.; Janin-Bussat, M.-C.; Bobaly, B.; Fekete, S.; Guillarme, D.; Beck, A. Determination of isoelectric points and relative charge variants of 23 therapeutic monoclonal antibodies. *J. Chromatogr. B* **2017**, *1065–1066*, 119–128. [CrossRef]

39. Erk, T. Gedig surface chemistry in SPR technology. In *Handbook of Surface Plasmon Resonance*; Richard, B.M.S., Ed.; The Royal Society of Chemistry: Cambridge, UK, 2017; pp. 173–220. ISBN 978-1-78801-028-3.

40. Glaser, R.W. Antigen-antibody binding and mass transport by convection and diffusion to a surface: A two-dimensional computer model of binding and dissociation kinetics. *Anal. Biochem.* **1993**, *213*, 152–161. [CrossRef]

41. León-Janampa, N.; Zimic, M.; Shinkaruk, S.; Quispe-Marcatoma, J.; Gutarra, A.; Le Bourdon, G.; Gayot, M.; Changanaqui, K.; Gilman, R.H.; Fouquet, E.; et al. Synthesis, characterization and bio-functionalization of magnetic nanoparticles to improve the diagnosis of tuberculosis. *Nanotechnology* **2020**, *31*, 175101. [CrossRef]

42. Darain, F.; Gan, K.L.; Tjin, S.C. Antibody immobilization on to polystyrene substrate—On-chip immunoassay for horse IgG based on fluorescence. *Biomed. Microdevices* **2009**, *11*, 653–661. [CrossRef]

43. Mariani, S.; Minunni, M. Surface plasmon resonance applications in clinical analysis. *Anal. Bioanal. Chem.* **2014**, *406*, 2303–2323. [CrossRef]

44. Puiu, M.; Bala, C. SPR and SPR imaging: Recent trends in developing nanodevices for detection and real-time monitoringof biomolecular events. *Sensors* **2016**, *16*, 870. [CrossRef]

45. Jiang, T.; Minunni, M.; Wilson, P.; Zhang, J.; Turner, A.P.F.; Mascini, M. Detection of TP53 mutation using a portable surface plasmon resonance DNA-based biosensor. *Biosens. Bioelectron.* **2005**, *20*, 1939–1945. [CrossRef]

46. Dubacheva, G.V.; Araya-Callis, C.; Geert Volbeda, A.; Fairhead, M.; Codée, J.; Howarth, M.; Richter, R.P. Controlling multivalent binding through surface chemistry: Model study on Streptavidin. *J. Am. Chem. Soc.* **2017**, *139*, 4157–4167. [CrossRef]

47. BioNavis MP-SPR Navi LayerSolver User Manual. Available online: http://www.bionavis.com/en/material-science/products/additional-options-and-consumables/software-mp-spr-navi/layersolver/ (accessed on 29 June 2020).

48. Baniukevic, J.; Kirlyte, J.; Ramanavicius, A.; Ramanaviciene, A. Comparison of oriented and random antibody immobilization techniques on the efficiency of immunosensor. *Procedia Eng.* **2012**, *47*, 837–840. [CrossRef]

49. Tsai, W.C.; Li, I.C. SPR-based immunosensor for determining staphylococcal enterotoxin A. *Sens. Actuators B Chem.* **2009**, *136*, 8–12. [CrossRef]

50. Frederix, F.; Bonroy, K.; Laureyn, W.; Reekmans, G.; Campitelli, A.; Dehaen, W.; Maes, G. Enhanced performance of an affinity biosensor interface based on mixed self-assembled monolayers of thiols on gold. *Langmuir* **2003**, *19*, 4351–4357. [CrossRef]

51. Kang, C.D.; Cao, C.; Lee, J.; Choi, I.S.; Kim, B.W.; Sim, S.J. Surface plasmon resonance-based inhibition assay for real-time detection of Cryptosporidium parvum oocyst. *Water Res.* **2008**, *42*, 1693–1699. [CrossRef] [PubMed]

52. Liang, H.; Miranto, H.; Granqvist, N.; Sadowski, J.W.; Viitala, T.; Wang, B.; Yliperttula, M. Surface plasmon resonance instrument as a refractometer for liquids and ultrathin films. *Sens. Actuators B Chem.* **2010**, *149*, 212–220. [CrossRef]

53. Mytych, D.T.; La, S.; Barger, T.; Ferbas, J.; Swanson, S.J. The development and validation of a sensitive, dual-flow cell, SPR-based biosensor immunoassay for the detection, semi-quantitation, and characterization of antibodies to darbepoetin alfa and epoetin alfa in human serum. *J. Pharm. Biomed. Anal.* **2009**, *49*, 415–426. [CrossRef]

54. Šípová, H.; Ševců, V.; Kuchař, M.; Ahmad, J.N.; Mikulecký, P.; Osička, R.; Malý, P.; Homola, J. Surface plasmon resonance biosensor based on engineered proteins for direct detection of interferon-gamma in diluted blood plasma. *Sens. Actuators B Chem.* **2012**, *174*, 306–311. [CrossRef]

55. Picca, R.A.; Manoli, K.; Macchia, E.; Sarcina, L.; Di Franco, C.; Cioffi, N.; Blasi, D.; Österbacka, R.; Torricelli, F.; Scamarcio, G.; et al. Ultimately sensitive organic bioelectronic transistor sensors by materials and device structures' design. *Adv. Funct. Mater.* **2019**, *30*, 1904513. [CrossRef]

56. Macchia, E.; Picca, R.A.; Manoli, K.; Di Franco, C.; Blasi, D.; Sarcina, L.; Ditaranto, N.; Cioffi, N.; Österbacka, R.; Scamarcio, G.; et al. About the amplification factors in organic bioelectronic sensors. *Mater. Horiz.* **2020**, *7*, 999–1013. [CrossRef]

MDPI

St. Alban-Anlage 66

4052 Basel

Switzerland

Tel. +41 61 683 77 34

Fax +41 61 302 89 18

www.mdpi.com

Sensors Editorial Office

E-mail: sensors@mdpi.com

www.mdpi.com/journal/sensors